ACTIVATED CARBON
COMPENDIUM

ACTIVATED CARBON COMPENDIUM

A collection of papers from the journal
Carbon 1996-2000

Edited by

HARRY MARSH

North Shields, UK

2001
ELSEVIER

AMSTERDAM – LONDON – NEW YORK – OXFORD – PARIS – SHANNON – TOKYO

ELSEVIER SCIENCE Ltd
The Boulevard, Langford Lane
Kidlington, Oxford OX5 1GB, UK

First edition 2001

Library of Congress Cataloging in Publication Data
A catalog record from the Library of Congress has been applied for.

British Library Cataloguing in Publication Data
A catalogue record from the British Library has been applied for.

ISBN: 0-08-044030-4

Transferred to digital printing 2005
Printed and bound by Antony Rowe Ltd, Eastbourne

FOREWORD

This compendium is intended to promote interest in and an understanding of activated carbons. Part-I is a brief introduction to considerations of porosity, adsorption and industrial applications. Part-II consists of papers published in CARBON to illustrate the wide-ranging topics and challenges presented by these activated carbons.

PART-I

Activated carbon is a porous carbon material, a char that has been subjected to reaction with gases, *e.g.* carbon dioxide or steam, sometimes with the addition of chemicals such as zinc chloride or phosphoric acid, before, during or after carbonization in order to increase its adsorptive properties. Activated carbons have a large adsorption capacity, especially for small molecules and are used for purification and separation of liquids and gases. Considerable *flexibility* exists in the manufacturing processes so allowing for activated carbon with a wide range of properties and adsorption capabilities. The carbons are used mainly in granular and powdered forms but also are manufactured in a textile (fibre) format.

What are active carbons?

Those materials of plant origin which do not fuse, form carbons (chars) on carbonization (heating in an inert atmosphere) and these can be activated to produce the activated carbon. For reasons of economics, including collection, transportation, sustainable supply and price, the industry that makes activated carbons essentially restricts its supplies to softwood, hard wood, peat, lignin, coal, and coconut shell. Speciality activated carbons are produced from carbon fibres and synthetic non-fusing polymers such as PAN and SARAN. Other possible sources are harvested only at the end of a growing season, or are scattered over entire countries, and cannot be transported regularly to the manufacturing site. Currently, globally, about 400,000 ton of activated carbons are manufactured annually requiring about one million ton of raw material. Worldwide, there are about 150 companies manufacturing activated carbons, the largest including Calgon, Norit, Nuchar, and Westvaco.

What is porosity?

Porosity, essentially, is composed of volume elements (called pores) of zero electron density existing within a solid, the volume elements having dimensions of molecules. Carbons offer a maximum in *flexibility* in terms of an infinite number of different pore size distributions (PSD). This arises because of the unique chemical and physical properties of carbon materials. The process of carbonization is the key to the understanding of this. The parent organic material must not melt or fuse. On carbonization, these cellulosic/lignitic materials lose small molecules such as water, carbon dioxide, and low molecular weight organics with the original solid state lattice, (dominantly composed of the residual carbon) remaining more or less intact. The positions vacated by the constituent atoms of the volatile decomposition products are the created porosity. This new carbon lattice which contains the porosity is quite disorganized but with further heating becomes more structured in terms of approximation to the hexagonal array of carbon atoms as found in the sheets or graphene layers which constitute single crystal graphite.

The use of high-resolution lattice-imaging transmission electron microscopy provides an insight into structure within activated carbons. Figure 1 is such a micrograph of a carbon from polyvinylidene chloride activated to 75 wt % burn-off. The micrograph is a two-dimensional vertical projection of the small sections of the defective graphene sheets. It shows their lack of planarity which prevents parallelism and creates the porosity. It also shows how the porosity can have an infinite number of shape and size.

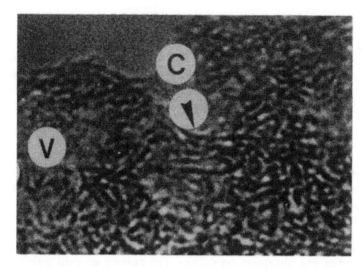

Figure 1. High-resolution lattice-imaging transmission electron micrograph of an activated
carbon, 73 wt % burn-off, from polyvinylidene chloride.
|_____| = 5 nm
Positions C and V illustrate shapes and sizes of porosity

There are three properties of active carbons of dominant importance in controlling their
adsorption characteristics, *viz*, (a) the distribution of pore sizes within the carbon, (b) the mean
pore size and (c) the composition of the surfaces which contain the porosity. Pore widths can
range from <1.0 nm to 50 nm with surface compositions varying from clean graphitic to
oxygenated surfaces exhibiting significant polarity. A nomenclature exists to describe size
ranges of porosity, as follows:

Ultra-microporosity:	< 0.5 nm diameter
Microporosity:	0.5 to 2.0 nm diameter
Mesoporosity	2.0 to 50 nm diameter
Macroporosity	> 50 nm diameter

These definitions are based on adsorption behaviour and cannot be considered as being
absolute.

What is the activation process?

When an organic precursor is carbonised a certain pore size distribution (PSD) is generated.
Such a carbon may not function well enough for a particular application, *e.g.* adsorption from
an aqueous medium. Should the precursor be carbonised with zinc chloride or phosphoric acid
(chemical activation) these act as dehydrating agents and modify the pyrolysis chemistry. As a
result, less carbon is lost with the volatiles and the resultant PSDs are different, and differ for
different activating agents. In addition the PSDs generated by chemical activation can be further
modified by physical activation. Further modifications (physical activation) by gasification with
steam or carbon dioxide selectively remove carbon atoms from within the walls of the original
porosity. Physical activation processes should be dominantly within the region of chemical
control of reaction rate to enable gasification to occur within the centres of the particles
(grains). Oxygen is not a suitable activating agent because it is too reactive a molecule and
would essentially react with the external surfaces of the particles without enhancing the PSDs.
Strong oxidizing agents introduce surface oxygen groups to the surfaces of the porosity. On the
other hand, carbon deposition techniques can be used to reduce the dimensions of entrances to
porosity.

How can PDSs be measured?

An activated carbon with an effective surface area of about 1000 m^2 g^{-1} contains about 5 x 10^{20} pores, all being different in some way but forming a continuous distribution of adsorption potential (energy). Direct measurement of such molecule-sized porosity is not an option and recourse has to be made to interpretive methods. There are many quite different methodologies available and these are listed in Table 1. This Table 1 shows the wide range of techniques used and the range of expertise needed in interpretations. This is an area of considerable research activity and forms the basis of many of the papers cited in Part-II.

Table 1.
Methodologies of Analyses of Porosity

Adsorption isotherms.

      Effective surface area (m^2 g^{-1}) based on nitrogen adsorption at 77 K (500-2500 m^2 g^{-1})

      Surface areas using other adsorbates (m^2 g^{-1})

      Volumes in micro-, meso- and macroporosity (0.5-2.5 cm^3 g^{-1})

      Isotherm constants (energetics) 'b' of the Langmuir equation and 'C' of the BET equation.

      Gradient of plot of log n_a versus log^2 p/p^o (Dubinin-Radushkevitch equation) as an indicator of pore (energy) size distribution and mean pore diameter.

      Use of the Dubinin-Astakhov, Dubinin-Izotova and Dubinin-Stoeckli equations.

      Distributions of characteristic energies of adsorption.

Fractal dimension calculations

Transmission Electron Microscopy.

Phase-contrast, lattice-imaging, high-resolution, transmission electron microscopy (TEM) to reveal structure.

Dark-field TEM, indicating stacking order of lamellar constituent molecules (LCM) or graphene layers.

Small angle X-ray scattering (SAXS) and

Small angle neutron scattering (SANS).

Molecular sieve experiments.

Pre-adsorption of n-nonane.

Scanning tunnelling microscopy and atomic force microscopy.

      To reveal atomic surface features.

Calorimetry.

      To measure the energetics and dynamics of adsorption and desorption..

^{13}C Nuclear Magnetic Resonance

X-ray photolectron spectroscopy (XPS)

Computer molecular modeling, to simulate structures using minimum energy considerations.

Dynamic breakthrough curves.

Particle size	(mm or sieve sizes)
Apparent density	(g cm^{-3})
Particle density	(g cm^{-3})
Hardness number	50-100
Abrasion number	
Ash content	(1-20 wt %)
CCl_4 activity	(35-125)
Butane working capacity (g 100 cm^{-3})	(4-14)
Iodine number	(500-1200)
Decolorizing index (Westvaco)	(15-25)
Molasses number (Calgon)	(50-250)
Molasses number (Norit)	(300-1500)
Heat capacity at 100 °C	(0.84-1.3 J g^{-1} K^{-1})
Thermal conductivity	(0.05-0.10 W m^{-1} K^{-1})

What is the adsorption process?

The process of physical adsorption is largely governed by van der Waals forces of interaction. In a dynamic process, the adsorbate enters (penetrates) the porous network to locate a position of minimum energy (the process of adsorption is exothermic). But it only stays there for an interval of time as short as 10^{-10} seconds before moving on to another site. And there are many sites. One gram of a typical porous activated carbon contains about 10^{20} sites for adsorption, this being a good statistical number. A photograph of the adsorption process, using a shutter speed of $<10^{-10}$ seconds, would show a certain distribution of adsorbate over the surface. A photograph taken 10^{-10} seconds later would show a different distribution. At positions of equilibrium the position is far from being static with a continuous rapid exchange of molecules between the adsorbed phase and the gas phase.

Surfaces of pores are not homogeneous. Sites which are most disorganized (quasi-crystalline) are energetically more favorable for adsorption. The smaller of porosity is energetically more favorable for adsorption. Polarity in the surface such as the presence of sulphur, nitrogen, phosphorus or oxygen is more energetically favorable for adsorption sites. The more energetically favourable of sites are occupied first and this is why heats of adsorption decrease with increasing extents of adsorption.

Figure 2 is a scanning electron micrograph of an etched silicon surface. Note the heterogeneous nature of the surface and the multiplicity of site structures. This surface had been contaminated with smoke from a cigar and it is interesting to note that all of the smoke particles adhered to one type of surface. Adsorption in carbons has similarities in that the adsorption sites are heterogeneous and an adsorbate entering into the porous system will adsorb initially and preferentially on sites of maximum adsorption potential, as does the smoke particle.

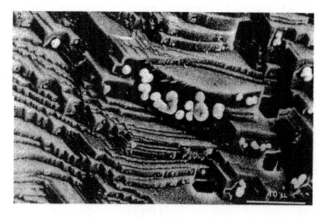

Figure 2. SEM micrograph of an etched surface of silicon contaminated
by smoke particles

Surface areas: fact or fiction?

In many of the papers cited in Part-II of this Compendium, results of surface area measurements are presented. However, such values need to be assessed with caution and may require an experienced researcher to make a judgment. The concept of 'internal surface area' within the mind cannot easily be converted to reality. Measurements of internal surface areas are based on interpretations of adsorption data and other data and this is a very subjective exercise. It is not possible, indeed certainly not recommended, that surface areas within porosity should be regarded with the same veracity as a weight or a volume. The problem is that different methods of assessment often provide quite different results. For example, a carbon which adsorbs nitrogen at 77 K whose isotherm is interpreted by the BET equation, has a surface area of 5000

$m^2 \ g^{-1}$. But if the Dubinin-Radushkevich (DR) equation is used the area is 1,000 $m^2 \ g^{-1}$. It happens that different methods examine different parts of the porosity. There is no purpose in entering into a debate as to which *'internal surface area'* is the correct one. In an adsorbing system, it is the *'effective surface area'* which is meaningful, that is the area which the adsorbent presents to the adsorbate.

Applications of active carbons

Producers of active carbons have a wide range of products. For example, Norit, in Europe and in North America, has an extensive website listing over 100 carbons with defined applications. The principal applications include (a) gas separation, (b) gas purification, (c) solvent recovery, (d) water treatment, (e) food and beverage processing, (f) chemical and pharmaceutical processing, (g) sewage treatment, (h) catalyst supports and (i) gold recovery.

Research into active carbons

Porosity in carbons presents researchers with two major challenges.   The first is its characterization in terms of PSDs and surface energetics; the second is the optimization of porosity in terms of efficiency of adsorption processes. The papers cited in Part-II of this compendium illustrate the far-reaching interests of researchers with the citations being grouped into:

A. **The Activation Process** looking at mechanisms of generation of porosity.
B. **Modification to Porosity** looking at how an initial porosity can be further improved.
C. **Properties of Activated Carbons** looking at relationships between structure and adsorption.
D. **Applications** which describes research programs in specific areas of applications.
E. **Theoretical** including suitability of theoretical adsorption equations, etc.

PART-II

PAPERS PUBLISHED BY CARBON

In the period 1996-2000 the journal CARBON, Volumes 34-38, published 257 papers specifically concerned with activated carbons. Figure 3 is a histogram of papers published in the years 1994-2000.

What emerges from an assessment of these 257 papers, on the one hand, is the *complexity* of analyses of porosity in such carbons and, on the other hand, is the great *flexibility* in methods of preparation to match properties to performance.

A critical awareness is encouraged, in some depth, of those associated with activated carbons of the wide-ranging properties of these materials and of the problems of structural analyses.

To this end, the publications in CARBON constitute a major database of understanding.

Of the 257 published papers, 27 have been selected by the Editor of this Compendium to represent the dominant research interests and points-of-view of researchers. There is little overlap in these points-of-view further illustrating the *complexity* and *flexibility* of manufacture and use of such materials.

The 27 papers are grouped into five sections:

(A) The Activation Process
(B) Modifications to Porosity
(C) Properties of Activated Carbons
(D) Theoretical
(E) Applications

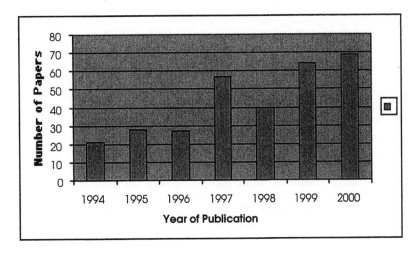

Figure 3. The variation of number of papers concerned with active carbons with year of publication in the Journal CARBON.

A. The Activation Process

(1)   N.R. Khalili, M. Campbell, G. Sandí and J. Golaś
        Production of micro- and mesoporous activated carbon from paper mill sludge.
        I Effect of zinc chloride activation
        Carbon 2000;**38(14):**1905-1915

Although phosphoric acid is shown to be the most environmentally sound chemical for activation processes, studies still use zinc chloride due to its effective activating capability. Khalili *et al.* (1) report on the production of micro and mesoporous activated carbons from paper mill sludge, using zinc chloride as the activating agent. Established theoretical methods were used to interpret isotherms of nitrogen adsorption. Amounts of $ZnCl_2$ used for chemical activation controlled the characteristics of the carbons, including surface area, micro- and mesoporosity and pore size distributions. For $ZnCl_2$/sludge mass ratios of 0.75 to 2.5 mesopore volumes increased by 600 %. $ZnCl_2$/sludge mass ratios of <1.0 and ≥ 1.5 are needed to obtain both micro-and mesoporous carbons. Surface areas in excess of 1,000 $m^2 g^{-1}$ are reported.

(2)   H. Teng, J.-A. Ho and Y.-F. Hsu
        Preparation of activated carbons from bituminous coals with CO_2 activation - influence
        of coal oxidation
        Carbon 1997;**35(2):**275-283

Teng *et al.* (2) carried out a study of the preparation of activated carbons from bituminous coals with CO_2 activation, the coals having been pre-oxidised. Three Australian coals were used of volatile contents 26.7, 31.6 and 33.1 wt %. For non-oxidised coals carbonised to 800 °C in an atmosphere of CO_2, to ~ 40 wt % burn-off, surface areas of resultant carbons were 436, 563, and 704 $m^2 g^{-1}$. For coals initially oxidised at 200 °C for 6 hours, surface areas of resultant carbons were 825, 819 and 800 $m^2 g^{-1}$. respectfully. Surface areas are dominantly contained within microporosity.

(3)   J.A.F MacDonald and D.F. Quinn
        Adsorbents for methane storage made by phosphoric acid activation of peach pits
        Carbon 1996;**34(9):**1103-1108

MacDonald and Quinn (3), starting from peach stones, prepared adsorbents (activated carbons), by activation with phosphoric acid, for the purpose of methane storage. The maximum BET surface area for the series of carbons produced was 1560 $m^2 g^{-1}$ for a carbon of heat treatment temperature (HTT) of 500 °C similar to other carbons prepared from lignocellulosic materials. Although low-temperature charring produces samples with high nitrogen surface areas and micropore volumes these samples are not suitable for methane adsorption. This is probably due to the high concentration of hydrogen and oxygen remaining on the surface,

(4)   H. Teng, T. - S. Yeh and L.-Y. Hsu
        Preparation of activated carbon from bituminous coal with phosphoric acid activation
        Carbon 1998;**36(9):**1387-1395.

Teng *et al.* (4) studied the preparation of activated carbons from an Australian Black Water bituminous coal (high volatile A) using phosphoric acid. No prior oxidation of the coal was made. Using 40 wt % of H_3PO_4 produced carbons with 510-550 $m^2 g^{-1}$ of surface mainly in microporosity. Further activation by carbon dioxide increased surface areas to 750 $m^2 g^{-1}$ mainly in microporosity. Phosphoric acid is not as effective as an activating agent as zinc chloride.

(5)   M. Molina-Sabio, F. Rodríguez-Reinoso, F. Caturla and M.J. Sellés
        Development of porosity in combined phosphoric acid-carbon dioxide activation
        Carbon 1996;**34(4):**457-462

Molina-Sabio *et al.* (5) write that phosphoric acid has a double role in the production of active carbons. It causes hydrolysis of the lignocellulose material and the subsequent extraction of some of its components. Also, the acid occupies a volume which inhibits the contraction of a particle during heat treatment so leaving a porosity when the carbon is extracted by washing. The main effect of gasification with carbon dioxide is the creation and widening of existing pores. Thus, a new family of activated carbons can be prepared (in this case from peach stones) using an initial activation with phosphoric acid and subsequent treatment with carbon dioxide. The family can range from totally microporous carbons to totally mesoporous carbons.

(6)   K.P. Gadkaree
Carbon honeycomb structures for adsorption applications
Carbon 1998;**36(7-8)**:981-989

Gadkaree (6) described activated carbon honeycomb structures based on synthetic precursors without the use of binders so making them more durable. They are thus an attractive alternative to the traditional packed bed system. A major advantage of the open structure is the low pressure drop but a disadvantage could be a poor adsorption performance. Gadkaree (6) found that in spite of the open structure, the honeycombs gave breakthrough curves similar to packed beds for butane, toluene, isopropanol, etc. However, this was not the case for low concentrations of small molecules such as aldehydes. This behaviour was explained in terms of the dynamic equilibrium between adsorption and diffusion of the adsorbate molecules into the honeycomb carbon walls.

B.   Modifications to Porosity

(7)   R. Moene, H.Th. Boon, J. Schoonman, M. Makkee and J.A. Moulijn
Coating of activated carbon with silicon carbide by chemical vapour deposition
Carbon 1996;**34(5)**:567-579

Moene *et al.* (7) coated active carbons with silicon carbide by chemical vapour deposition in order to improve the oxidation resistance and mechanical strength of such carbons. Such carbons have applications as catalyst support materials rather than as adsorbents. Silicon carbide deposition was achieved from $CH_4/SiCl_4$ mixtures and also from CH_3SiCl_3. In both cases oxidation resistances improved and side crushing strengths improved by 1.7 and 1.4 respectively. Residual surface areas were 176 and 530 m^2 g^{-1} such carbons being considered suitable for catalyst support applications. This paper incorporates a theoretical study, which evaluates a chemical vapour infiltration design chart, which correlates initial Thiele moduli with the porosity after deposition.

(8)   S. Wang and G.Q. Lu
Effects of acid treatments on the pore and surface properties of Ni catalyst supported on activated carbon
Carbon 1998;**36(3)**:283-292

Wang and Lu (8), with interests in Ni catalyst supported activated carbons, studied the effects of acid treatments on pore and surface properties. A Calgon activated carbon was treated with HCl, HNO_3 and HF and changes in properties were monitored by adsorption of nitrogen at 77 K, temperature programmed desorption (TPD) and X-ray photoelectron spectroscopy (XPS). The adsorption capacity of Ni^{2+} on the carbon supports increased with acid treatments. Acid treatment results in a more homogenous distribution of the nickel salt in the carbon. Catalyst activity correlates closely with the chemical composition of the carbon surface.

(9)   A. Oya, R. Horigome, D. Lozano-Castello and A. Linares-Solano
Change of porous structure and surface state of charcoal as a result of coating with thin carbon layer
Carbon 1999;**37(9)**:1499-1502

In a Letter-to-the-Editor Oya *et al.* (9) describe a study whereby a charcoal (from a Japanese cypress wood) is coated with a thin carbon layer (from Kureha petroleum pitch) to change its porous structure as well its surface chemistry. The study has two objectives, (a) to increase the specific surface area and (b) to suppress the formation and release of fine carbon particles when in use. The coated charcoal was carbonized at 900 °C under nitrogen for 30 min and then activated in steam for 5 min at 900 °C to a burn-off of 10.8 wt %. For the pristine charcoal, before activation, the specific surface area increased from 13 to 428 m^2 g^{-1} for the carbon-coated charcoal. Similarly, after activation, the specific surface area increased from 13 to 428 $m^2 g^{-1}$ for the carbon-coated charcoal. The coating promoted a decrease in macropore volumes and an increase in meso-pore volumes. The technique is effective in fixing the fine carbon particles on the charcoal surface.

C.   Properties of Activated Carbons

(10)   Y. Suzin, L.C. Buettner and C.A. LeDuc
Characterizing the ignition process of activated carbon
Carbon 1999;**37**(2):335-346

Properties of activated carbons at elevated temperatures, specifically the spontaneous ignition temperature (SIT) and the point of initial oxidation (PIO) are of significance in a number of unit operations. Suzin *et al.* (10) attempted to devise standard methods to compare various carbons. The SIT was determined by the ASTM and the thermal analysis (TG and DSC) methods. The PIO was determined by effluent CO_2 concentration analysis (TG and DSC) and temperature profiling. For three base carbons and three impregnated carbons these methods do not have general utility, some are not easily correlated and there is evidence of the validity of two reaction regimes of interest. The simplest method that provides the most consistently conservative estimate of reaction commencement is by measuring the PIO with the temperature profiling technique.

(11)   F. Haghseresht, G.Q. Lu and A.K. Whittaker
Carbon structure and porosity of carbonaceous adsorbents in relation to their adsorption properties
Carbon 1999;**37**(9):1491-1497

Adsorption properties of carbons (HTT 600 °C) from acid-washed coal waste materials were studied (Haghseresht *et al.* (11)) by nitrogen, phenol and para-nitrophenol adsorption, by benzene adsorption and Raman spectroscopy and ^{13}C-NMR spectrometry. The coal was given a pretreatment with nitric acid varying acid concentration, residence time and reaction temperature. The ^{13}C-NMR work indicated that as the severity of the oxidation conditions increased, the ratio of aromatic to aliphatic carbon, from the ratio of the Raman D-bands G-bands, affects both the pore structure and the adsorption properties of the resultant adsorbents. It was demonstrated that as the ratio of D-band to G-band increased, the adsorption capacity of the resulting chars for organic compounds increased. Furthermore, the increase in the extent of the disorganized carbon also indicates less space between the clusters of aromatic compounds, hence creating micropores.

(12)   R. Diduszko, A. Swiatkowski and B.J. Trznadel
On surface of micropores and fractal dimension of activated carbon determined on the basis of adsorption and SAXS investigations
Carbon 2000;**38**(8):1153-1162

As indicated in Part-I the concept of 'surface area' of an activated carbon is fraught with difficulties in the sense that experimental values are a function of the method used. One explanation of this is that different methods may not be examining equivalent parts of the surface. This study of Diduszko *et al.* (12) addresses this problem. Three samples of activated hardwood were examined by (a) adsorption of benzene, using the Dubinin-Izotova (DI) and Dubinin-Stoeckli (DS) equations, and (b) (SAXS) small angle scattering of X-rays. Results of the SAXS study are higher than obtained from adsorption methods because of access of X-rays to the finer pores inaccessible to the benzene adsorbate molecule. A better agreement existed between the SAXS and DS equations. Relationships between fractal dimensions and adsorption data are a matter for further clarification.

D.   Applications

(13)   S. Sircar, T.C. Golden and M.B. Rao
       Activated carbon for gas separation and storage
       Carbon 1996;**34(1)**:1-12

Sircar *et al.* (13) assess the use of activated carbon for gas separation and storage. Activated carbons offer a broad spectrum of pore size distributions and surface chemistry for adsorption of gases, which are being used to design practical pressure swing and thermal adsorption processes for separation and purification of gases. Separation and purification of gas mixtures by adsorption have become a major unit operation in the chemical and petrochemical industries as shown by the number of worldwide patents issued every year (average about 300). Of the several adsorbents used, including zeolites, activated carbons, alumina, silica gels, and polymeric adsorbents, activated carbons play a major role. *Flexibility* in production of adsorbents with specific properties has allowed the development of pressure swing adsorption (PSA) and thermal swing adsorption (TSA) separation techniques. In addition to trace-impurity removal, solvent vapor removal and recovery and air separation, separations are possible of hydrogen-hydrocarbon mixtures, as well gas drying by PSA using a carbon with a hydrophobic surface.

(14)   J.A. Meidl
       Responding to changing conditions: how powdered activated carbon systems can provide the operational flexibility necessary to treat contaminated groundwater and industrial wastes
       Carbon 1997;**35(9)**:1207-1216

The use of activated carbon for wastewater and contaminated groundwater treatment is increasing throughout the world as a greater awareness becomes evident of this planet's limited supply of water. Meidl (14) discusses how powdered activated carbon systems provide the operational *flexibility* necessary to treat contaminated waters. The use of powdered activated carbon technology (PACT) is increasing because of superior performance and operational *flexibility* that powdered activated carbons can provide. By merging biological and physical treatment into a single process step, the system is able to buffer toxic loads which might otherwise impair a biological system and reduce the amount of carbon otherwise needed by a adsorption treatment system. This contribution by Meidl (7) describes several case studies in some detail.

(15a,15b,15c)   W.H. Lee and P.J. Reucroft
       Vapor adsorption on coal- and wood-based chemically activated carbons.
       Part I. Surface oxidation states and adsorption of H_2O
       Carbon 1999;**37(1)**:7-14
       Vapor adsorption on coal- and wood-based chemically activated carbons.
       Part II. Adsorption of organic vapors
       Carbon 1999;**37(1)**:15-20
       Vapor adsorption on coal- and wood-based chemically activated carbons.
       Part III. NH_3 and H_2S adsorption in the low relative pressure range
       Carbon 1999;**37(1)**:21-26

Lee and Reucroft (15a,15b,15c) published assessments of the adsorption of H_2O, organic vapors and NH_3 and H_2S in the low relative pressure range using coal- and wood-based chemically activated carbons. Eight carbons were prepared from Illinois bituminous coals and white oaks using KOH and H_3PO_4 as chemical activating agents. X-ray photoelectron spectroscopy (XPS) evaluated surface chemical compositions. Generally, high surface area carbons had low concentrations of surface oxygen complexes and low surface area carbons had high concentrations of surface oxygen complexes. These data correlated with temperatures of formation of the carbons. The high amounts of water vapor adsorption on the lower surface area carbons can be explained by the higher concentrations of surface oxygen complexes. The α_0 values of the Dubinin-Serpinsky equation correlated well with the total oxygen concentration and the percentage of carbon oxygen groups such as C-O and C=O obtained from XPS.

Using adsorption data of CCl_4 and acetone, Lee and Reucroft (15a,15b,15c) indicate that the coal-based KOH activated carbons have narrower micropores and more uniform micropore size distributions than the wood-based H_3PO_4 activated carbons. From the use of NH_3 and H_2S in the low relative pressure range these authors conclude that the adsorption capacity of NH_3 and H_2S was independent of the degree of surface area development. This is a most important statement for those working with the removal of contaminants at low relative concentrations. *The relevance to surface areas determined by nitrogen adsorption over the full relative pressure range is minimal.*

(16)   I. Mochida, Y. Korai, M. Shirahama, S. Kawano, T. Hada, Y.Seo, M. Yoshikawa and
       A. Yasutake
       Removal of SO_x and NO_x over activated carbon fibers
       Carbon 2000;**38(2):**227-239

Mochida *et al.* (16) review methodologies for removal of SO_x and NO_x from the combustion products of fossil fuels. The need for such activity is obvious. In Japan some 1800 wet desulfurization units of high efficiency are operating. These units remove about 95 % of SO_2 but only about 50 % of SO_3 from emissions, the latter being attributable for 'Yokkaichi asthma'. These authors (16) consider that higher packing densities of activated carbon fibers need to be developed, including control over the shape and size of the fiber. The use of a thinner paper in the filters may moderate the pressure drop in the packed bed. Process design is concerned with how to combine the de-SO_x and de-NO_x processes and how to contact SO_2, NO, O_2 and H_2O with the ACF.

E.   Theoretical

(17)   K.A. Sosin, D.F. Quinn and J.A.F. MacDonald
       Changes in PSD of progressively activated carbons obtained from their supercritical methane isotherms
       Carbon 1996;**34(11):**1335-1341

Pore size distributions (PSD) in activated carbons are of crucial importance but are difficult to quantify. Sosin *et al.* (17) monitored changes in the PSD of carbons prepared from polyvinylidene chloride (Saran) by pyrolysis to 700 °C, followed by activation in nitrogen/steam at 800°C (55 wt % yield) and in carbon dioxide at 850 °C (35 wt % yield). Methane isotherms, obtained at 25 °C at pressures up to 4.1 MPa (41 atmospheres) were interpreted in terms of PSD, total micropore and mesopore volumes as well as total internal surface area (TISA). The authors conclude that the effects of steam activation of Saran carbon are very similar to those of carbon dioxide. Further, they consider that both activating agents act in existing pores only and no new pores are being created in the process. Considering methane storage, it is further concluded that the effects of activation are rather small and that any gain in storage capacity (uptake by unit volume of carbon) is too small to compensate for the loss of carbon.

(18)   N. Vahdat
       Theoretical study of the performance of activated carbon in the presence of binary vapor mixtures
       Carbon 1997;35(**10-11):**1545-1557

Industrial situations where purifications are needed often involve binary or more complex mixtures. Laboratory studies on the other hand are most likely to involve only one adsorbate. Studies involving mixtures of adsorbates require continuous analysis of the adsorptives and this adds to the complexity and cost of the experimental rigs. Vahdat (18) undertook a theoretical evaluation of a binary mixture of adsorbates. The effect of the adsorption isotherm of pure compounds on the breakthrough curves was evaluated by varying Langmuir constants for each compound. An increase in the adsorption capacity of the more strongly adsorbed component causes an increase in the breakthrough time of both compounds. The effect of the adsorption capacity of the weakly adsorbed compound on the other hand is to increase its breakthrough time and to reduce the breakthrough time of the other compound. The effect of low rate on breakthrough times of the two components of a binary mixture is similar to that of the pure compounds. Both breakthrough times vary linearly with the reciprocal of flow rate.

(19)   C.R. Clarkson, R.M. Bustin and J.H. Levy
    Application of the mono/multilayer and adsorption potential theories to coal methane adsorption isotherms at elevated temperature and pressure
    Carbon 1997;**35(12)**:1689-1705

As mentioned in the Introduction of this Compendium, the most frequently used theoretical adsorption equations are those of Langmuir, BET, and the D-R (Dubinin-Radushkevich) and (D-A) Dubinin-Astakhov equations. Clarkson *et al.* (19) with interests in methane adsorbed in wet coal (as with coal-bed methane) looked for the optimum equation to be applied to adsorption isotherm data. The conceptual model for pore filling differs for the four equations and therefore the study of Clarkson *et al.* (19) has important implications for the modeling of equilibrium isotherm data in coal. The three-parameter Dubinin-Astakhov equation yielded the best curve-fit to the high pressure (>0.101 MPa) methane experimental data, but the two-parameter Dubinin-Radushkevich and BET equations are better than the Langmuir equation.

(20)   P.D. Paulsen, B.C. Moore and F.S. Cannon
    Applicability of adsorption equations to argon, nitrogen and volatile organic compound adsorption on to activated carbon
    Carbon 1999;**37(11)**:1843-1853

Paulsen *et al.* (20) studied the applicability of adsorption equations to the adsorption of argon, nitrogen, methylisobutylketone (MIBK) and m-zylene. The activated carbons used were from coconut shell, peat, lignite and a bituminous coal. Two theoretical equations were tested, namely a Modified Freundlich (MF) equation and the Dubinin-Astakhov (DA) equation. The MF and DA equations both performed well in the relative pressures between 10^{-5} and 0.1. For argon adsorption in the range 10^{-5} to 10^{-4} the DA equation performed slightly better. When nitrogen was used as the adsorbate for two of the activated carbons, the MF equation performed better than the DA equation over the pressure range studied. The adsorption of argon on to a coconut-based carbon at the boiling point of argon was quite similar to the polytherm-predicted adsorption of both MIBK and m-zylene. This uniformity of normalized adsorption behavior offers a promising opportunity to approximate the adsorption of a lightly tested volatile organic compound by normalizing it to the extensively tested MIBK on the same carbon.

(21)   D.D. Do and K. Wang
    A new model for the description of adsorption kinetics in heterogeneous activated carbons
    Carbon 1998;**36(10)**:1539-1554

Do and Wang (21) used an Ajax activated carbon and seven adsorbates to propose a new model describing the adsorption kinetics in this heterogeneous activated carbon. The model incorporates pore diffusion, surface diffusion and finite mass interchange rate between the gas and adsorbed phase. The driving force for surface diffusion is taken as the chemical potential gradient and the mass interchange is described by Langmuir kinetics. Surface heterogeneity is attributed to the micropore size distribution obtained through the use of the Lennard-Jones potential theory. Predictions were found to be excellent in general, In particular, the model predicted the long tail behaviour observed in the desorption mode, something which other models were unable to account for.

(22)   J.P. Barbosa Mota, A.E. Rodrigues, E. Saatdjian and D. Tondeur
    Dynamics of natural gas adsorption storage systems employing activated carbon
    Carbon 1997;**35(9)**:1259-1270

Barbosa Mota *et al.* (22) undertook a study of the dynamics of natural gas (95 % methane) adsorption storage systems employing activated carbon. A storage performance indicator for adsorbed natural gas is volume of gas adsorbed per volume of carbon. It is considered that the theoretical maximum storage capacity predicted by molecular simulation is 209 v v^{-1} for a monolithic carbon but this is reduced to 146 v v^{-1} for pelletized carbon. Natural gas, compressed to 20.7 MPa has an equivalent of 240 v v^{-1}. Experimental carbons have storage capacities in the region of 100 to 150 v v^{-1}. The authors consider that a major limitation to the effective use of adsorbed methane as a fuel source is the filling times, that is, the resistance to gas flow and this is where the dynamics of the process must be understood.

(23)   A. Burian, A. Ratuszna, J.C. Dore and S.W. Howells
        Radial distribution function analysis of the structure of activated carbons
        Carbon 1998;**36(11):**1613-1621

Rather than rely on interpretations of adsorption isotherms to determine pore size distributions, another method is their direct measurement. Burian *et al.* (23) carried out a radial distribution function analysis of wide-angle neutron diffraction data to investigate the structure of activated carbons prepared by the carbonization of phenol-formaldehyde resin and activation in carbon dioxide to 14 and 32 wt % burn-off. The neutron scattering experiments were carried out using the pulsed neutron facility in the U.K. Rutherford Appleton Laboratory. The derived model gave a satisfactory fit to the experimental data assuming the graphite-like arrangement within the single layer and the para-crystalline-type distortion of the two-dimensional hexagonal lattice. Simulations indicate that on average only about four of these layers are associated. This open structure leads to surface areas for the three carbons of 580, 840 and 1000 m^2 g^{-1}.

(24)   N.R. Khalili, M. Pan and G. Sandí
        Determination of fractal dimensions of solid carbons from gas and liquid phase
        adsorption isotherms
        Carbon 2000;**38(4):**573-588

According to Khalili *et al.* (24) activated carbons are by far the most frequently used adsorbents for solvent recovery, gas refining, air purification, exhaust desulfurization, deodorization, and gas separation and recovery. This study of fractal dimensions used two granular activated carbons from NORIT peat-based and coal-based, a carbon from peat suitable for solvent recovery, a carbon suitable for water treatment prepared by the steam activation of a coal, and a honeycomb-type of carbon derived from a natural Montmorillonite clay. The adsorptions of nitrogen at 77 K and of a labelled phenanthrene in aqueous solution were studied. A modified BET equation and fractal Frenkel-Halsey-Hill (FHH) models were used to estimate surface fractal dimensions. Fractal geometry is a mathematical tool that deals with complex systems that have no characteristic length scale. Results suggested the existence of van der Waals forces of adsorption, and the presence of microporosity with the possibility that some porosity may be 'smooth'.

(25)   S.W. Rutherford and D.D. Do
        Adsorption dynamics of carbon dioxide on a carbon molecular sieve 5A
        Carbon 2000;**38(9):**1339-1350

Rutherford and Do (25) measured batch adsorption of carbon dioxide on a carbon molecular sieve (Takeda 5A) and report that no molecular sieving action occurs but instead that micropore diffusion is rate limiting the adsorption dynamics. The gas phase diffusion is a combined Knudsen and viscous mechanism.

CONTENTS

(A) The Activation Process

Activated Carbon Compendium
H. Marsh (Editor)

Production of micro- and mesoporous activated carbon from paper mill sludge
I. Effect of zinc chloride activation

Nasrin R. Khalili[a,*], Marta Campbell[a], Giselle Sandi[b], Janusz Golaś[c]

[a]*Department of Chemical and Environmental Engineering, Illinois Institute of Technology, 10 W. 33rd Street, Chicago, IL 60616, USA*
[b]*Argonne National Laboratory Chemistry Division, Argonne, IL 60439, USA*
[c]*The Faculty of Mining Surveying and Environmental Engineering, Director of International School of Technology, University of Mining and Metallurgy, Al. Mickiewicza 30, 30-059 Cracow, Poland*

Received 1 July 1999; accepted 20 December 1999

Abstract

A series of micro- and mesoporous activated carbons were produced from paper mill sludge using a modified carbonization methodology. N_2-adsorption isotherm data and mathematical models such as the D–R equation, α_s-plot, and MP and BJH methods were used to characterize the surface properties of the produced carbons. Results of the surface analysis showed that paper mill sludge can be economically and successfully converted to micro- and mesoporous activated carbons with surface areas higher than 1000 m^2/g. Activated carbons with a prescribed micro- or mesoporous structure were produced by controlling the amount of zinc chloride ($ZnCl_2$) used during chemical activation. Pore evolvement was shown to be most affected by the incremental addition of $ZnCl_2$. Increasing the $ZnCl_2$ to sludge ratio from 0.75 to 2.5 resulted in a 600% increase in the mesopore volume. $ZnCl_2$ to sludge ratios less than 1 and greater than 1.5 resulted in the production of micro- and mesoporous carbons, respectively. At a $ZnCl_2$ to sludge ratio of 3.5, an activated carbon with a predominantly (80%) mesoporous structure was produced. The calculated D–R micropore volumes for activated carbons with the suggested microporous structure were in good agreement with those obtained from the α_s method, while estimated micropore volumes from the α_s method deviated markedly from those obtained from the D–R equation for carbons with a predominantly mesoporous structure. © 2000 Elsevier Science Ltd. All rights reserved.

Keywords: A. Activated carbon; B. Carbonization; Activation; C. Adsorption; D. Adsorption properties

1. Introduction

Although activated carbons have been extensively used as adsorbents, catalysts and catalyst supports in a variety of industrial and environmental applications (i.e. purification processes, recovery of chemical products, and removal of organic and metals), their adsorption capability and catalytic activity are shown to be largely controlled by their surface characteristics [1,2]. For example, carbons used for the adsorption of gases and vapors should include pores with effective radii considerably smaller than 16–20 Å [2], while activated carbons with developed transitional porosity in the range of 20 to 500 Å have been shown to be significant adsorbents for removal of coloring impurities from liquid phase systems [3]. The existing relationship between the surface properties of activated carbon and its effectiveness as an adsorbent or catalyst accentuates the importance of developing methodologies to produce activated carbons with specific surface properties.

Conventionally, activated carbon is produced from wood, peat, coal, and wastes of vegetable origin (e.g. nutshells, fruit stones). Today, one promising approach for the production of cheap and efficient activated carbon is the reuse of waste sludge, such as biosolids produced at municipal or industrial wastewater treatment facilities. The usage of waste sludge is especially important due to its

*Corresponding author. Tel.: +1-312-567-3534; fax: +1-312-567-8874.
E-mail address: khalili@iit.edu (N.R. Khalili).

mass production and resulting occupation of valuable landfill space. Studies conducted more recently by Walhof [4] and Lu et al. [5] showed that activated carbons can be produced from municipal wastewater treatment sludge. Walhof [4] developed a procedure by which biosolids produced at two municipal wastewater treatment facilities in Chicago were successfully converted to activated carbons with surface areas of 600–1000 m^2/g. She also investigated the possibility of producing activated carbons from wastewater sludge generated at paper mill facilities, but failed to optimize the carbon production processes for this type of sludge.

The combination of the chemical and physical activation processes leads to the production of activated carbon with specific surface properties. Chemical activation involves impregnation of the raw material with chemicals such as phosphoric acid [6], potassium hydroxide [7], and zinc chloride [8–10]. Although, phosphoric acid is shown to be the most environmentally sound chemical for the activation processes, most studies have used zinc chloride due to its effective activating capability [11–13]. The common feature of these impregnants is their ability for carbonization and therefore, development of a desired pore structure [11]. The degradation of cellulose material and the aromatization of the carbon skeleton upon $ZnCl_2$ treatment result in the creation of the porous structure. Caturla et al. [11] showed that zinc chloride acts as a dehydrating agent that promotes the decomposition of carbonaceous material during the pyrolysis process, restricts the formation of tar, and increases the carbon yield. Essentially, vacant interstices in the carbon matrix are formed upon extensive post pyrolysis washing of the pores.

The extent of chemical activation can significantly alter the characteristics of the produced carbons. According to Caturla, at high $ZnCl_2$ concentrations, some $ZnCl_2$ remains in the external part of the carbon particles and widens the porosity by a localized decomposition of the organic matter. This process results in the enhancement of the meso- and macropore formation. Gonzalez-Serrano et al. [14] tested different chemical activation ratios (i.e. $ZnCl_2$/ raw material) and showed that activated carbon with a porous structure consisting of narrow microporosity can be obtained by using a $ZnCl_2$ to sample ratio of less than 1. Ahmadopour and Do [15] described pore evolution with respect to the activating agents and showed that micropore formation is predominant when the $ZnCl_2$ to precursor mass ratio is less than one. They also observed that at impregnation ratios in the range of 1–2, creation and widening of the micropores take place simultaneously; when the impregnation ratio is greater than 2, pore widening becomes the dominant mechanism and mesopores are formed.

According to Torregrosa-Macia et al. [12], for a given industrial application, a solid adsorbent with a relatively wide pore size distribution can be obtained solely through a chemical activation process. Physical activation can further enhance the adsorbent's pore structure due to a partial oxidation of the carbonized material by gases such as CO/CO_2 or steam [16,17].

Analysis of the surface physical properties of the carbon includes determination of the total surface area, extent of microporosity, and characterization of the pore size distribution. Nitrogen adsorption isotherms are commonly used for these types of surface analyses [18–21]. The measured relative pressure and adsorbed volume of nitrogen gas are commonly used in various mathematical models (i.e. BET model) to calculate the monolayer coverage of nitrogen adsorbed on the adsorbent surface, while the characteristics of the pore structure and pore distribution can be identified from comparative plots such as 't' and 'α_s'-plots [22–25]. The extent of the microporosity is commonly evaluated by applying low-pressure isotherm data to the Dubinin–Radushkevich (DR) equation [22,26]. The pore size distribution and the corresponding surface areas of the mesoporous adsorbents are traditionally calculated from the hysteresis loop according to the BJH theory, which is based on the Kelvin equation [24,25].

The aforementioned surface analyses were used in this study to evaluate the effect of production processes on the physical characteristics of the produced carbons.

The significant feature of sludge-based activated carbon that makes it a unique and a particularly economical adsorbent is that it can be produced from waste materials such as paper mill sludge. A current problem faced by pulp mills is the generation of an excessive amount of sludge during the paper making process and secondary treatment of wastewater. For environmental and ecological reasons, the innocuous disposal of these sludges has become immensely important. Optimization of the processes involved with the conversion of paper mill sludge to activated carbon provides an innovative, environmentally safe, and economically feasible solution to the problem of sludge management at paper mill facilities. Sludge to carbon conversion processes can significantly reduce the sludge volume produced in the paper mill industry, eliminate the need for further treatment of sludge, reduce the cost of hauling and landfilling the sludge, and reduce transportation costs.

The emphasis of this study was to optimize processes involved with the production of activated carbons with prescribed surface properties (micro- or mesoporous structure) and specific end uses from paper mill sludge, and to examine the applicability of the surface analysis methodologies (i.e. α_s, MP methods) for examining surface properties of the produced carbons according to the method of production. The higher purity (when compared to biosolids), negative cost, high rate of production, and strong carbonaceous structure of paper mill sludge makes it useful as a precursor for carbon production.

2. Experimental

2.1. Production of activated carbons from paper mill sludge

A previous study conducted by Walhof [4] showed that both sewage sludge and sludge produced at paper mill wastewater treatment facilities can be successfully converted to activated carbons with surface areas between 600 and 1000 m^2/g. Since the preliminary results indicated that the chemical composition of paper mill sludge is more uniform and deficient in impurities, the optimization process was conducted to produce activated carbons with specific surface properties from paper mill sludge.

Raw sludge was first dried in an oven at 110°C for 24 h, then crushed mechanically using a paint-mixing machine. Crushing provided smaller particles with increased surface area and also enabled more efficient chemical activation of the raw material. Samples were sieved after mechanical crushing to obtain particle sizes smaller than 600 μm. This particle size range was found to be the most suitable for the chemical activation process that was performed using six different $ZnCl_2$ to dried sludge mass ratios of 0.75, 1, 1.5, 2, 2.5, and 3.5. To ensure a complete reaction between $ZnCl_2$ and sludge particles, slurries of the sludge and zinc chloride were mixed at 85°C for 7 h. After chemical activation, samples were dried at 110°C for 24–36 h. The time required for drying varied depending on the amount of zinc chloride used for the activation process (i.e. higher impregnation ratios required longer drying time). After drying, the sludge was crushed again into a fine powder. As was proposed by Walhof [4], chemically activated samples were exposed to light and humidity (L&H) for about 22 h to enhance the development of the pore structure during pyrolysis.

Pyrolysis of the chemically activated and L&H-treated sludge was carried out in an inert environment (70 ml/min flow of nitrogen) at 800°C for 2 h. Upon completion of the pyrolysis, the sample was removed from the reactor and crushed using a mortar and pestle. Pyrolysis was followed by rinsing using 500 ml of 1.2 M HCl, and 500 ml of distilled water to remove excess zinc chloride and residual inorganic matter. Upon drying, samples were transferred to 20-ml vials for storage prior to conducting the physical activation process.

During the physical activation, chemically activated, dried and crushed sludge was heated for 2 h at 800°C in a mixture of carbon monoxide and carbon dioxide (75% CO and 25% CO_2). Upon completion of the physical activation, the produced carbons were characterized according to their surface properties.

2.2. Characterization of the produced activated carbons

Characterization of the produced carbons included de-termination of the surface area, extent of the micro- and mesoporosity, and pore size distribution. To estimate the surface area (a primary measure for the assessment of the produced carbons), the Brunauer, Emmet and Teller (BET) model was applied to the N_2-adsorption data at a relative pressure of 0.05 to 0.2, where monolayer coverage of nitrogen molecules is assumed to be complete. To conduct calculations using the BET model, it was assumed that the surface area, A, is related to the monolayer capacity by the simple equation $A = n_m a_m L$, where a_m is the average area occupied by a molecule of the adsorbate in the completed monolayer and L is the Avogadro constant, taking 0.162 nm^2 as the cross-sectional area of a nitrogen molecule [23]. The surface areas of the produced activated carbons were also evaluated from the α_s-plots following the procedure proposed by Carrott et al. [18]. The extent of the micro- or mesoporosity was determined from N_2-adsorption data using the DR equation, and a comparative plot such as the α_s plot [22–25]. The α_s-plots were constructed following a procedure described in Webb and Orr [19] using the standard data suggested in Carrott et al. [18] as a reference carbon. The α_s method involved plotting the amount of nitrogen gas adsorbed at each relative pressure against the reduced adsorption, α_s, determined for a non-porous reference material having a surface structure similar to the carbons under investigation.

Since porosity causes deviation of the α_s-plot from linearity, it was possible to evaluate the external surface area of the produced carbons from the slope of the plot, and the micropore capacity by back-extrapolation of the plot to $\alpha_s = 0$ (the intercept of the linear segment of the α_s-plots with y-axis representing the adsorbed volume of N_2 gas allowed for estimation of the micropore volumes).

The D–R micropore volume was determined from the Dubinin–Radushkevich equation:

$$\log_{10}W = \log_{10}W_0 - B(T/\beta)^2 \log_{10}^2\left(\frac{p_0}{p}\right). \tag{1}$$

In this equation, W represents the volume adsorbed, W_0 is the micropore volume, B is the structural constant, β is the similarity coefficient, T is the temperature, and p_0/p is the inverse of the partial pressure [23].

A plot of $\log_{10}W$ versus $\log^2 (p_0/p)$ allowed determination of the micropore volume, W_0. While the measured adsorbed volume (W) at each relative pressure (p/p_0) was used in the calculations, an intercept of a straight line applied to the low-pressure isotherm data ($p/p_0 < 0.04$) provided an estimate of the micropore volume, W_0. The mesopore volume was calculated by subtracting the micropore volume (W_0) (obtained from the DR equation) from the total pore volume (i.e. the amount adsorbed at $p/p_0 = 0.98$).

The micropore analysis (MP method, Mikhail et al. [19]) which is one of the first methods developed for the analysis of microporous solids, was used to determine

6

micropore size distributions [20,21]. This method uses a *t*-curve that is constructed by plotting the adsorbed volume of the nitrogen gas versus thickness, *t*, which represents the average thickness of the adsorbed layer. The *t* values were calculated as a function of p/p_0 for the adsorbed nitrogen gas at 78 K using the Harkins and Jura equation:

$$t = \left[\frac{13.99}{0.034 - \log(p/p_0)} \right]^{1/2}. \qquad (2)$$

The pore size distributions of the produced carbons were determined by manually drawing a series of tangent lines in a contiguous range of *t*-values. Knowing the surface area of the pores and the corresponding values of *t*, the size and the volume of the micropores were estimated.

The mesopore volume distribution was calculated according to the BJH theory [23,24]. The BJH method is based on the Kelvin equation, which relates the relative pressure of nitrogen in equilibrium with the porous solid to the size of the pores where capillary condensation takes place. The pore size radii covered by the BJH calculations ranged from 17 to 3000 Å.

3. Results and discussion

3.1. N₂-adsorption isotherms

For ease of comparison, the produced carbons are identified as Zn_xC, where *x* represents the mass ratio of the zinc chloride to the dried sludge, ranging from 0.75 to 3.5. The Brunauer, Deming, Deming, and Teller (BDDT) theory, the basis of the modern IUPAC classification, was used in this study to characterize the N₂-adsorption isotherms. The analysis of the N₂-adsorption isotherms shown in Fig. 1 provided an approximate assessment of the pore size distributions. As shown, Fig. 1a and b had very similar shapes. The nitrogen uptake was significant only in the low-pressure region (region 1) where $p/p_0 < 0.2$. At higher relative pressure (region 2: $p/p_0 > 0.2$), no further adsorption was observed and the adsorption curve reached equilibrium at $p/p_0 \approx 0.2$. According to the IUPAC classification, these curves resemble the type I isotherm which represents microporous solids having a relatively small external surface area (e.g. activated carbons, molecular sieve zeolites). According to Warhust et al. [26], in these materials, the limiting uptake is controlled by the accessible micropore volume rather than by the internal surface area.

Fig. 1c and d showed that an increase in the amount of ZnCl₂ resulted in the production of activated carbons with N₂ adsorption isotherms containing three characteristic regions: (1) when compared to the plots (a) and (b) the initial nitrogen uptake elevated at $p/p_0 < 0.2$, (2) at $0.2 < p/p_0 < 0.8$, the knee of the isotherms became more open, the plateau started at higher relative pressures and the

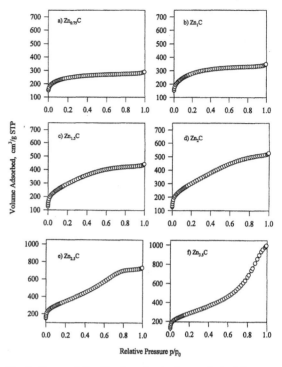

Fig. 1. Representative N₂-adsorption isotherms obtained for carbons treated with different zinc chloride to sludge mass ratios: (a) 0.75, (b) 1, (c) 1.5, (d) 2, (e) 2.5, (f) 3.5.

slope generated in the range of $p/p_0 > 0.2$ increased with increased amounts of zinc chloride, and (3) both plots reached equilibrium at $p/p_0 \approx 0.9$. The observed characteristics indicated a significant development of the mesoporosity and production of carbons with isotherms different from those identified by the IUPAC classification (plots a and b).

Fig. 1e and f also indicated that increasing the ZnCl₂ to sludge ratios to 3.5 markedly affected and continuously changed the shape of the isotherms in the mesopore region. As shown, $Zn_{2.5}C$ and $Zn_{3.5}C$ isotherms contain a second knee in the meso- and macropore region indicating the occurrence of the important process of pore widening. In fact increasing *x* resulted in the production of carbons with adsorption curves of type IV. As reported by Aranovich and Donohue [25], this type of isotherm is traditionally attributed to the mesoporous solids. A characteristic feature of this group of isotherms is the hysteresis loop, which is associated with capillary condensation occurring in mesopores and with the limiting uptake occurring at high relative pressure [26].

Fig. 2 shows the adsorption–desorption isotherms obtained for $Zn_{0.75}C$, $Zn_{1.5}C$, and $Zn_{3.5}C$. As shown, when the zinc to carbon ratio was increased by a factor of about 4, isotherms changed from a type I with a characteristic H₄

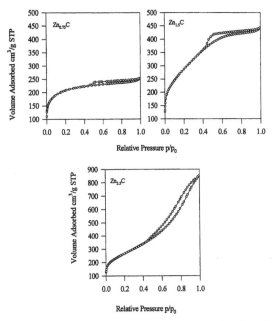

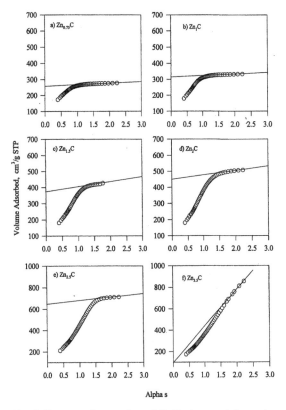

Fig. 2. The N_2-adsorption–desorption isotherms for $Zn_{0.75}C$, $Zn_{1.5}C$, and $Zn_{3.5}C$.

hysteresis loop, to a type IV with an H_3 hysteresis loop. Since the H_3 and H_4 hysteresis loops are often associated with the existence of slit-shaped pores [26], the different nature of the loops suggested the presence of pores with different structures (shapes).

Review of the N_2-adsorption isotherm data suggested that an increase in the amount of $ZnCl_2$ used for chemical activation affects both the extent of the mesoporosity and the characteristics of the produced carbons.

3.2. Pore structure analysis

The extent of the microporosity and the evolution of the pore structure via chemical activation were evaluated by identifying the micro- and mesopore volumes and pore volume distributions. Two empirical methods, α_s-plot and DR equation, were used to estimate the micropore volumes. The α_s-plots were constructed by following the procedure described in Webb and Orr [19] using the standard data suggested in Carrott [18] as a reference carbon (the reference material was a nonporous ungraphitized carbon). Fig. 3 presents α_s-plots constructed for Zn_xC carbons. The y-axis represents the amount of nitrogen adsorbed, and the x-axis represents the α_s values (the ratio of the amount adsorbed by non-porous standard carbon at each relative pressure and the amount adsorbed at $p/p_0 = 0.4$). Since a linear α_s-plot represents a non-porous carbon, the observed deviation from linearity suggested the presence of micro- and mesoporous structures. The volumes of the micropores for all carbons were identified by the

Fig. 3. Representative α_s-plots of $ZnCl_2$ impregnated samples with different zinc chloride to sludge mass ratios: (a) 0.75, (b) 1, (c) 1.5, (d) 2, (e) 2.5, and (f) 3.5.

extrapolation of the high-pressure branch to the adsorption axis.

The α_s-plots of $Zn_{0.75}C$ and Zn_1C deviated from linearity in the low α_s region ($p/p_0 < 0.4$). At high-pressure region where p/p_0 exceeds 0.4, α_s-plots approached linearity and became parallel to the α_s axis. These characteristics suggested that $Zn_{0.75}C$ and Zn_1C are carbons with predominantly microporous structures.

As is indicated in Fig. 3c–f, α_s-plots of carbons activated with an increased amount of zinc chloride had elevated slopes at higher p/p_0 regions indicating formation of the mesoporous structure. For $Zn_{3.5}C$ carbon, the slope of the curve increased notably in $p/p_0 \geq 0.4$, specifying a significant uptake of N_2 in the region that represents mesoporous structure. Due to the dominant contribution of the mesopores, the $Zn_{3.5}C$ micropore volume was significantly lower than that estimated for the other carbons (Table 1). These results suggested that addition of zinc chloride resulted in the transformation of the micro- to mesopores through the process of pore widening.

The calculated micropore volumes from α_s-plots (Table 1) showed that an increase in the amount of $ZnCl_2$ not only enhances microporosity, it also promotes formation of

Table 1
Micropore volume of the produced carbons

Impregnation ratio	$V_{(0)DR}^a$ (cm^3/g)	$V_{(0)\alpha_s}^a$ (cm^3/g)
$Zn_{0.75}C$	0.361	0.400
Zn_1C	0.377	0.485
$Zn_{1.5}C$	0.393	0.595
Zn_2C	0.398	0.657
$Zn_{2.5}C$	0.486	0.990
$Zn_{3.5}C$	0.396	0.155

[a] Gas volumes were multiplied by 0.001547 to obtain liquid volume.

the mesoporous structure as was indicated by the greater slope of the linear branch of α_s plots. Increasing the $ZnCl_2$/dried sludge mass ratio from 0.75 to 2.5 increased the micropore volumes from 0.40 cm^3/g to 0.99 cm^3/g.

To better understand the impact of $ZnCl_2$ activation and to confirm the results of the α_s-plots, the common method of micropore analysis proposed by Dubinin and Radushkevich [23] was applied to the low relative pressure data ($p/p_0 < 0.04$). The analysis of the D–R plots (Fig. 4)

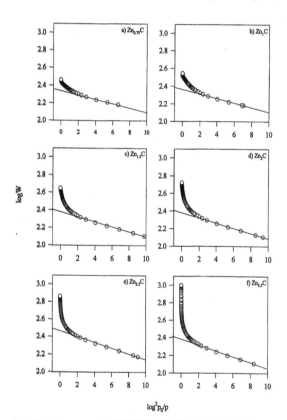

Fig. 4. Representative D–R plots obtained for carbons treated with different zinc chloride to sludge mass ratios: (a) 0.75, (b) 1, (c) 1.5, (d) 2, (e) 2.5, (f) 3.5.

showed that the shape of the curves is clearly dependent on the degree of $ZnCl_2$ activation. The deviation from linearity and upward turn at $p/p_0 \geq 0.04$ indicated formation of the multilayer and capillary condensation in the mesopores. The extent of the deviation, however, was influenced by the amount of $ZnCl_2$ used for impregnation. As is shown, formation of the mesopores and their contribution to the adsorption by capillary condensation were positively affected by the amount of $ZnCl_2$.

The contrast between the estimated micropore volumes from the D–R equation and α_s plots was more prominent for carbons produced using a higher amount of $ZnCl_2$. At high impregnation ratios, the α_s micropore volume reached a value of 0.99 cm^3/g versus 0.486 cm^3/g calculated from the D–R equation. The observed differences between the D–R and α_s micropore volumes were related to the characteristics and limitations of these methods: (1) since α_s-plots of highly activated carbons ($ZnCl_2$/sludge ratios ≥ 1.5) barely reached a plateau, it was assumed that these carbons contained both primary and secondary micropores and that pore filling had occurred simultaneously in both groups of pores [13]. The range of the α_s used for determination of the micropore volumes, therefore, became smaller and shifted more to the right (i.e. to the macropore region) and resulted in the estimation of micropore volumes that included the volume of super micropores and very small mesopores (primary and secondary micropores), (2) due to the mesoporous filling, D–R plots exhibited strong upward deviation. Although the shape of the plots depended on the amount of $ZnCl_2$ used for impregnation, the changes did not affect the range of the data used for determination of the micropore volume ($\log^2 p/p_0 > 2$). Therefore, the D–R equation correctly estimated the volume of the narrow micropores, which were filled below a relative pressure of 0.04.

The estimated DR and α_s micropore volumes were comparable for carbons activated with a low amount of $ZnCl_2$ ($x \leq 1.5$). However, since the estimated α_s micropore volumes for carbons activated with a higher amount of $ZnCl_2$ ($x \geq 1.5$) included the total micropore volume and volume of the small mesopores, their values were higher than the D–R micropore volumes that were a measure of the volume of the narrow micropores only. These results suggested that although the aforementioned methods of the surface analysis are significant and meaningful, they have some limitations for the analysis of the carbons with enhanced mesoporous structure.

3.3. Analysis of the pore size distribution

The micro- and mesopore size distributions were evaluated using micropore analysis (MP) and Barret, Joyner, and Halenda (BJH) methods, respectively [20,23]. The BJH method that is based on the Kelvin equation relates the relative pressure of the nitrogen in equilibrium with the porous solid to the size of the pores where capillary condensation takes place (Eq. (3)). The highly idealized BJH model assumes that porous solids contain cylindrical capillaries that are closed at one end and have uniform cross sections:

$$RT \ln \frac{p}{p_0} = -2\sigma \frac{V_M}{r_K}. \tag{3}$$

In this equation, p is the pressure at which a cylindrical pore of Kelvin radius r_K is filled, σ is the surface tension, V_M is the molar volume, p_0 is the saturation pressure of the liquid adsorbate at temperature T, and R is the gas constant. By substituting for the values of different constants using the data presented in Table 2, it was possible to calculate the core radius (Kelvin radius) of the liquid in the capillaries:

$$r_K (\text{Å}) = 4.14 \log \frac{p}{p_0}. \tag{4}$$

Since adsorption occurs naturally on the walls of the mesopores, r_K values do not correspond to the true pore radius r_p. Therefore, to estimate r_p accurately, r_K was corrected using the calculated thickness of the adsorbed multilayer, t, as follows:

$$r_p (\text{Å}) = r_K + t\left(\frac{p}{p_0}\right). \tag{5}$$

The t as a function of p/p_0 was evaluated for nitrogen at 78 K from the Harkins and Jura equation [20,21]:

$$t = \left[\frac{13.99}{0.034 - \log(p/p_0)}\right]^{1/2}. \tag{6}$$

The N_2-adsorption data and calculated values of r_p were used to construct the BJH graphs and to determine the pore size distributions for the Zn_xC carbons.

Analysis of the BJH graphs presented in Fig. 5 suggested a systematic dependence of the mesoporosity development on the degree of chemical activation. The peaks representing the predominant pore sizes were shifted to the right and became wider according to the amount of $ZnCl_2$ used for activation. The suggested pore diameter for the microporous carbons; $Zn_{0.75}C$ and Zn_1C (Fig. 5a and b) was smaller than 20 Å, while activated carbons produced with $ZnCl_2$/sludge ratio up to 2.5 contained a fairly narrow mesoporous structure (up to 100 Å). The largest impregnation ratio used in this study ($Zn_{3.5}C$) produced carbon with a wide pore size distribution (Fig. 5f) and the

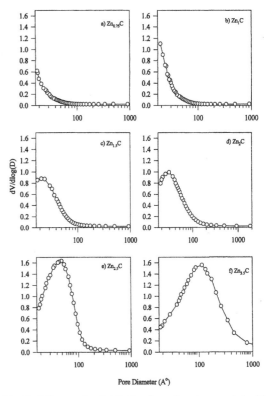

Fig. 5. BJH pore size distribution for carbons impregnated with different zinc chloride to sludge mass ratios: (a) 0.75, (b) 1, (c) 1.5, (d) 2, (e) 2.5, and (f) 3.5.

most frequently occurring pore diameter of about 100 Å. In addition, as is shown in Table 3, the percentage of the total pore volume occupied by the micropores decreased as the pore sizes were increased.

Since the Kelvin equation did not provide an apparent pore size distribution for the microporous carbons ($Zn_{0.75}C$ and Zn_1C) [23], the MP method was used to establish the formation and evolution of the pores in microporous solids [20].

Micropore analysis by the MP method was based on the t-curve in which the adsorbed volume of the adsorbate was plotted versus the thickness of adsorbed layer, t, starting from $t = 3.5$ Å. The thickness, t, was calculated using the Harkins and Jura equation (Eq. (6)). The pore size

Table 2
Constants used in calculations of pore size distribution

Characteristic	Constant
Surface tension of nitrogen	8.855 mN/m
Molar volume	34.6 cm^3/mole
Normal boiling point of N_2	77.3 K
Ideal gas constant	8.31×10^7 erg/mol K
Conversion factor from STP to liquid volume	0.001547

Table 3
Contribution of micropore volume in the total pore volume

Impregnation ratio	$V_{(0)DR}$ (cm^3/g)	V_{total} (cm^3/g)	% of micropore volume	Estimated pore size (Å)
Zn$_{0.75}$C	0.361	0.446	81	<20
Zn$_1$C	0.377	0.534	71	<20
Zn$_{1.5}$C	0.393	0.679	58	25
Zn$_2$C	0.398	0.809	59	30
Zn$_{2.5}$C	0.486	1.128	43	30
Zn$_{3.5}$C	0.396	1.529	26	100

distributions for the microporous carbons were extracted by manually drawing a series of tangent lines in a contiguous range of t-values. Using the surface area of the filled pores and the corresponding values of t, the size and volume of the micropores were calculated using Eq. (7):

$$V_p = \left[\frac{(S_n - S_{n+1}) \cdot (t_n + t_{n-1})}{2} \right] \cdot 15.47. \qquad (7)$$

S_n and S_{n+1} represented the surface areas obtained from the slopes of tangents n. $n+1$, t_{n+1} and t_n represented the thickness at points $n+1$ and n, and 15.47 was the constant converting gas volume at STP to liquid volume and Å units to cm. The micropore size distributions for the two microporous carbons were determined using estimated values of V_p and the mean hydraulic radius, $t_n + t_{n+1}/2$ (Fig. 6).

As Fig. 6 shows, increasing the amount of ZnCl$_2$ resulted in the enhancement of the micropore volume (i.e. formation of the new pores) and widening of the existing micropores. The most frequently occurring pores were those with a radius of 3.5 Å and 4 Å for Zn$_{0.75}$C and Zn$_1$C, respectively.

Fig. 7 summarizes the results of the pore size analysis and illustrates the relationship between the ZnCl$_2$ impregnation ratio and the estimated mesopore volume and average pore width. As is shown, the micropore formation and the enlargement of the existing micropores significantly affect the estimated values of the pore volume and pore width. The most significant changes in mesopore volume and average pore diameter were observed for Zn$_{3.5}$C (Table 4).

These results contradicted the Gonzalez-Serano et al. [14] study that suggested the impregnation ratio does not have a significant impact on the macropore size distribution. In fact, it was observed that an excess amount of ZnCl$_2$ could significantly enhance the process of pore widening and formation of the mesoporous structure (by promoting externally located devolatilization processes). The analysis of the micro- and mesopore size distribution using BJH and MP methods also suggested that the constitution of the ZnCl$_2$ activation has a significant impact on the evolution of the micro- and mesoporous structure.

3.4. Surface area analysis

The surface area of the Zn$_x$C activated carbons was determined from the BET equation and α_s plots. The BET equation was applied to the N$_2$-adsorption data at relative pressures ranging from 0.05 to 0.2, where completion of the monolayer is assumed. The α_s specific surface area was calculated from Eq. (7) using the adsorbed volume of the nitrogen gas, V, and α_s value calculated at $p/p_0 = 0.4$ [18,19]:

$$A_{\alpha_s} = 2.86 \frac{V}{\alpha_s}. \qquad (8)$$

The estimated BET and α_s surface areas presented in Table

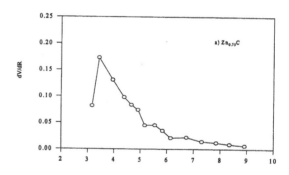

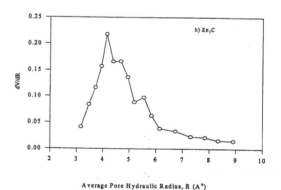

Average Pore Hydraulic Radius, R (Å)

Fig. 6. Micropore size distributions for Zn$_{0.75}$C and Zn$_1$C.

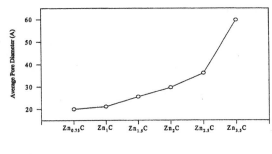

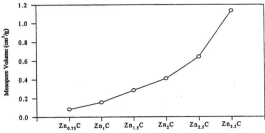

Fig. 7. Effect of chemical ratio on mesopore volume and pore width of $ZnCl_2$ treated carbons.

Table 4
Effect of chemical activation on the development of the mesopore volume and pore width

Sample	Mesopore volume (cm^3/g)	Average diameter[a] (Å)
$Zn_{0.75}C$	0.084	19.93
Zn_1C	0.157	21.06
$Zn_{1.5}C$	0.286	25.46
Zn_2C	0.411	29.62
$Zn_{2.5}C$	0.641	36.13
$Zn_{3.5}C$	1.133	59.70

[a] Average diameter calculated as $4V/A$ by BET.

5 and Fig. 8 showed that the surface area of the produced carbons increases with increasing amounts of $ZnCl_2$ used for the activation up to $x = 2.5$. The maximum BET surface area of 1249 m^2/g was obtained for $Zn_{2.5}C$, but this value decreased due to the transformation of the micro- to mesopores at $x = 3.5$.

For carbons activated with $ZnCl_2$/sludge mass ratio <1, the BET surface areas deviated markedly from those obtained from the α_s method (Fig. 8). These results agreed well with the Gregg and Sing [23] results that showed that

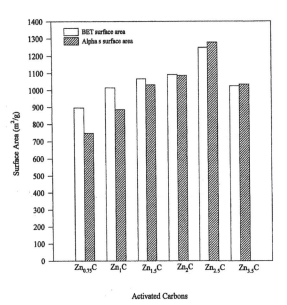

Fig. 8. Comparison of total surface area obtained from BET and α_s method for carbons impregnated with different amount of zinc chloride.

BET surface area is significantly higher than the α_s surface area for solids containing mainly micropores since the building up of the monolayer is affected by the presence of the neighboring surfaces. Their results also suggested that the mechanism of the adsorption in very fine pores is controlled mainly by the pore filling processes rather than the surface coverage. The calculated BET and α_s surface areas converged for carbons activated with a higher amount of zinc chloride and developed micro- and mesoporous structure.

Fig. 9 shows that calculated surface areas and D–R micropore volumes follow a very similar pattern. This similarity suggested that micropores make the largest contribution to the calculated specific surface areas and confirmed the results of the micropore volume analysis. This indicated that the D–R model provides a true estimation of the extent of microporosity.

4. Conclusions

The results of this study showed that paper mill sludge can be successfully converted into activated carbon with

Table 5
BET and α_s surface areas of carbons treated with different $ZnCl_2$ ratio

Sample	$C_1Zn_{0.75}$	C_1Zn_1	$C_1Zn_{1.5}$	C_1Zn_2	$C_1Zn_{2.5}$	$C_1Zn_{3.5}$
BET surface area (m^2/g)	895	1015	1067	1095	1249	1025
α_s surface area (m^2/g)	744	885	1031	1086	1278	1034

12

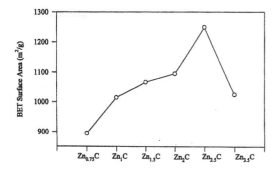

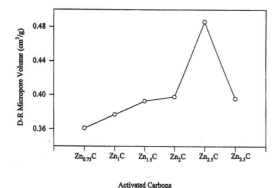

Activated Carbons

Fig. 9. Effect of $ZnCl_2$ impregnation ratio on: (a) BET surface area and (b) D–R micropore volume.

specific surface properties. It was found that the amount of $ZnCl_2$ used for chemical activation controls the characteristics of the carbons including surface area, micro- and mesoporosity, and pore size distributions. Increasing the amount of $ZnCl_2$ used for chemical activation ($ZnCl_2$/sludge mass ratio from 0.75 to 2.5) resulted in a 600% increase in the mesopore volume. Detailed surface analyses using different mathematical models suggested that: (1) the $ZnCl_2$ to sludge ratios required to obtain micro- and mesoporous carbons are <1 and ≥ 1.5, respectively; (2) a single theoretical approach cannot provide complete information about the surface structure of the produced activated carbons; (3) for highly activated carbons with an extended pore size distribution, the α_s method estimates total micropore volume (volumes used for both primary and secondary filling processes); and, (4) the calculated D–R micropore volumes for activated carbons with suggested microporous structure were in good agreement with those obtained with the α_s method, while estimated micropore volumes from the α_s method deviated markedly from those obtained with the D–R equation for carbons with predominant mesoporous structures.

Acknowledgements

Work at Argonne National Laboratory was performed under the auspices of the Office of Basic Energy Sciences; Division of Chemical Sciences, US Department of Energy, under contract number W-31-109-ENG-38. We would also like to acknowledge the Kosciuszko Foundation for the partial support they provided for the graduate student working on this project.

References

[1] Bansal RC, Donnet J, Stoeckli F. Active carbon, Marcel Dekker Inc., 1988.
[2] Boppart S, Ingle L, Potwora RJ, Rester DO. Understanding activated carbons. Chem Process 1996;79–84.
[3] Walker PL. Chemistry and physics of carbon, Ann Arbor, Michigan: Book Demand, 1996.
[4] Walhof LK. Procedure to produce activated carbon from biosolids, Chicago, IL: Illinois Institute of Technology, 1998, MS thesis.
[5] Lu GQ. Preparation and evaluation of adsorbents from waste carbonaceous materials for SO_x and NO_x removal. Environ Prog 1995;15(1):12–8.
[6] Molina-Sabio M, Rodriguez-Reinoso F, Caturla F, Selles MJ. Porosity in granular carbons activated with phosphoric acid. Carbon 1995;33(8):1105–13.
[7] Otowa T, Nojima Y, Miyazaki T. Development of KOH activated high surface area carbon and its application to drinking water purification. Carbon 1997;35(9):1315–9.
[8] Ibarra JV, Moliner R, Palacios JM. Catalytic effects of zinc chloride in the pyrolysis of Spanish high sulfur coals. Fuel 1991;70:727.
[9] Hourieh MA, Alaya MN, Youssef AM. Carbon dioxide adsorption and decolorizing power of activated carbons prepared from pistachio shells. Adsorpt Sci Technol 1997;15(6).
[10] El-Nabarawy TH, Mostafa MR, Youssef AM. Activated carbons tailored to remove different pollutants from gas stream and from solution. Adsorpt Sci Technol 1997;15(1):61–8.
[11] Caturla F, Molina-Sabio M, Rodriguez-Reinoso F. Preparation of activated carbon by chemical activation with $ZnCl_2$. Carbon 1991;29(7):999–1007.
[12] Torregrosa-Macia R, Martin-Martinez JM, Mittelmeijer-Hazeleger MC. Porous texture of activated carbons modified with carbohydrates. Carbon 1997;35(4):447–53.
[13] Balci S, Dogu T, Yücel H. Characterization of activated carbon produced from almond shell and hazelnut shell. J Chem Tech Biotech 1994;60:419–26.
[14] Gonzalez-Serano E, Cordero T, Rodriguez-Mirasol J, Rodriguez JJ. Development of porosity upon chemical activation of Kraft lignin with $ZnCl_2$. Ind Eng Chem Resources 1997;36:4832–8.
[15] Ahmadpour A, Do DD. The preparation of activated carbon from macadamia nutshell by chemical activation. Carbon 1997;12:1723–32.
[16] Molina-Sabio M, Gonzalez MT, Rodriguez-Reinoso F,

Sepulveda-Escribano A. Effect of steam and carbon dioxide activation in the micropore size distribution of activated carbon. Carbon 1996;34(4):505–9.

[17] Rodriguez-Reinoso F, Molina-Sabio M, Gonzalez MT. The use of steam and CO_2 as activating agents in the preparation of activated carbons. Carbon 1995;33(1):15–23.

[18] Carrott PJM, Roberts RA, Sing KSW. Standard nitrogen adsorption data for nonporous carbons. Carbon 1987;25(6):769–70.

[19] Mikhail RS, Brunauer S, Bodor EE. Investigations of a complete pore structure analysis. I. Analysis of micropores. J Colloid Interface Sci 1968;26:45–53.

[20] Harkins WD, Jura G. An adsorption method for the determination of the area of a solid without the assumption of a molecular area, and the area occupied by nitrogen molecules on the surfaces of solids. J Chem Phys 1943;11:431.

[21] Aranovich G, Donohue M. Analysis of adsorption isotherms: lattice theory predictions, classification of isotherms for gas–solid equilibria, and similarities in gas and liquid adsorption behavior. J Colloid Interface Sci 1998;200:273–90.

[22] Lippens BC, de Boer JH. Studies of pore systems in catalysts. V. The t method. J Catal 1965;4:319–23.

[23] Gregg SJ, Sing KSW. Adsorption, surface area and porosity, Academic Press, 1982.

[24] Barret EP, Joyner PB, Halenda P. The determination of pore volume and area distribution in porous substances. I. Computations from nitrogen isotherms. J Am Chem Soc 1951;73:373–80.

[25] Sing KSW, Everett DH, Haul RAW, Moscou L, Pierotti RA, Rouquerol J, Siemieniewska T. Reporting physisorption data for gas/solid systems with special reference to the determination of surface area and porosity. Pure Appl Chem 1985;57(4):603–19.

[26] Warhurst AM, Fowler GD, McConnachie GL, Pollard SJT. Pore structure and adsorption characteristics of steam pyrolysis carbons from moringa oleifera. Carbon 1997;35(8):1039–45.

PREPARATION OF ACTIVATED CARBONS FROM BITUMINOUS COALS WITH CO_2 ACTIVATION—INFLUENCE OF COAL OXIDATION

H. Teng,* J.-A. Ho and Y.-F. Hsu

Department of Chemical Engineering, Chung Yuan Christian University, Chung-Li 32023, Taiwan

(*Received 22 May 1996; accepted in revised form 6 September 1996*)

Abstract—In this study activated carbons were prepared from three Australian bituminous coals. The influence of coal oxidation on the carbonization and activation processes and the structure of the resulting activated carbons were explored. The oxidation was conducted at 200°C in O_2 for 6 hours. During the oxidation the amount of oxygen uptake decreased with the O/C atomic ratio of the raw coal. The carbonization of the oxidized coal exhibited a broader volatile evolution with respect to temperature, and the resulting char from the oxidized coal had a larger surface area. The oxidation of the coal precursors resulted in increases in the specific surface area and pore volume of the resulting activated carbons. For the carbons from the oxidized coals, the activation rate in CO_2 was higher and the structure more accessible for CO_2 penetration. © 1997 Elsevier Science Ltd. All rights reserved

Key Words—A. Activated carbon, B. activation, oxidation, D. surface properties.

1. INTRODUCTION

Coal is a commonly used material for producing activated carbon [1]. Among coal-based activated carbons, bituminous products are in greater demand since they have greater density, hardness and abrasion resistance, and are more durable [2]. Consequently, an extensive study on the preparation and the properties of activated carbons from bituminous coals is needed. Physical activation, consisting of carbonization of the raw material followed by gasification of the resulting char [3–5], is the preparation process investigated in this study. As for the properties, pore structure, in terms of surface area and pore volume, is an important characteristic of activated carbons. In general, activated carbons with both a high surface area and porosity, allowing large amounts of adsorption, are desirable.

It is well known [1,6] that upon carbonization a certain portion of the carbonaceous material in bituminous coals will traverse through a plastic phase. The caking portion will present a much more ordered, less porous and less reactive structure than its non-caking counterpart. The occurrence of caking may affect the pore development of coals upon carbonization by inducing the collapse of pores. However, introduction of oxygen in the coal structure can convert a caking coal from thermoplastic to thermosetting, because oxygen functional groups, such as carbonyl, carboxyl and hydroxyl [7,8], appear to play a major role in promoting the cross-linking reactions between the aromatic and the hydro-aromatic building blocks of the coal.

In the preparation of activated carbon, structure changes during the activation step strongly depend on the initial textural characteristics of the char [6]. It has been reported that preliminary oxidation of bituminous coals can influence not only the properties of the chars from carbonization, but also that of the activated carbons by subsequent activation [1,4,6,9]. This technique has also been employed in producing activated carbon from waste tires to increase the surface area [10]. However, scarcely any fundamental studies on the application of preliminary oxidation to produce activated carbons from bituminous coals with CO_2 activation have been reported previously.

In the present study, Australian bituminous coals have been chosen as precursors of activated carbon. Within the above scope this study is devoted to the kinetics of carbonization and activation and the changes in surface area and porosity in activated carbons, employing different preparation procedures. This report describes the influences of preliminary oxidation of the raw coals on the behaviors during carbonization and activation and on the development of the surface structure of the resulting activated carbons.

2. EXPERIMENTAL

2.1 Coal characteristics

Three Australian bituminous coals, Black Water (BW), Gregory (GG) and Mt. Thorley (MT), were used as the starting material. The proximate and ultimate analyses of the raw coals are shown in Table 1. The contents of C, H, N, O and S elements in the ultimate analysis were determined by an elemental analyzer (Heraeus, CHN–O–RAPID). The data shown in Table 1 have been normalized to constitute a sum of 100%. The as-received coals were

---

*Corresponding author.

Reprinted from *Carbon* **35** (2), 275-283 (1997)

Table 1. Coal analysis (wt%)

	Black Water (BW)	Gregory (GG)	Mt. Thorley (MT)
Ultimate			
Carbon	83.2	81.7	78.9
Nitrogen	3.7	4.0	3.3
Hydrogen	5.1	5.4	5.4
Oxygen	7.4	8.1	11.8
Sulfur	0.6	0.8	0.6
O/C atomic ratio	0.067	0.074	0.11
Proximate			
Moisture	2.7	2.4	3.1
Volatile matter	26.7	31.6	33.1
Fixed carbon	61.8	57.9	55.0
Ash	8.9	8.2	8.8

crushed and sieved to a particle size of 210–300 μm before being treated.

The analytical results shown in Table 1 reveal that the major difference in the bituminous coals in the ultimate analysis is the oxygen content, which has a great impact on the properties of the coals; e.g. in the proximate analysis the volatile matter increases and the fixed carbon decreases with the oxygen content of the coals. The O/C atomic ratio, representing a normalized property of each coal shown in Table 1, is employed instead of the oxygen content in the following discussion of the oxygen impact throughout this report. Table 1 includes the value of the O/C atomic ratio, showing that the trend of the ratio with the coal type is the same as that for the oxygen content.

2.2 Sample preparation

2.2.1 Carbonization. Pyrolyses of as-received coals were performed in a thermogravimetric analyzer (TGA, DuPont TGA 51) under a stream of CO_2. The samples were heated at 30°C/min from room temperature to maximum heat treatment temperatures in the range of 800–950°C.

2.2.2 Coal oxidation. The oxidation of the coals was implemented in the TGA under a stream of O_2 at 200°C for a period of 6 hours. Following the oxidation process, the oxidized samples were carbonized, also in CO_2, at 30°C/min from 200°C to the maximum heat treatment temperature.

2.2.3 Activation. Following the carbonization process the char samples were gasified, also in the TGA, in a stream of CO_2 at the maximum heat treatment temperature. Activated carbons with various degrees of burn-off were prepared.

2.3 Sample characterization

Specific surface areas and pore volumes of the samples were determined by gas adsorption. An automated adsorption apparatus (Micromeritics, ASAP 2000) was employed for these measurements. Adsorption of N_2, as a probe species, was performed at 77 K. Before any such analysis the sample was

degassed at 300°C in a vacuum at about 10^{-3} Torr. Surface areas and micropore volumes of the samples were determined from the application of the BET [11,12] and Dubinin–Radushkevich (D–R) equations [12–14], respectively, to the adsorption isotherms at relative pressures between 0.06 and 0.2.

In type I isotherms [11,12], the amount of N_2 adsorbed at pressures near unity corresponds to the total amount adsorbed at both micropores (filled at low relative pressures) and mesopores (filled by capillary condensation at pressures above 0.2); and, consequently the subtraction of the micropore volume (from the D–R equation) from the total amount (determined at $p/p_0 = 0.98$ in this case) will provide the volume of the mesopore [15].

A Hitachi S-4000 scanning electron microscope was used to study the structural features of the carbon surface.

3. RESULTS AND DISCUSSION

3.1 Oxygen treatment

Both oxidized and unoxidized coals were used to produce activated carbons in this study. As stated in the experimental section, the oxidation of the coals was implemented in O_2 at 200°C, and the oxygen uptake process was monitored by the TGA system. Figure 1 shows an example of oxygen uptake of the BW coal during oxidation. One can observe from the figure that the sample mass shows a rapid increase at the beginning of oxidation at 200°C and levels off after 6 hours of oxidation. The other coals show a similar behavior in the oxygen treatment. Obviously, after 6 hours of oxidation the coals were saturated with oxygen at the treatment temperature.

The amounts of oxygen uptake for different coals are shown in Table 2. The data show that the amount

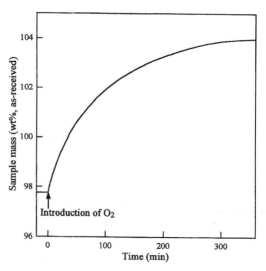

Fig. 1. Mass uptake during the course of oxidation of the BW coal. The process was implemented at 200°C and a oxygen pressure of 1 atm.

Table 2. Amounts of mass uptake from the coals during 6 hours of oxidation at 200°C and the O/C atomic ratios of the oxidized coals. The amounts are represented as the weight fractions of the respective coals on a dry basis

	BW	GG	MT
Amount of oxygen uptake (wt% of dry coal)	4.4	4.0	3.4
O/C atomic ratio of oxidized coal	0.13	0.14	0.17

of oxygen uptake decreases with the O/C ratio of the raw coal; i.e. the BW coal has the highest amount of oxygen uptake, and the MT has the lowest. The results reveal that the oxygen atoms originally retained in the raw coals may occupy the sites which are available for oxygen chemisorption during the treatment.

Since the major evolution from the coals at temperatures lower than 200°C is moisture, the O/C atomic ratio of the oxidized coals can be calculated by assuming that little gasification occurs during the oxidation. Both the oxygen originally retained and that chemisorbed during treatment were included in the calculation. The O/C atomic ratios are also shown in Table 2. The results reveal that the trend of the O/C ratio of the oxidized coal with the coal type is the same as that for the unoxidized coal, although the amount of oxygen uptake shows an opposite trend. The influences of the different types of oxygen, either originally retained or introduced during treatment, on the preparation of activated carbon are further discussed in the following sections.

In this study the process for the preparation of activated carbons principally consists of two consecutive steps: carbonization and activation. How the operating conditions of the steps affect the development of the surface structures of the activated carbons are shown, and the results are discussed separately.

3.2 Carbonization

In most situations for the production of activated carbon, carbonization has been implemented under inert conditions, such as in nitrogen [1,9,15,16]. However, carbonization may be implemented in mildly oxidizing atmospheres, such as carbon dioxide, to improve the quality of the pyrolytic products [10]. The present researchers' previous study [17] demonstrates that a gaseous environment (CO_2 or N_2) during carbonization has a negligible effect on the volatile evolutionary behavior of the carbonization process and the surface structure of the resulting activated carbon. The carbonizations of the coal samples were implemented in a CO_2 environment in this study.

The carbonization processes of the oxidized and unoxidized coals were monitored and the results were compared. Figure 2 shows the comparison of the carbonization behaviors of the oxidized and unoxidized BW coals in CO_2. One can observe from

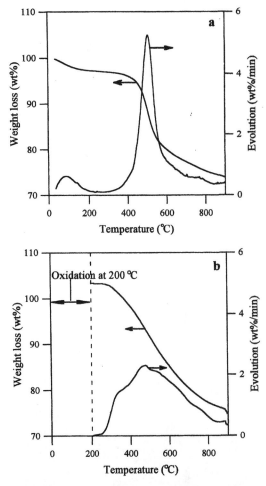

Fig. 2. Weight loss and evolution rate during carbonization of BW coal in CO_2: (a) unoxidized BW coal and (b) oxidized BW coal (wt% on an as-received basis; heating rate: 30°C/min; maximum heat treatment temperature: 900°C).

Fig. 2(a) that the carbonization of the unoxidized coal can be approximately described by two lumps: one for the moisture and the other for the volatiles. Tar is probably a predominant product of devolatilization for a significant part of the process, so that a sharp peak for volatile evolution is observed. However, as shown in Fig. 2(b), the carbonization of the oxidized coal shows a broader volatile evolution with respect to temperature. In a study of waste tire carbonization [10], it has been shown that oxygen treatment leads to a decrease in tar yield, and, on the other hand, an increase in oxygen-containing gases (such as H_2O, CO, CO_2 and SO_2), gaseous fuels (such as CH_4 and C_2H_4), and char. Since the increase in char yield due to oxygen treatment is insignificant (<3%) in this study, the broader evolution during devolatilization of the oxidized coals can probably be attributed to the increased evolution in gases, and the corresponding decrease in tar yield.

It has been reported in a previous study [17] that

surface areas of chars from carbonization (up to 900°C) of the unoxidized coals are 10 m²/g for MT coal, 2 m²/g for GG and less than 1 m²/g for BW. The surface area of the resulting char is an increasing function of the O/C atomic ratio of the starting materials. The implication from that study is that increasing the oxygen content of the raw coals by oxidation may also affect the development of surface area during carbonization. It was found in this study that the BET surface areas of the chars from the oxidized BW and MT coals are 122 and 116 m²/g, respectively, which are much larger than those from the unoxidized coals. Obviously, introduction of oxygen at 200°C to the coals can effectively promote cross-linking reactions between the coal structures.

Figure 3 shows the scanning electron micrograph (SEM) of the surface structures of the chars coming from the carbonization of the oxidized and unoxidized BW coals. The micrographs reveal that the char from the unoxidized BW coal has a fairly smooth surface, indicating that the unoxidized coal has passed through a plastic phase during carbonization to restructure the surface, which is expected to be coarse with the escape of tars there from. On the other hand, the char from the oxidized BW coal has a very rough surface, indicating that the oxidized

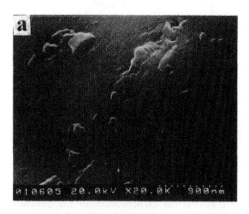

Fig. 3. SEM of the chars from unoxidized and oxidized BW coals: (a) from unoxidized BW coal, and (b) from oxidized BW coal.

coal is less thermoplastic during carbonization, hence the caking of the carbonaceous aggregates was hindered.

The above results indicate that the oxygen functional groups, either originally retained in the coal or introduced by oxidation, play an important role in reducing the caking of the bituminous coals, thus resulting in an increase in the surface areas. However, it appears that the functional groups resulting from oxidation at 200°C are more efficient in cross-linking the coal structures, since the chars from the oxidized BW and MT coals have similar surface areas; whereas the O/C ratio of the oxidized MT coal is higher than that of the oxidized BW coal. This aspect necessitates a more systematic study.

3.3 Activation

Following the carbonization of the coals, the chars obtained are subject to activation in CO_2 at the maximum heat treatment temperature, in order to increase the surface areas and pore volumes of the carbons. Changes in pore surface area and pore volume during activation are monitored and controlled to produce various activated carbons in this study. Variations in surface area and pore volume with carbon burn-off during activation are normally taken as the essential characteristics of a particular activation process [18].

It was found that the adsorption isotherms of N_2 on the activated carbons are typical of microporous carbons (type I). The adsorption isotherms are employed to deduce the BET surface area and the micropore and mesopore volumes. The average pore diameter can be determined according to the surface area and total pore volume (the sum of the micropore and mesopore volumes), if the pores are assumed to be parallel and cylindrical. This parameter is very useful when comparing the porous texture of activated carbon.

The developments of the surface area, pore volume, pore size distribution and average pore diameter as functions of the extent of burn-off are shown in Tables 3 and 4 for the samples from the unoxidized and oxidized coals, respectively. In the tables the values of the surface area and pore volume enclosed in parentheses are determined on a dry-ash-free basis. One can observe from Table 3 that for the carbons from the unoxidized BW coal the surface area and pore volume generally increase upon activation to a maximum value at a burn-off level around 50% and begin to decrease with further activation. The decrease in specific surface area at higher levels of burn-off can be attributed to the presence of significant ash, and to the breaking through of pore walls, which also results in a decrease in the number of micropores [19]. The trend of the surface area on a dry-ash-free basis with the burn-off level and the decrease in the micropore proportion can support the above argument. As for the total pore volume, the decrease at higher levels of burn-off can be only

Table 3. Variations in the surface area, pore volume, pore size distribution and average pore diameter with the levels of burn-off for the activated carbons from the unoxidized coals. The data are presented on a dry basis; values determined on a dry-ash-free basis are parenthesized

Burn-off (%)	BET SA (m^2/g)	Pore volume (cm^3/g)	Pore size distribution		Average pore diameter (Å)
			Micro (%)	Meso (%)	
BW carbon					
0	0.149 (0.183)	ND*	ND	ND	ND
11	165 (192)	0.081(0.097)	96	4	19.7
25	247 (296)	0.12 (0.14)	99	1	19.1
35	360 (436)	0.19 (0.23)	90	10	21.2
51	370 (496)	0.21 (0.28)	84	16	22.8
64	282 (432)	0.17 (0.26)	79	21	24.3
GG carbon					
40	435 (543)	0.23 (0.29)	78	22	21.1
MT carbon					
41	545 (704)	0.28 (0.36)	92	8	20.7

*Not detectable.

Table 4. Variations in the surface area, pore volume, pore size distribution and average pore diameter with the levels of burn-off for the activated carbons from the oxidized coals. The data are presented on a dry basis; values determined on a dry-ash-free basis are parenthesized

Burn-off (%)	BET SA (m^2/g)	Pore volume (cm^3/g)	Pore size distribution		Average pore diameter (Å)
			Micro (%)	Meso (%)	
BW carbon					
0	122 (140)	0.057(0.066)	100	0	18.8
10	278 (323)	0.13 (0.15)	99	1	19.1
23	438 (521)	0.20 (0.25)	100	0	18.5
37	665 (825)	0.32 (0.40)	98	2	19.3
52	969 (1308)	0.47 (0.63)	99	1	19.4
63	1035 (1552)	0.52 (0.78)	96	4	20.1
75	963 (1867)	0.48 (0.93)	96	4	20.0
GG carbon					
41	656 (819)	0.33 (0.41)	95	5	19.8
MT carbon					
40	625 (800)	0.32 (0.41)	91	9	20.7

partially attributed to the increasing presence of the ash, since the pore volume on a dry-ash-free basis still shows a decreasing trend upon activation at high burn-off levels. The major reason for this can be the gasification of the exterior portion of the carbon upon activation, which results in the disappearance of the exterior pores [17]. Figure 4(a) shows the SEM of the carbon from the unoxidized BW coal at a high burn-off level. The sample has a deficient and eroded surface, indicating that the occurrence of gasification on the exterior portion of the carbon is significant.

Table 4 shows the variations in surface structures with the extent of burn-off for the activated carbons from the oxidized coals. Upon comparison with the data in Table 3, it is seen that oxidation of the coal precursors results in increases in the specific surface area and pore volume of the resulting activated carbons. The influence on the surface structures by

the oxidation increases with the amount of oxygen uptake during oxidation; that is, the BW has the most significant influence and the MT has the least. Apart from the surface area and pore volume, one can observe from the tables that an increase in the volume proportion of micropores and a corresponding decrease in average pore diameter result from the oxidation of the coals. Obviously the preliminary oxidation destroys the caking properties to facilitate the development of a microporous structure in the char during carbonization [7,9].

The data in Table 4 reveal that the carbon from the oxidized BW coal has its highest values of BET surface area and pore volume at about 60% carbon burn-off, while it is about 50% for the sample coming from the unoxidized BW coal. This behavior demonstrates that the influences of the breaking of micropore walls and the erosion of the exterior portion of

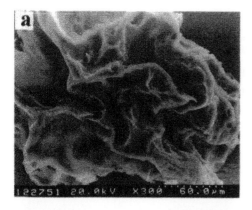

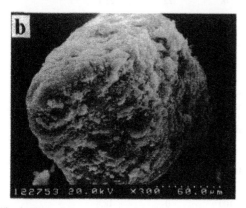

Fig. 4. SEM of the activated carbons from unoxidized and oxidized BW coals: (a) from unoxidized BW coal at 64% char burn-off, and (b) from oxidized BW coal at 63% char burn-off.

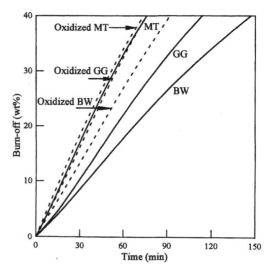

Fig. 5. Rate of activation of the chars from different coals (unoxidized represented by solid lines and oxidized by dashed lines) at 900°C in CO_2.

a carbon particle are less significant for the carbons from the oxidized coal. This interpretation can be supported by the results in Table 4, showing that both the BET surface area and the pore volume, on a dry-ash-free basis, of the carbon coming from the oxidized BW coal increase upon activation, even at a burn-off level as high as 75%. Figure 4(b) shows that the carbon from the oxidized BW coal still has no significant change on exterior surface structure or morphology, even at a high burn-off level, thereby indicating that gasification occurs mainly in the interior of the particle to increase the surface area and pore volume at high burn-off levels.

Oxidation also affects the activation process. Figure 5 shows the rate of activation of the chars resulting from the respective coals at 900°C. The graph shows how the degree of activation varies with the activation time. For the unoxidized samples the activation rate of the resulting char in CO_2 increases with the O/C atomic ratio of the coal precursor; that is, the MT char has the highest activation rate and the BW char has the lowest. With the oxidation of the coals, the resulting chars become more reactive in CO_2 than those from the unoxidized coals, as revealed in Fig. 5. The increase in the activation rate

in CO_2 due to the oxidation is an increasing function of the amount of oxygen uptake during the treatment; that is, the BW char has the highest level of increase in activation rate due to the oxidation, and the MT has the least. The increase in activation rate can be, at least partially, attributed to the increase in the surface area of the oxidized sample.

The temperature dependence of the reactivites of the chars in CO_2 at various levels of burn-off was determined. Since the gasification was performed within a narrow temperature range (800–950°C), the apparent activation energy was assumed to be constant in this range. The representative results are shown in Fig. 6, illustrating a change in apparent activation energy with the extent of char burn-off. Since the average pore diameter varies with the change of burn-off level, the results indicate that pore diffusion resistance is an important factor in determining the gasification rate. Also, the apparent activation energies shown in Fig. 6 are high, in comparison with that of a reaction controlled by diffusion. Obviously, the gasification of the solid carbons occurs in Zone II or in the transition region between Zone II and Zone I [20,21].

Figure 7 shows the influence of burn-off level on the apparent activation energy for the gasification of the chars coming from the BW and MT coals. Data for the unoxidized carbons show that the activation energy for the gasification of the BW char decreases with the degree of activation, while that of the MT increases. For a carbon gasification occurs in Zone II or in the transition region between Zone II and Zone I, the apparent activation energy generally increases with a decrease in pore diffusion resistance. The increase in the activation energy with the level of burn-off for the MT carbon demonstrates that the pores are widened upon activation to lower the

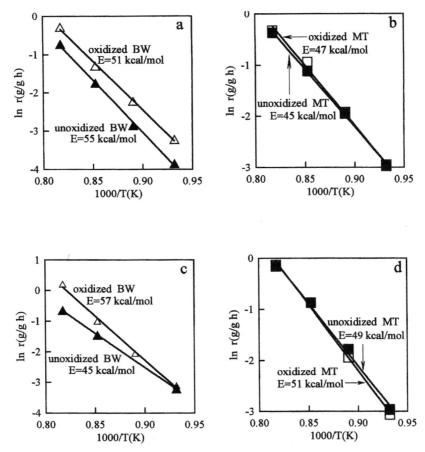

Fig. 6. Temperature dependence of the reactivity of the chars in CO_2. The chars are from different coals and at different extents of burn-off: (a) from oxidized and unoxidized BW coals at 2.5% char burn-off; (b) from oxidized and unoxidized MT coals at 2.5% char burn-off; (c) from oxidized and unoxidized BW coals at 33% char burn-off; and (d) from oxidized and unoxidized MT coals at 33% char burn-off.

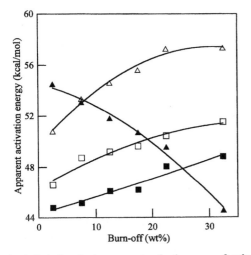

Fig. 7. Variations in the apparent activation energy for the gasification of the chars from different coals: (▲) unoxidized BW; (△) oxidized BW; (■) unoxidized MT; (□) oxidized MT.

diffusion resistance [17]. The decrease in the BW carbon implies a strong diffusion limitation for pores created in the caking portion of the carbon. The SEM of the BW samples in Fig. 4(a) demonstrates the occurrence of gasification on the exterior portion of the particle at a high burn-off level, indicating a strong pore resistance. Since the MT coal has a higher oxygen content, the above results may suggest that the cross-linked structure induced by the oxygen is more accessible for CO_2 penetration during activation.

As for the carbons from the oxidized coals, both of the activation energies for the gasification of the BW and MT carbons increase with the level of burn-off. The results for the MT carbon are as expected, since the oxygen content of the coal precursor is increased by the oxidation, and the activation energy for the gasification of the carbon from the unoxidized MT coal has already shown an increasing trend with the level of burn-off. However, it is interesting to observe that the trend of activation energy with carbon burn-off for the BW sample was inversed due to the oxidation. The results indicate that the oxygen introduced during oxidation can also induce a structure which is accessible for CO_2 penetration during activation. This conclusion can be supported by the SEM of the BW sample in Fig. 4(b), in which the sample particle still has no significant change on exterior surface structure or morphology at a high burn-off level, indicating that the gasification occurs mainly in the pores of the particles.

In a previous study [17] for the production of activated carbons from the unoxidized coals, it was found that the average pore diameter increases as the activation temperature was elevated, and there is no obvious trend for the variations in surface area and pore volume with respect to the activation temperature. The effect of the activation temperature on the development of the surface structure of the chars coming from the oxidized coals was explored in this study. Table 5 shows the variations in the surface area, pore volume and average pore diameter with the activation temperature ranging from 800 to 950°C. Unlike the results for the carbons from the unoxidized coals, there is no obvious temperature effect on the average pore diameter of the carbons coming from the oxidized BW and GG coals. It has been stated that due to the increased pore resistance at higher temperatures pore widening is the predominant mechanism at high temperatures (gasification is so fast that diffusion becomes the slow step; the gasification occurs on the outer portion of the pores before CO_2 molecules diffuse deeply inside, and, thus, pore widening happens) while pore deepening is favored by low temperatures (the rates of diffusion and gasification are comparable, and, thus, pore deepening takes place) [17]. Since oxidation of the coals can lower the pore diffusion resistance for the resulting chars, it is not surprising to observe that the activation temperature has no significant effect on the average pore diameter of the carbons from the oxidized BW and GG coals. For the MT sample, since the influence from the oxidation is less significant, it is reasonable to observe that the average pore diameter still increases with the activation temperature.

Table 5 also shows that for the carbons from the oxidized GG and MT coals there is no obvious trend for the variations in surface area and pore volume with respect to the activation temperature. The results are similar to those for the carbons from the unoxidized coals. However, for the carbons from the oxidized BW coal the surface area and pore volume decrease with an increase in the activation temperature, especially for temperatures above 900°C. The increase in surface area and pore volume of the BW carbon from the oxidation is related to the formation of cross-links between carbonaceous aggregates. Therefore, at higher temperatures it is concluded that the cross-links break, with a consequent rearrangement of carbonaceous aggregates and the collapse of pores. This effect of the activation temperature on the surface structure of the carbons from the oxidized BW coal, which has the maximum oxygen uptake during oxidation, may suggest that the cross-links induced from the oxygen functional groups introduced during the oxidation are less thermally stable than those from the oxygen originally retained in the coals.

4. CONCLUSIONS

This study has demonstrated that oxidation of bituminous coal precursors has a significant influence on the surface structure of the resulting chars and the activated carbons upon CO_2 activation.

During the oxidation, the amount of oxygen uptake decreases with the O/C atomic ratio of the raw coals. However, the trends of the O/C ratio with the coal type for the oxidized and unoxidized coals are the same.

The carbonization of the oxidized coals exhibits a broader volatile evolution with respect to temperature than that of the unoxidized coal, indicating an increased evolution in gases at the expense of a reduced tar yield. The resulting char from the oxidized coals has a much larger surface area than that from the respective unoxidized coal, indicating that the oxidation effectively promotes the cross-linking reaction during carbonization.

The oxidation of the coal precursors results in increases in the specific surface area and pore volume of the resulting activated carbons, whereas the average pore diameter of the carbons decreases. The rate of activation in CO_2 is higher for the carbon from an oxidized coal than that from a corresponding unoxidized coal. The increase in the activation rate due to the oxidation can be, at least partially, attributed to an increase in surface area. Also, the oxidation of the coals can produce a char structure which is more accessible for CO_2 penetration during activa-

Table 5. Effects of the activation temperature on the BET surface area, pore volume and average pore diameter of the activated carbons from the oxidized coals

Activation temperature (°C)	BET SA (m^2/g)	Pore volume (cm^3/g)	Average pore diameter (Å)
BW carbon (~36% burn-off)			
800	689	0.34	19.6
850	673	0.32	18.9
900	665	0.32	19.3
950	520	0.25	19.3
GG carbon (~40% burn-off)			
800	605	0.29	19.5
850	652	0.32	19.8
900	655	0.33	19.8
950	587	0.29	19.9
MT carbon (~41% burn-off)			
800	622	0.29	18.7
850	640	0.33	20.4
900	625	0.32	20.7
950	657	0.35	21.6

tion; and, therefore the carbon gasification by CO_2 occurs mainly in the pores of the particles.

The effect of activation temperature on the average pore diameter is less significant for the carbons from the oxidized coals than those from the unoxidized coals. This behavior can be explained by the reduced pore diffusion resistance of the carbon due to the oxygen treatment. The data also show that the cross-links induced from the oxygen introduced during the oxidation are less thermally stable than those from the oxygen originally retained in the coals.

Acknowledgements—This research was supported by the National Science Council and the China Steel Corp., both in Taiwan, under Project NSC 85-2214-E-033-003 and Contract No. RE85601, respectively.

REFERENCES

1. M. J. Muñoz-Guillena, M. J. Illán-Gómez, J. M. Martín-Martínez, A. Linares-Solano and C. Salinas-Martínez de Lecea, *Energy Fuels* **6**, 9 (1992).
2. M. Greenbank and S. Spotts, *Water Technology* **16**, 56 (1993).
3. R. C. Bansal, J. B. Donnet and H. F. Stoekcli, *Active Carbon*. Marcel Dekker, New York (1988).
4. T. Wigmans, *Carbon* **27**, 13 (1989).
5. J. Laine and S. Yunes, *Carbon* **30**, 601 (1992).
6. E. M. Suuberg, In *Chemistry of Coal Conversion* (Edited by R. H. Schlosberg), Chapter 4, pp. 67–119. Plenum Press, New York (1985).
7. T. Alvarez, A. B. Fuertes, J. J. Pis, J. B. Parra, J. Pajares and R. Menéndez, *Fuel* **73**, 1358 (1994).
8. M. A. Serio, P. R. Solomon, E. Kroo, R. Bassilakis, R. Malhotra and D. F. McMillen, *ACS Div. of Fuel Chem. Prepr.* **35**, 61 (1990).
9. T. A. Centeno and F. Stoeckli, *Carbon* **33**, 581 (1995).
10. H. Teng, M. A. Serio, M. K. Wójtowicz, R. Bassilakis and P. R. Solomon, *Ind. Eng. Chem. Res.* **34**, 3102 (1995).
11. S. J. Gregg and K. S. W. Sing, *Adsorption, Surface and Porosity*. Academic Press, London (1982).
12. S. Lowell and J. E. Shields, *Powder Surface Area and Porosity*, 3rd edn. Chapman and Hall, London (1991).
13. H. F. Stoeckli, *Carbon* **28**, 1 (1990).
14. P. J. M. Carrott, M. M. L. Ribeiro Carrott and R. A. Roberts, *Colloids and Surfaces* **58**, 385 (1991).
15. F. Rodríguez-Reinoso, M. Molina-Sabio and M. T. González, *Carbon* **33**, 15 (1995).
16. G. Kovacik, B. Wong and E. Furimsky, *Fuel Proc. Technol.* **41**, 89 (1995).
17. H. Teng, J.-A. Ho, Y.-F. Hsu and C.-T. Hsieh, *Ind. Eng. Chem. Res.* accepted (1996).
18. G. Q. Lu, *Fuel* **73**, 145 (1994).
19. P. L. Walker, Jr and A. Almagro, *Carbon* **33**, 239 (1995).
20. P. L. Walker, Jr, F. Rusinko and L. G. Austin, *Advances in Catalysis* **11**, 133 (1959).
21. M. F. R. Mulcahy and I. W. Smith, *Rev. Pure and Appl. Chem.* **19**, 81 (1969).

ADSORBENTS FOR METHANE STORAGE MADE BY PHOSPHORIC ACID ACTIVATION OF PEACH PITS

J. A. F. MacDonald, and D. F. Quinn

Department of Chemistry and Chemical Engineering, Royal Military College of Canada, Kingston,
ON K7K 5L0, Canada

(*Received 11 October 1995; accepted in revised form 16 April 1996*)

Abstract—Powdered peach pit was impregnated with various concentrations of phosphoric acid and then heated to temperatures between 350 and 900°C. The resulting chars were characterized by nitrogen isotherms at 77 K and mercury porosimetry. Methane isotherms were measured gravimetrically at 298 K to 3.5 MPa.

Although many of the chars had relatively high 77 K nitrogen BET surface areas and micropore volumes, they consistently adsorbed less methane at 298 K than expected. A linear relationship between surface area and methane uptake for many different carbons had previously been observed [1]. With increasing heat treatment temperature, these chars followed more closely the predicted relationship. From reflectance IR absorption and temperature-programmed desorption, it is apparent that these chars cannot be considered as only carbon because of their high oxygen and hydrogen content. Only at temperatures above 700°C do they begin to show the methane uptake expected from their surface area and micropore volume.

The adsorption potential of the char pore towards methane at 298 K appears to be lower than that of a truly carbonaceous adsorbent. Alternatively, the adsorption potential to nitrogen at 77 K may be greater for these chars than the adsorption potential of a microporous carbon.

The chars produced in this study by phosphoric acid treatment of peach pit do not adsorb sufficient methane to be considered as suitable adsorbents for natural-gas storage applications. Copyright © 1996 Elsevier Science Ltd

Key Words—Methane storage, phosphoric acid activation.

1. INTRODUCTION

The production of an economical carbon adsorbent for natural-gas storage has been actively pursued in recent years. The ideal adsorbent would be in the form of a dense monolith with limited macropore volume, and a high adsorption capacity for methane. Carbons have been prepared from a vast number of precursor materials using almost every conceivable method of activation [2–5].

It appears, however, that the use of phosphoric acid as an activating agent has not yet been vigorously exploited in the production of a methane-adsorbent carbon. In complete agreement with Laine *et al.* [6], it was discovered that little research on the activation process is available in the scientific literature. One process using sawdust has been described [7], and the relatively low temperatures (~500°C) required in conjunction with the recovery of the acid make the process commercially viable.

The positive aspects of phosphoric acid activation were demonstrated by workers such as Laine *et al.* [6], who obtained a coconut-shell carbon with a surface area of about 2000 m^2 g^{-1} after impregnation of the raw material with concentrated acid and heating to a maximum of 500°C. At lower acid concentrations, the optimum temperature for maximum surface area was about 450°C. The production of high surface areas at low temperatures was also confirmed by the work of Jagtoyen, Derbyshire and

co-workers [8,9] using white oak and coal, and by Botha and McEnaney [10] using a South African coal.

Included in the limited quantity of older literature mentioning phosphoric acid activation is a summary of work by Berl [11] that used 1000°C for the dehydration of cellulose- and lignin-containing material with phosphoric acid. Elevated temperatures of greater than 800°C were also advocated by Urbain [12] in the carbonization of vegetable matter to produce an active carbon. The procedure described included the distillation and recovery of phosphorous from the carbon at elevated temperatures.

In this study, chars were prepared using peach pit dust that was impregnated with phosphoric acid and heated to a wide variety of temperatures, usually under an inert atmosphere. Peach pits have been shown to be a good precursor for carbons suitable for hydrocarbon adsorption [4,13]. These carbons have been activated using zinc chloride and air. A study using phosphoric acid as the activating agent has not been carried out on peach pits, but high surface areas have been reported for other precursor materials [7]. This study was undertaken with the prospect of producing a high surface area, high bulk density carbon in a single pyrolysis step at relatively low temperature, which would be suitable as an adsorbent for use in natural-gas storage.

In this study, "char" has been used to refer to the materials produced by heat treatment with phosphoric acid. This is used to deliberately distinguish them from carbons which, by comparison, have much lower oxygen and hydrogen content.

2. EXPERIMENTAL

The precursor material, peach pits, was ground to a fine powder (<150 μm) and soaked in phosphoric acid solutions where the solution volume and acid concentration were both varied. The soak time of the acid impregnation before charring also varied, but in most cases long soak times (>24 hours) were used. Heating of the mixtures took place under an atmosphere of dry nitrogen in a tube furnace equipped with a quartz tube. (The exception was one sample which was heated in air.) Hold times at the uppermost temperature varied but always exceeded 1 hour. After being cooled under nitrogen and then removed from the furnace, samples were repeatedly washed with copious amounts of hot water until pH was neutral, and dried in an oven at 110°C prior to characterization.

Attempts were made to press some peach pit/phosphoric acid mixtures into pellets and form monoliths before the heat treatment. Most of these efforts failed, although in some instances samples were produced as extrudates. Often, attempts to form pellets resulted in the elimination of some of the acid from the sample.

After charring, dried samples were examined for their adsorption capacity for nitrogen at 77 K, using a Micromeritics ASAP 2000 instrument, following degassing to ~ 2 μmHg at 110°C. The nitrogen isotherms were fitted to both the BET equation [14] to determine a surface area, and the Dubinin–Radushkevich equation [15] for the calculation of a micropore volume. Methane adsorption capacity was measured gravimetrically using a Sartorius 4436 high-pressure microbalance described elsewhere [3]. A Quantachrome Autoscan-60 mercury porosimeter was used to characterize the samples for macropore volumes, external surface areas and densities.

Temperature-programmed desorption (TPD) analyses were carried out using a Dataquad quadrupole mass Spectrometer (Spectromass Inc.) connected to a desorption system using a procedure described previously [16]. This system does not have the ability to distinguish between the CO, CO_2, H_2 and water desorbed from the surface of the carbon, or that liberated from the sample as primary degradation products of the charring. Since in their preparation some samples had not been heated above 400°C, a substantial fraction of the total oxygen content must be the result of further thermal degradation during the TPD analysis of partially decomposed cellulosic material.

Infrared spectra were measured at room temperature using a Niclolet 510P FT-IR equipped with a diffuse reflectance accessory. The samples used were prepared as a 5% w/w KBr mixture.

3. RESULTS AND DISCUSSION

A typical example of surface areas versus charring temperature for the phosphoric acid peach pit samples is shown in Fig. 1. The peach pit chars used in this FOS 6 series were produced using concentrated phosphoric acid (85%) and a high acid-to-precursor ratio; 2 ml acid for each gram of dry peach pits. The residence time for the acid to contact the powdered peach pits at room temperature was longer than two weeks, after which the mixture had a black, tarry consistency. The maximum BET surface area for this series was 1560 m^2 g^{-1} at a temperature of 500°C, quite similar to that found by other workers using different lignocellulosic precursor materials. Between 500 and 700°C the surface area showed a decrease, but increased again above 700°C. Three samples were pyrolysed to 900°C at different rates and using different hold times at the maximum temperature. The variation of the surface areas for the 900°C carbons, from 1465 to 1522 m^2 g^{-1}, was not considered sufficiently different to be significant. A constriction of mesopores, narrowing them into micropores on heat treatment, may contribute to the increase in the measured surface area between samples heated to 700 and 900°C. Table 1 shows both the macropore and mesopore volume of the sample heated to 900°C to be less than that of the sample heated to 700°C.

Figure 2 shows the methane adsorbed per gram of sample at 3.4 MPa (500 psia) and 25°C for each of the FOS 6 series as a function of charring temperature. It is immediately obvious that the maximum adsorption does not occur at the same temperature as for nitrogen (Fig. 1). Indeed, the sample charred to 600°C had a lower surface area than that charred at 500°C, but adsorbed significantly more methane.

Previously a linear correlation was established between the D–R micropore volume, calculated from nitrogen isotherms at 77 K, of a series of activated peach pit carbons and methane uptake at 25°C and 3.4 MPa [5]. The carbons in the previous study had

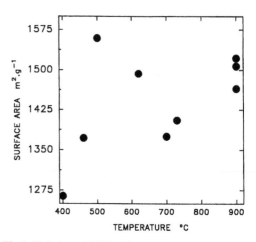

Fig. 1. Variation of BET surface area, from nitrogen isotherms at 77 K, with preparation temperature for FOS 6 series of chars.

Table 1. Characterization of macro- and mesopore volumes of some phosphoric-acid activated peach pit chars, including external surface areas and bulk densities from mercury porosimetery, and D–R micropore volumes calculated from nitrogen isotherms at 77 K

Sample	Maximum temperature (°C)	Bulk density (g ml^{-1})	Macropore volume (ml g^{-1})	Mesopore volume (ml g^{-1})	Micropore volume (ml g^{-1})	External area (m^2 g^{-1})
FOS 1	475	0.44	0.50	0.78	0.53	260
FOS3/T1	475	0.48	1.01	0.33	0.58	139
FOS 3/T2	425	0.41	1.01	0.56	0.49	238
FOS 4/T1	475	0.46	0.71	0.54	0.78	255
FOS 4/T2	425	0.64	0.91	0.43	0.79	223
FOS 4/T2B	700	0.54	0.94	0.15	0.58	117
FOS 6/T1	400	0.60	0.94	0.44	0.56	209
FOS 6 EXTRUDED	425	0.57	0.37	0.76	0.61	299
FOS 6/T2	460	0.42	0.84	0.78	0.61	259
FOS 6/T2 EXTRUDED	460/AIR	0.54	0.40	0.53	0.66	255
FOS 6/T4	620	0.36	0.60	1.39	0.66	553
FOS 6/T5	700	0.40	0.68	1.14	0.60	543
FOS 6/T7C	900	0.41	0.44	0.96	0.66	489

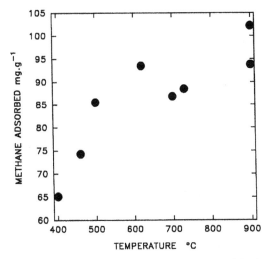

Fig. 2. Variation of adsorbed methane at 3.4 MPa and 25°C, with preparation temperature for FOS 6 series of chars.

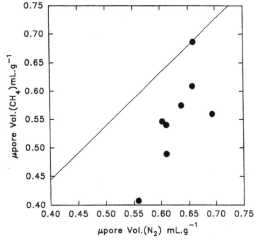

Fig. 3. Variation of micropore volumes from methane isotherms at 298 K with Dubinin micropore volumes from nitrogen isotherms at 77 K and previously determined relationship.

all been pyrolysed to 750°C before they were subjected to the air cyclic activation. The observation shown in Fig. 3, where the line represents the previous correlation for peach pit carbons, shows that this series of phosphoric-acid activated peach pit chars does not fit that correlation. The one result which lies on the line represents a sample heated to 900°C for ~4 hours.

In an attempt to gain insight into the composition of the chars their DRIFT spectra were measured. These spectra, in the region between 400 and 2000 cm^{-1}, are shown in Fig. 4 for the FOS 6 series of phosphoric acid samples charred at temperatures from 400 to 700°C. Three dominant bands in Fig. 4 are of interest. The 1707 cm^{-1} band suggests a strong carbonyl presence in chars heated to temperatures below 500°C. This C=O stretching vibration disappears at higher temperatures. The assignment of this band to a carbonyl group is consistent with examples in the literature [17–21] which encompass a wide variety of different types of carbons such as those

obtained from oxidised polyfurfuryl alcohol, cellulose, oxidised Saran chars, and carbon blacks.

No consensus has been reached on the assignment of the strong IR active band at 1585 cm^{-1}. Morterra and Low [22] explain that it is always present in the spectra of carbons, and tend to favour an IR active aromatic C=C vibration to this band. This band remains present in all the FOS 6 series chars shown in Fig. 4 regardless of the charring temperature. Starsinic et al. [20] reported that the intensity of this band was dependant on the amounts of oxygen surface structures. In this series of phosphoric-acid activated chars, the strength of both the 1585 and 1220 cm^{-1} bands decreases with increased charring temperature, initially linearly with temperature then flattening (see Fig. 5). This suggests that above a heat treatment temperature of 600°C, the char has become a more truly carbonaceous matrix. The band in the 1220 cm^{-1} region has been assumed to be characteris-

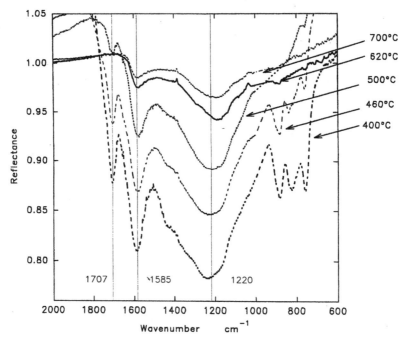

Fig. 4. DRIFT spectra on FOS 6 series of phosphoric acid activated chars heated to 400, 460, 500, 620 and 700°C.

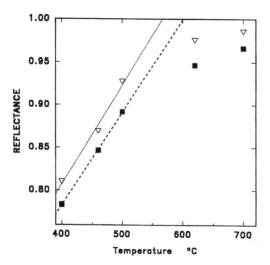

Fig. 5. Intensity of specific IR bands (∇ 1585 cm^{-1}, $\blacksquare$ 1220 cm^{-1}) for the FOS 6 series of chars versus pyrolysis temperature.

tic of a C–O single bond, although an undisputed assignment may not be possible due to an aromatic P=O stretch in the IR region of 1190 cm^{-1} [9]. Although analysis of these samples for phosphorous content may have been desirable in this study, facilities to do this were not available. The P–O stretching mode at 985 cm^{-1} was not observed in any of the spectra. The bands between 900 and 675 cm^{-1} are characteristic of aromatic C–H out-of-plane bending [23]. Further evidence of a high concentration of non-carbon atoms on the surface of these chars is shown in Fig. 6. The TPD carried out on these

samples shows that the total [H], as both H_2 and H_2O, desorbed from the chars decreases as the charring temperature increases. Not unexpectedly, these samples also desorbed significant quantities of CO and CO_2 on TPD.

Figure 7 shows the methane adsorption at 25°C and 3.4 MPa (500 psia) versus Dubinin micropore volume, calculated from nitrogen isotherms at 77 K, on a series of 9 commercial carbons. The line representing the predicted adsorption was reported elsewhere [24]. These commercial carbons were chosen for their diversity, as they come from a variety of sources and countries and span a wide micropore range. They are, in decreasing order of micropore volume, AX-21, known to be a KOH-activated carbon, Sutcliffe Speakman GS-60, Barnabey Cheney MI, GMS-70 from the California Carbon Company, Kureha BAC, Darco Vapure, CNS 196 from A.C. Carbone, Calgon BPL and Norit EX6. The methane adsorbed at 25°C and 500 psia fits well with the predicted adsorption for most of these carbons, deviating slightly for the lowest adsorption. These carbons, in general, have little or no volatile matter.

The same line predicting methane adsorption from nitrogen micropore volume is shown in Fig. 8 along with the results for many phosphoric-acid activated peach pit samples. The expected methane adsorption on the samples produced at low temperatures (<620°C) was predominantly below the predicted value, while the samples pyrolysed at higher temperatures appeared to fit the relationship previously determined for the commercial carbons. For the FOS 6 series of phosphoric acid peach pit chars, the samples charred below 620°C attain less than 80% of

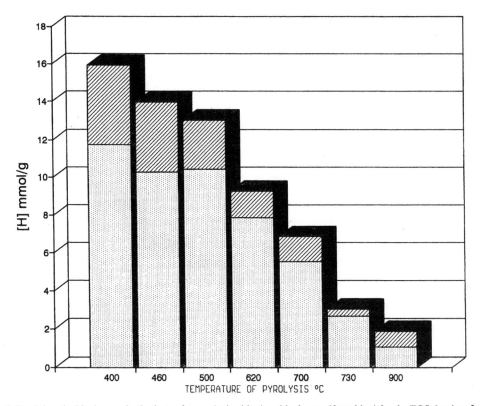

Fig. 6. Total desorbed hydrogen, in the form of water (striped bar) and hydrogen (dotted bar) for the FOS 6 series of chars pyrolysed at the given temperatures.

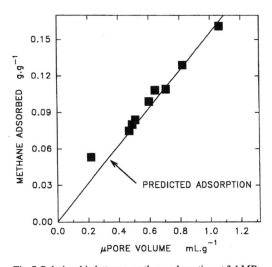

Fig. 7. Relationship between methane adsorption at 3.4 MPa and 25°C and Dubinin micropore volume from nitrogen isotherms at 77 K for a series of commercial carbons. The line represents previous correlation for predicted adsorption.

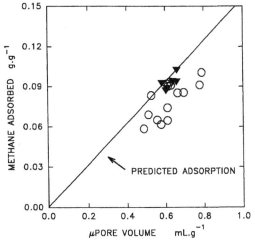

Fig. 8. Relationship between predicted methane adsorption at 3.4 MPa and 25°C and Dubinin micropore volume from nitrogen isotherms at 77 K for peach pit chars. ▼ pyrolysed to 620°C or greater, ○ heated to <620°C. The line represents previous correlation for predicted adsorption.

the predicted methane adsorption, while the same mixture pyrolysed to higher temperatures reaches 90% or more of the predicted value. The highest sample in this series had been heated to 900°C for a longer period of time (4 hours compared to 1 hour for the other samples), and reached the predicted methane adsorption.

The chars produced in this study all had fairly large macropore and mesopore volumes as determined by mercury porosimetry and their bulk densities varied in the range 0.40–0.64 g ml^{-1}. Some typical examples are summarised in Table 1.

In general, the phosphoric-acid prepared chars

have too large macro- and mesopore volumes to be considered as desirable adsorbents for the volumetric storage of methane. In these pores the methane density is not much higher than that of the gas phase, and so their volume does not contribute to enhancing the methane storage capacity. Extrusion did help to reduce some of the unwanted macropore volume (see Table 1).

Further work is needed to determine whether high-density monoliths can be prepared from phosphoric-acid activated carbons. These carbons must be pyrolysed to temperatures above 620°C to create a true carbonaceous surface which is more attractive for methane adsorption at 25°C. Although low-temperature charring produces samples with high nitrogen surface areas and micropore volumes these samples are not suitable for methane adsorption. This is probably due to the high concentration of hydrogen and oxygen remaining on the surface.

REFERENCES

1. J. T. Mullhaupt, W. E. BeVier, K. C. McMahon, R. A. Van Slooten, I. C. Lewis, R. A. Greinke, S. L. Strong, D. R. Ball and W. E. Steele, *Proc. Carbon*, Essen, Germany, p. 387 (1992).
2. I. C. Lewis, R. A. Greinke and S. L. Strong, *Extended Abstracts 21st Biennial Conf. Carbon*, p. 490 (1993).
3. D. F. Quinn and J. A. F. MacDonald, *Carbon* **30**, 1097 (1992).
4. J. A. F. Holland and D. F. Quinn. unpublished report on 'The Development of Natural Gas Adsorbents', presented to AGLARG (1988).
5. J. A. F. MacDonald and D. F. Quinn, *J. Porous Mater.* **1**, 43 (1995).
6. J. Laine, A. Calafat and M. Labady, *Carbon* **27**, 191 (1989).
7. R. C. Bansal, J. B. Donnet and F. Stoeckli, *Active Carbon*. Marcel Dekker, New York (1988).
8. M. Jagtoyen and F. Derbyshire, *Carbon* **31**, 1185 (1993).
9. M. Jagtoyen, M. Thwaites, J. Stencel, B. McEnaney and F. Derbyshire, *Carbon* **30**, 1089 (1992).
10. F. D. Botha and B. McEnaney, *Adsorption Sci. Technol.* **10**, 181 (1993).
11. E. Berl, *Trans. Faraday Soc.* **34**, 1040 (1938).
12. E. Urbain, U.S. Pat. No. 1 610 399 (1926).
13. F. Caturla, M. Molina-Sabio and F. Rodriguez-Reinoso, *Carbon* **29**, 999 (1991).
14. S. Brunauer, P. H. Emmett and E. Teller, *J. Am. Chem. Soc.* **60**, 309 (1938).
15. M. M. Dubinin, E. D. Zaverina and L. V. Radushkevich, *Zh. Fiz. Khim.* **21**, 1351 (1947).
16. S. S. Barton, M. J. B. Evans and J. A. F. MacDonald, *Carbon* **29**, 1099 (1991).
17. J. Zawadzki, *Carbon* **16**, 491 (1978).
18. R. A. Friedel and L. J. E. Hofer, *J. Phys. Chem.* **74**, 2921 (1970).
19. J. M. O'Reilly and R. A. Mosher, *Carbon* **21**, 47 (1983).
20. M. Starsinic, R. L. Taylor, P. L. Walker and P. C. Painter, *Carbon* **21**, 69 (1983).
21. C. Morterra, M. J. D. Low and A. G. Severdia, *Carbon* **22**, 5 (1983).
22. C. Morterra and M. J. D. Low, *Carbon* **21**, 283 (1983).
23. R. M. Silverstein, G. C. Bassler and T. C. Morrill, *Spectrometric Identification of Organic Compounds*. Wiley, New York (1981).
24. N. D. Parkyns and D. F. Quinn, *Porosity in Carbons* (Edited by J. W. Patrick), Ch. 11. Edward Arnold, London (1995).

PREPARATION OF ACTIVATED CARBON FROM BITUMINOUS COAL WITH PHOSPHORIC ACID ACTIVATION

HSISHENG TENG,* TIEN-SHENG YEH and LI-YEH HSU

Department of Chemical Engineering, National Cheng Kung University, Tainan 70101, Taiwan,

(Received 5 August 1997; accepted in revised form 10 March 1998)

Abstract—Activated carbons were prepared from an Australian bituminous coal in this study. The preparation process consisted of phosphoric acid impregnation followed by carbonization in nitrogen at 400–600°C for 1–3 hours. The results reveal that the surface area and pore volume of the resulting carbons increase with the chemical ratio, H_3PO_4/coal. Within the ranges of carbonization temperature and time, the chemically activated carbon prepared from carbonization at 500°C for 3 hours was found to have maximum surface area and pore volume values. Physical activation with CO_2 of the initially H_3PO_4 activated carbon was examined and the results suggest that the combined activation is suitable for producing high porosity carbons with a high proportion of mesoporosity. © 1998 Elsevier Science Ltd. All rights reserved.

Key Words—A. Activated carbon, B. activation, B. carbonization, C. adsorption, D. surface properties.

1. INTRODUCTION

Activated carbons, with their high porosity, are extensively used in industrial purification and chemical recovery operations. Most types of industrial activated carbons are produced from naturally occurring carbonaceous materials like coal, petroleum, peat, wood and other biomass. Because of its availability and cheapness, coal is the most commonly used precursor for activated carbon production [1–3]. Among coal-based carbons, bituminous coal products are in greater demand since they have greater density, hardness and abrasion resistance, and are more durable than other coal-based carbons [4]. In the present study, an Australian bituminous coal was chosen as a precursor of activated carbon.

The high adsorptive capacities of activated carbons are associated with their internal porosity and are related to properties such as surface area, pore volume and pore size distribution. Generally, activated carbons are mainly microporous, but in addition to micropores they contain meso and macropores, which are very important in facilitating access of the adsorbate molecules to the interior of carbon particles [5]. In practical terms, the type of porosity is dictated by the type of raw material employed; however, the method of activation is another parameter which may influence the final pore size distribution [6]. Basically, there are two different processes for the preparation of activated carbon: physical and chemical activation. Physical activation involves the carbonization of a carbonaceous precursor followed by activation of the resulting char in the presence of some mildly oxidizing gases such as carbon dioxide or steam [1,3,7]. The other method, chemical activation, consists of carbonization at a

relatively low temperature (e.g. 400–700°C) in the presence of a dehydrating agent (e.g. $ZnCl_2$, KOH and H_3PO_4). These chemical reagents may promote the formation of cross-links, leading to the formation of a rigid matrix, less prone to volatile loss and volume contraction upon heating to high temperatures [8,9].

In previous studies [10,11], the preparation of activated carbons from Australian bituminous coals through physical activation (CO_2 gasification) has been extensively studied. Because of the low reaction rate between the activating agent and the char, a high temperature (> 800°C) was necessary for activation. The results have shown that activated carbons with high surface area can be obtained only at a high extent of char burn-off. Therefore, the product yield is low for the preparation of activated carbons from these bituminous coals through physical activation. Some authors have studied the combination of chemical and physical activation, attempting to obtain high surface area carbons with high yields [1,12].

Among the numerous dehydrating agents for chemical activation, zinc chloride and phosphoric acid are the most widely used chemical agents [1,13–15]. However, the use of phosphoric acid is preferred and the use of zinc chloride has declined due to problems of environmental contamination with zinc compounds. The preparation and the properties of activated carbons from phosphoric acid activation have not been thoroughly compared with those from physical activation, when bituminous coal is used as the starting material. However, it should be pointed out that chemical activation with H_3PO_4 is only practised with lignocellulosic materials [8]. Previous studies have shown that the role H_3PO_4 in activation of biomass is to accelerate the cleavage of bonds between biopolymers (principally cellulose and

---

*Corresponding author.

Reprinted from *Carbon* **36 (9)**, 1387-1395 (1998)

lignin), followed by recombination reactions in which a rigid cross-linked solid is formed [13,16]. Since the greater portion of coal structure derives from coalification of biomass, it has been suggested that the response to H_3PO_4 activation decreases with the degree of coalification [16], as evidenced by the fact that the surface areas of carbons produced by H_3PO_4 activation decrease with increased rank of the starting materials. Another drawback in using coals as the precursors is the increased ash content of the carbon products [2,8]. This can be attributed to the interactions between phosphorus and the coal mineral matter, forming species such as $FePO_4$, $Al(PO_3)_3$ and $Si_2P_2O_7$ [8]. High ash content is undesirable for granular activated carbon, since it reduces the mechanical strength of the carbon. Apparently, preparation of activated carbon from bituminous coal by chemical activation with H_3PO_4 does not represent a practical approach in industrial applications.

This study is devoted to investigating the effects of different parameters on the preparation of activated carbon from bituminous coals by H_3PO_4 activation. Subsequent activation of the chemically activated carbon with CO_2 was also studied. These results, including the effects of the H_3PO_4 impregnation conditions, behavior during carbonization and pore structures of the resulting activated carbon, are then compared with those obtained by physical activation of the same raw material, using CO_2 activation.

2. EXPERIMENTAL

An Australian bituminous coal, Black Water (BW, high volatile A in ASTM classification), was used as the starting material. The proximate and ultimate analyses of the raw coal are shown in Table 1. The as-received coal was ground and sieved to 210–300 μm, before being treated.

Chemical activation was performed with H_3PO_4. The experimental procedure used in the activation process was as follows: 20 g of as-received coal were mixed, by stirring, with 100 g of an aqueous solution that contained 0, 20, 40 or 85% of H_3PO_4 by weight. The chemical ratios of activating agent/coal were then 0.0, 1.0, 2.0 or 4.25, respectively. Mixing was performed at 50 or 85°C for 1–3 hours. After mixing, the coal slurry was subjected to vacuum drying at 100°C for 24 hours. The resulting samples were then carbonized in a horizontal cylindrical furnace (25 mm i.d.) in an N_2 atmosphere, with a flow rate of

100 ml min^{-1}. The samples were heated at 30°C min^{-1} from room temperature to carbonization temperatures in the range of 400–600°C and then held at that temperature for 1–3 hours before cooling under N_2. After cooling the carbonized products were leached by mixing with distilled water at 150 ml g^{-1} char, followed by filtration of the mixtures. Leaching was carried out for several times until the pH value of the water–char mixture was above 6. The leached products were then dried by vacuum at 50°C for 24 hours.

In order to have a better understanding in the carbonization process, a thermogravimetric analyzer (TGA, DuPont TGA 51, New Castle DE) was employed to monitor the volatile evolution behavior, also in an N_2 atmosphere. A sample of 30–50 mg was used for each TGA analysis. It was found that the chemical and surface properties of the carbon obtained from carbonization in the TGA apparatus were similar to those of the carbon prepared from the cylindrical furnace, although the sizes of the samples carbonized in these two systems were different.

Physical activation with CO_2 of an initially H_3PO_4 activated carbon was also examined. Gasification with CO_2 was performed at 900°C for various extent of carbon burn-off. Since physical activation takes place at a much higher temperature (900°C versus 400–600°C) a loss of material is expected during heat-up, carried out at 30°C min^{-1} in nitrogen, up to 900°C, due to further carbonization of the sample. It should be noted that the carbon heat treated in nitrogen at 900°C was referred to as a 0% burn-off carbon for the combined activation. The further carbonization and the gasification of chemically activated carbon were studied by the TGA system.

Specific surface areas and porosities of the samples were determined by gas adsorption. An automated adsorption apparatus (Micromeritics, ASAP 2000, Norcross, GA) was employed for these measurements. The adsorption of N_2, as a probe species, was performed at -196°C. Before such analysis the sample was degassed at 300°C in a vacuum of ca 10^{-3} Torr. Surface areas and micropore volumes of the samples were determined from the application of the Brunauer–Emmett–Teller (BET) and Dubinin–Radushkevich (D–R) equations, respectively, to the adsorption isotherms at relative pressures between 0.06 and 0.2. The amount of N_2 adsorbed at relative pressures near unity corresponds to the total amount adsorbed in both micropores and mesopores; and, consequently the subtraction of the micropore volume (from the D–R equation) from the total amount (determined at $p/p_0 = 0.98$ in this case) will provide the volume of the mesopores [17]. The average pore diameter can be determined according to the surface area and total pore volume (the sum of the micropore and mesopore volumes), if the pores

Table 1. Analysis of BW bituminous coal

Ultimate (wt%, dry-ash-free basis)		Proximate (wt%, as-received)	
Carbon	83.2	Moisture	2.7
Nitrogen	3.7	Volatile matter	26.7
Hydrogen	5.1	Fixed carbon	61.8
Oxygen	7.4	Ash	8.9
Sulfur	0.6		

are assumed to be cylindrical and have no intersection.

A Hitachi S-4000 scanning electron microscope was used to study the structural features of the carbon surface.

3. RESULTS AND DISCUSSION

The preparation of activated carbon by chemical activation principally consists of two consecutive steps: impregnation and carbonization. It has been reported that the ratio of reagent to precursor and the conditions of impregnation and carbonization are important factors to determining the properties of the resulting activated carbon. These factors were extensively examined in the present study. The pore structures of the carbons prepared by chemical activation were then compared with those obtained by physical activation with CO_2 [10]. The development of porosity in combined H_3PO_4–CO_2 activation was also studied. These results are discussed separately as follows.

3.1 Carbonization behavior

As stated in Section 2, carbonization was carried out under an N_2 atmosphere. The volatile evolution during carbonization of the H_3PO_4 treated samples was monitored by TGA and the results were compared to those of untreated coals. These samples were heated from 30 to 800°C at a heating rate of 30°C min^{-1}. Fig. 1 shows a comparison of the carbonization behaviors of the BW samples. One can observe from Fig. 1(a) that the carbonization process for the untreated sample can be approximately described by the evolution of moisture and volatile. Tar is probably a predominant product of the devolatilization process, so that a sharp peak for volatile evolution is observed. In a blank test, the carbonization of the samples treated with 0% H_3PO_4 (i.e. pure water) at 85°C for 3 hours showed similar results as shown in Fig. 1(a), indicating that the structure of the BW coal was little affected by contacting with water.

For the carbonization of the sample treated with 40% H_3PO_4 at 85°C for 3 hours, as shown in Fig. 1(b), a strong evolution of volatiles occurs below 300°C. The composition of the evolved matter can be water and, possibly, be carbon oxides and various volatile hydrocarbons. It has been reported [8] that H_3PO_4 accelerates the bond cleavage reactions, leading to the early evolution of volatiles. Comparing the results of Fig. 1(b) with those of Fig. 1(a), one can observe that the evolution of tar, which mostly occurs between 400 and 600°C, has been suppressed for the H_3PO_4 treated samples. This is consistent with other published findings [8,16] and with the proposition that a more highly cross-linked structure is developed after acid treatment, which is less prone to volatile loss. After losing water, the condensed phosphoric acid contains not only phosphoric acid but a mixture of polyphosphoric acids, including predominant

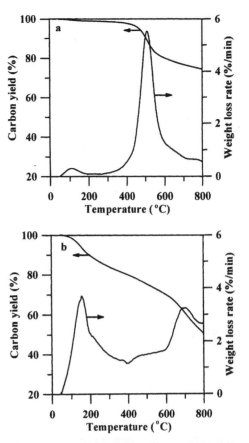

Fig. 1. Carbon yield and weight loss rate during carbonization of the different BW samples, using a heating rate of 30°C min^{-1}: (a) untreated BW coal (wt% on an as-received untreated coal basis); (b) BW coal treated with 40% H_3PO_4 at 85°C for 3 hours, having an H_3PO_4/coal weight ratio of 2.0 (wt% on an as-received treated coal basis).

species such as H_3PO_4, $H_4P_2O_7$ and $H_5P_3O_{10}$ and some others in lower proportion (e.g. $H_{n+2}P_nO_{3n+1}$) [13]. In a blank test, in which an aqueous solution of 85% H_3PO_4 was heated up from room temperature to 800°C in the TGA, it was found that *ca* 25% of weight loss occurred between 100 and 400°C, representing water loss upon heating. Another major weight loss, accounting for *ca* 65% of the H_3PO_4 solution, occurred between 600 and 800°C. Obviously, the polyphosphoric acids underwent decomposition or evaporation in this temperature range. Therefore, these polymerized acids could account for a significant part of the evolution of the H_3PO_4 treated sample during the heat-up from 600 to 800°C, as shown in Fig. 1(b). The existence of polyphosphates has also been suggested by other workers [8] studying the preparation of carbons from a bituminous coal by phosphoric acid activation.

3.2 Effects of the temperature and time of impregnation

The conditions, including temperature and time, under which the H_3PO_4 impregnation carried out is considered to play an important role in determining the pore structure of the resulting carbons [13]. To explore the effects of the impregnating conditions coal samples were added to a 40% H_3PO_4 solution stirred at different temperatures (50 and 85°C), and maintaining the mixing for different periods of time (1 or 3 hours). The H_3PO_4 treated samples were then carbonized at 500°C for 2 hours. The surface properties of the resulting carbons are shown in Table 2. It can be seen that the differences in the results of Table 2 are small. Therefore, the effects of the temperature and time of impregnation on the properties of the carbons are minor, on the basis of the present study. The results reveal that the impregnating temperature has little impact on the properties of the resulting carbons. However, the surface area and pore volume of the carbons were slightly increased due to extending the impregnating time from 1 to 3 hours, indicating that the time for impregnation should be taken into account in the preparation process to ensure access of the H_3PO_4 to the interior of the coal.

3.3 Effects of the ratio of chemical reagent to coal and carbonization time

It has been stated in the previous section that prior to carbonization the BW samples with different chemical ratios (H_3PO_4/coal) were prepared by treating the coal with different concentrations of H_3PO_4 solutions. The ratio can have a significant influence on the development of pore structure [8,13]. The time of carbonization with chemical agents has also been reported to be important in determining the pore structure of the resulting carbons [1]. The effects of the ratio and the time of carbonization on the pore structure of the resulting carbons were explored, and the results are shown in Fig. 2. The carbons were prepared by H_3PO_4 treatments carried out at 85°C for 3 hours, followed by carbonization at 500°C. As shown in Fig. 2, the surface area and total pore volume of the carbons increase with the chemical ratio at constant carbonization time. The results are consistent with the findings of other workers [13] who suggested that increasing the amount of phosphorus leads to an increase in the volumes of micro and mesopores. The variations of the volumes in micro and mesopores with ratio are shown in Fig. 3, showing that the volumes increase with the amount

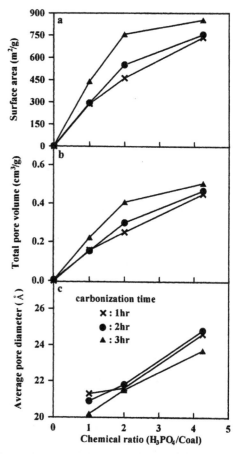

Fig. 2. Effects of the ratio of H_3PO_4 to coal and the carbonization time on (a) surface area, (b) total pore volume and (c) average pore diameter of carbons prepared by carbonization at 500°C. (Prior to the carbonization, the BW coal was treated with H_3PO_4 at 85°C for 3 hours.)

of H_3PO_4 impregnated. The experimental data in Fig. 2(c) also show an increase in the average pore diameter with increasing chemical ratio, showing that the development of porosity is also accompanied by a widening of the porosity as the amount of H_3PO_4 is increased.

While maintaining a constant chemical ratio, Fig. 2 shows that the surface area and pore volume of the resulting carbons increase with the time for carbonization at 500°C. The increase in the surface area and pore volume with carbonization time indicates that prolonged heat treatment is required for the full development of porosity at 500°C. The development of porosity is also accompanied by a narrowing of the porosity since the average pore diameter is smallest for the carbonization of 3 hours, as revealed in Fig. 2. It can be seen from Fig. 3 that the micropore volume generally increases with the carbonization time, whereas the carbonization time has little influence on the mesopore volume. However, as the carbonization temperature was increased to 600°C, it was found that the surface area and pore volume

Table 2. Properties of carbons prepared by impregnating the BW coal with a 40% H_3PO_4 solution under different conditions, followed by carbonization at 500°C for 2 hours

Impregnating temperature (°C)	Impregnating time (hour)	BET Surface area (SA) ($m^2 g^{-1}$)	Pore volume ($cm^3 g^{-1}$)
50	1	517	0.301
50	3	536	0.303
85	1	510	0.276
85	3	551	0.300

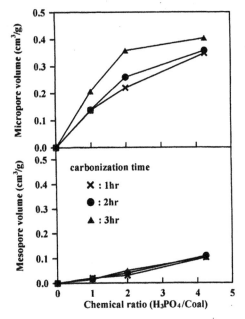

Fig. 3. Variations of the micro and mesopore volumes with the ratio of H_3PO_4 to coal and the carbonization time for carbons prepared by carbonization at 500°C. (Prior to the carbonization, the BW coal was treated with H_3PO_4 at 85°C for 3 hours.)

of the resulting carbons decreased with the time for carbonization (the data are not shown). Since the char structure has already been well developed at this temperature, the increase in carbonization time simply causes the break of the cross-links between carbon structures, with a consequent collapse of pores. Obviously, the time for carbonization affects the porosity of the resulting carbons, and the influence varies with the carbonization temperature.

3.4 Effect of the carbonization temperature

One of the advantages presented by the chemical activation of coals compared to physical activation is the lower temperature of the process. It has been reported that "phosphoric acid treatment of a bituminous coal accelerates structural alteration at lower temperatures than are realized by thermal treatment" [8]. "These alterations involve the loss of hydrogen, sulfur and oxygen and accelerated aromatization", leading "to rupture of weaker linkages in the coal structure and the early development of a rigidly cross-linked product through the formation of new and stronger linkages. This process is accompanied by development of a mainly microporous structure". "There is evidence that pore structure development can be tailored by varying the final heat treatment temperature".

Previous studies have shown that the maximum surface area developed with H_3PO_4 activation occurs at a carbonization temperature of 450°C for the preparation of carbons from biomass [16,18,19] and lignite [20], and 550°C from bituminous coal [8].

The temperature range of carbonization employed in the present study was 400–600°C. The variations of surface area, pore volume and average pore diameter with the carbonization temperature are shown in Table 3, in which the activations were carried out by treating the coal with a solution of 85% H_3PO_4 at 85°C for 3 hours, followed by carbonization at different temperatures for 3 hours. It can be seen from the results that the surface area, pore volume and average pore diameter of the resulting carbons pass through a maximum at a carbonization temperature of ca 500°C. This trend is consistent with the findings from previous studies [8,16,20]. The results show that the carbons are mainly microporous, but with a non-negligible mesoporosity. The increase in porosity by raising the carbonization temperature from 400 to 500°C can be attributed to the release of tars from the cross-linked framework generated by acid treatment at lower temperatures [21]. For carbonization temperatures above 500°C, the N_2 adsorption capacity decreases with increasing temperature. The increase in temperature from 500 to 600°C may induce not only a weight loss, but also a shrinkage in carbon structure, leading to a reduction, as well as a narrowing, in porosity [1,8,12,16]. The structural contraction above 500°C suggests that "the cross-links formed at lower temperature have reached their limit of thermal stability" [16]. The "extensive contraction upon thermal treatment collapses the porous structure and renders most of the porosity inaccessible to nitrogen".

3.5 Effect of combined activation with CO_2

The process of combined activation—physical after chemical—has already been tested for peach stones previously activated with H_3PO_4, and later gasified with CO_2 [12]. The main effect of gasification with CO_2 is similar to that described for other carbons, namely the creation and widening of the existing pores. However, since activation with H_3PO_4 may produce very different pore size distributions, a common CO_2 activation process, following the chemical activation, may produce very different effects [12]. The preparation of activated carbon from bituminous coal using combined activation has never been reported in literature. It would be of interest to compare the properties of carbons prepared by partial gasification in CO_2 and by combined activation.

In the combined activation process, a chemically activated carbon, which was prepared by treating the coal with 40% H_3PO_4 at 85°C for 3 hours, followed by carbonization at 500°C for 3 hours and leaching to remove H_3PO_4, was further activated by gasification with CO_2 at 900°C for different periods of time. In this way, final carbons covering a wide range of burn-off were obtained. It has been stated that since CO_2 gasification took place at a much higher temperature (900°C versus 500°C) a weight loss was expected during the heating stage up to 900°C in N_2. The carbon heat treated at 900°C in N_2 was

Table 3. Effects of the carbonization temperature on the properties of carbons prepared by H_3PO_4 activation (prior to the carbonization, which was lasted for 3 hours at the carbonization temperature, the BW coal was impregnated with a solution of 85% H_3PO_4 at 85°C for 3 hours)

Carbonization temperature (°C)	BET SA ($m^2 g^{-1}$)	Pore volume ($cm^3 g^{-1}$)	Pore size distribution		Average pore diameter (Å)
			Micro (%)	Meso (%)	
400	458	0.237	92	8	21.7
500	854	0.507	80	20	23.7
600	549	0.316	82	18	23.0

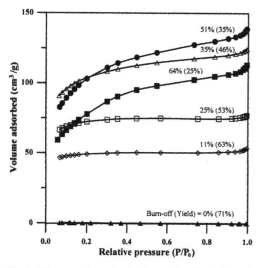

Fig. 4. Nitrogen adsorption isotherms for physically activated carbons with different extents of burn-off in CO_2 at 900°C. (Prior to the activation, the BW coal was carbonized in N_2 by heating at 30°C min^{-1} from room temperature to 900°C.)

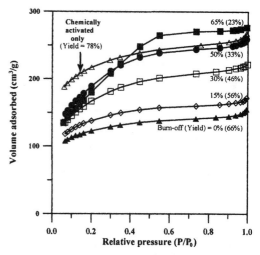

Fig. 5. Nitrogen adsorption isotherms for carbons prepared from the combined activation—physical after chemical— with different extents of burn-off in CO_2 at 900°C. (The chemical activation consisted of treating the coal with 40% H_3PO_4 at 85°C for 3 hours, followed by carbonization at 500°C for 3 hours and leaching to remove H_3PO_4.)

designated as the final carbon with 0% burn-off. The extent of burn-off represents the fraction of carbon removed by reaction with CO_2.

As a comparison, Fig. 4 shows the isotherms of the carbons obtained by physical activation with CO_2 at 900°C. The isotherms in Fig. 4 are typical of microporous carbons [10]. The adsorptive capacity of the activated carbon generally increases upon gasification except for the highest burn-off (64%), indicating a decrease in pore volume at high burn-off levels. It can also be seen from Fig. 4 that at low levels of burn-off the activated carbon is mainly microporous, and as the activation proceeds there is a widening of porosity due to increases in supermicro-porosity and mesoporosity, as inferred from the opening of the knee of the isotherm and the higher slope of the plateau [2].

Nitrogen adsorption isotherms for the carbons obtained by the CO_2 activation of the chemically treated carbon are shown in Fig. 5. It is important to explain the effects of increasing the temperature of the chemically activated carbon to 900°C before considering the effect of CO_2 activation at this temperature. As shown in Fig. 5, increasing the temper-

ature to 900°C produces a decrease in adsorption capacity and a slightly sharper knee of the isotherm, indicating that the heat treatment produces not only a weight loss, but also an ordering of the structure leading to a narrowing of the microporosity [12]. Activation with CO_2 increases the amount of N_2 adsorption, especially in the range of high relative pressures. The shape of the isotherm also changes with the extent of burn-off. The opening of the knee of the isotherm with activation in CO_2 indicates the widening of the porosity. A comparison of the knees of the isotherms shown in Figs. 4 and 5 also reflects that the carbons prepared from the combined activation possess a higher fraction of mesopores than those obtained by physical activation to a similar burn-off level.

The adsorption isotherms shown in Figs. 4 and 5 were employed to calculate the BET surface area and pore volumes, and the results are shown in Table 4. As expected from the adsorption isotherms, the surface area, pore volume and average pore diameter generally increase upon activation for the carbons from both processes. It should be noted that the surface areas of the carbons from both series increase

Table 4. Properties of carbons prepared by different activation processes

Burn-off* (%)	Yield[†] (%)	BET SA ($m^2 g^{-1}$)	Pore volume ($cm^3 g^{-1}$)	Pore size distribution		Average pore diameter (Å)
				Micro (%)	Meso (%)	
Physical activation						
0	71	0.15	ND[‡]	ND	ND	ND
11	63	165	0.081	96	4	19.7
25	53	247	0.12	99	1	19.1
35	46	360	0.19	90	10	21.2
51	35	370	0.21	84	16	22.8
64	25	282	0.17	79	21	24.3
Combined activation—physical after chemical						
Chemical[§]	78	756	0.408	88	12	21.5
0	66	427	0.234	87	13	21.9
15	56	483	0.260	89	11	21.3
30	46	597	0.339	84	16	22.7
50	33	687	0.395	83	17	23.0
65	23	657	0.436	71	29	26.5

*Percentage of the char carbonized at 900°C.
[†]Percentage of the original coal.
[‡]Not detectable.
[§]A chemically activated carbon prepared by treatment with 40% H_3PO_4 at 85°C for 3 hours, followed by carbonization at 500°C for 3 hours and leaching to remove H_3PO_4.

upon activation to a maximum value at burn-off levels *ca* 50% and begin to decrease with further activation. However, the pore volume does not show a decrease at high extent of burn-off for the combined activation series. This can be explained by the fact that the removal of carbon atoms at high extent of burn-off results in the elimination of pore walls, leading to a decrease in surface area. For the physically activated carbons, the pore volume decreases with activation at high extent of burn-off, indicating that gasification of the exterior surface of the carbons dominates [10].

When comparing the properties of the corresponding carbons (similar extent of burn-off or product yields) from these two processes, one can observe from Table 4 that the surface area, pore volume and average pore diameter of the carbon from the combined activation process are larger than those of the carbon from the physical activation process. The high porosity and the large pore size of the high burn-off (65%) carbon from the combined activation suggest that treatment with H_3PO_4 followed by CO_2 activation is suitable for producing high porosity carbons with a high proportion of mesoporosity (as high as 30%). As considering the activated carbon yield, the results in Table 4 show that, in comparison with the physical activation, high porosity carbons can be produced from the combined activation at a relatively high carbon yield, which is considered as a merit in a commercial process.

3.6 Comparison of the external surfaces of the resulting carbons

Scanning electron micrographs (SEMs) of the external structures of BW carbons prepared from by the chemical and combined activations are compared in Fig. 6. Figure 6(a) shows the structure of a chemi-

cally activated carbon prepared by treatment with a solution of 40% H_3PO_4 at 85°C for 3 hours, followed by carbonization at 500°C for 3 hours. The surface properties of this carbon has been presented in Table 4. The external surface of the chemically activated carbon, as shown in Fig. 6(a), is full of cavities. The reason for the formation of the cavities on the H_3PO_4 activated carbon is not clear. According to the micrograph, it seems that the cavities resulted from the removal of the phosphoric and polyphosphoric acids during preparation, leaving the space previously occupied by the acids. The carbonization temperature for chemical activation was too low to cause the agglomeration of the char structure. Figure 6(b) shows the SEM of the carbon prepared by partially gasifying the chemically activated carbon in CO_2 to 15% burn-off. Since the gasification temperature was high, 900°C, caking and agglomeration occurred on char structures and thus resulted in chars having no obvious cavities on the external surface, as revealed in Fig. 6(b). The structure of the carbon from the combined process with 65% burn-off is revealed in Fig. 6(c), showing that the carbon from the combined activation still shows no significant change in exterior surface structure or morphology, even at a high burn-off level, thereby indicating the gasification occurs mainly in the interior of the particles to increase the pore volume.

Examples of the SEM of physically activated carbons are shown in Fig. 7. The micrographs reveals that the activated carbon at a low burn-off levels has an intact external structure, indicating that most of the pore development occurs inside the particles. At a high burn-off level, the sample has a deficient and eroded surface, indicating that the occurrence of gasification of the external surface is significant, possibly due to a strong resistance for CO_2 diffusion

38

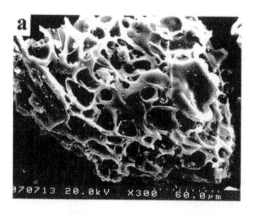

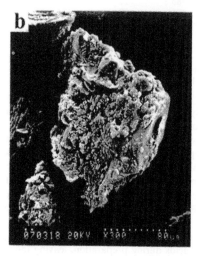 

Fig. 6. SEMs of carbons prepared by (a) chemical activation with carbonization at 500°C for 3 hours, corresponding to 78% yield; (b) combined activation with 15% burn-off in CO_2, corresponding to 56% yield; (c) combined activation with 65% burn-off in CO_2, corresponding to 23% yield.

in micropores [10]. The erosion of the exterior portion of a char particle upon activation should result in the disappearance of the exterior pores; this explains why both the surface area and the pore volume decrease with the degree of activation at high burn-off levels.

4. SUMMARY

This study has demonstrated that H_3PO_4 is a suitable activating agent for the preparation of high porosity carbons from bituminous coal. The surface area and the pore volume are larger for the carbon prepared from chemical activation with H_3PO_4 than that from physical activation with CO_2.

The carbonization behavior of the H_3PO_4 treated coal was compared with that of the untreated coal. For the carbonization of the H_3PO_4 treated sample, the release of volatiles is suppressed, indicating that a more highly cross-linked structure is developed

after acid treatment, which is less prone to volatile loss.

Investigations of the impregnating conditions have shown that the time for impregnation should be taken into account to ensure the access of H_3PO_4 to the interior of the coal. The surface area and pore volume of the resulting carbons increase with the chemical ratio, H_3PO_3/coal. The development of porosity is also accompanied by a widening of the porosity as the amount of H_3PO_4 is increased.

The carbonization time also affects the porosity of the resulting carbons, and the influence varies with the carbonization temperature. Within the ranges of carbonization temperature and time in the present study, the chemically activated carbon prepared by carbonization at 500°C for 3 hours was found to have maximum values of surface area and pore volume.

Physical activation with CO_2 of the previously H_3PO_4 activated carbon results in a decrease in

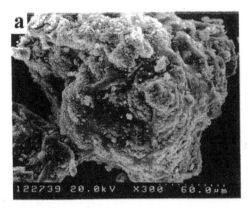

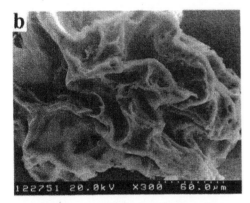

Fig. 7. SEMs of the physically activated carbons with (a) 11% burn-off, corresponding to 63% yield and (b) 64% burn-off, corresponding to 25% yield, in CO_2.

activation at high burn-off levels, gasification occurs mainly in the interior of the particles for the combined activation, whereas the occurrence of gasification of the external surface is significant for the physical activation.

Acknowledgements—This research was supported by the National Science Council of Taiwan, through Project NSC 86-2214-E-033-008.

porosity at low burn-off levels. The porosity of the resulting carbon increases upon further activation in CO_2. The results suggest that the combined activation—physical after chemical—is suitable for producing high porosity carbons with a high proportion of mesoporosity.

A scanning electron microscopic study shows that the external surface of a chemically activated (H_3PO_4 treated) carbon is full of cavities. For CO_2

REFERENCES

1. Ahmadpour, A. and Do, D. D., *Carbon*, 1996, **34**, 471.
2. Muñoz-Guillena, M. J., Illán-Gómez, M. J., Martín-Martínez, J. M., Linares-Solano, A. and Salinas-Martínez de Lecea, C., *Energy Fuels*, 1992, **6**, 9.
3. Bansal, R. C., Donnet, J. B. and Stoekcli, H. F., *Active Carbon*. Marcel Dekker, New York, 1988.
4. Greenbank, M. and Spotts, S., *Water Technology*, 1993, **16**, 56.
5. Ahmadpour, A. and Do, D. D., *Carbon*, 1995, **33**, 1393.
6. Laine, J. and Yunes, S., *Carbon*, 1992, **30**, 601.
7. Wigmans, T., *Carbon*, 1989, **27**, 13.
8. Jagtoyen, M., Thwaites, M., Stencel, J., McEnaney, B. and Derbyshire, F., *Carbon*, 1992, **30**, 1089.
9. Illán-Gómez, M. J., García-García, C., Salinas-Martínez de Lecea, C. and Linares-Solano, A., *Energy Fuels*, 1996, **10**, 1108.
10. Teng, H., Ho, J.-A., Hsu, Y.-F. and Hsieh, C.-T., *Ind. Eng. Chem. Res.*, 1996, **35**, 4043.
11. Teng, H., Ho, J.-A. and Hsu, Y.-F., *Carbon*, 1997, **35**, 275.
12. Molina-Sabio, M., Rodríguez-Reinoso, F., Caturla, F. and Sellés, M. J., *Carbon*, 1996, **34**, 457.
13. Molina-Sabio, M., Rodríguez-Reinoso, F., Caturla, F. and Sellés, M. J., *Carbon*, 1995, **33**, 1105.
14. Caturla, F., Molina-Sabio, M. and Rodríguez-Reinoso, F., *Carbon*, 1991, **29**, 999.
15. Kirubakaran, C. J., Krishnaiah, K. and Seshadri, S. K., *Ind. Eng. Chem. Res.*, 1991, **30**, 2411.
16. Jagtoyen, M. and Derbyshire, F., *Carbon*, 1993, **32**, 1185.
17. Rodríguez-Reinoso, F., Molina-Sabio, M. and González, M. T., *Carbon*, 1995, **33**, 15.
18. Laine, J., Calafat, A. and Labady, M., *Carbon*, 1989, **27**, 191.
19. Solum, M. S., Pugmire, R. J., Jagtoyen, M. and Derbyshire, F., *Carbon*, 1995, **33**, 1247.
20. Toles, C., Rimmer, S. and Hower, J. C., *Carbon*, 1996, **34**, 1419.
21. Ibarra, J. V., Moliner, R. and Palacios, J. M., *Fuel*, 1991, **70**, 727.

Activated Carbon Compendium
H. Marsh (Editor)

41

DEVELOPMENT OF POROSITY IN COMBINED PHOSPHORIC ACID–CARBON DIOXIDE ACTIVATION

M. Molina-Sabio, F. Rodríguez-Reinoso,* F. Caturla, and M. J. Sellés,
Departamento de Química Inorgánica, Universidad de Alicante, E-03080 Alicante, Spain

(*Received* 30 *January* 1995; *accepted in revised form* 11 *September* 1995)

Abstract—Five activated carbons with different porosity, prepared by chemical activation of peach stones with phosphoric acid, have been further activated in a carbon dioxide gas flow at 825°C for different periods of time to cover a wide range of burn-off. The porosity of all activated carbons was determined by adsorption of N_2 (77 K), CO_2 (273 K) and n-C_4H_{10} (273 K). The main effect of gasification with carbon dioxide is similar to that described for other carbons, namely the creation and widening of existing pores, the predominance of one or another being a function of burn-off. However, since activation with phosphoric acid may produce very different pore size distributions, a common carbon dioxide activation process may produce very different effects, which range from the development of only micropores to the development of only mesopores, thus enhancing the differences among the initial chemically activated carbons. Copyright © 1996 Elsevier Science Ltd

Key Words—Activated carbon, porosity, physical activation, chemical activation.

1. INTRODUCTION

Chemical activation with phosphoric acid of cellulosic and lignocellulosic materials [1–4] and some types of coal [5,6] may lead to activated carbons with highly developed porosity and, consequently, with the adsorption properties expected in competitive activated carbons. It is well known that one of the variables having a larger effect on the porosity of the final carbon is the degree of impregnation. This has been shown in previous work [7] on the phosphoric acid activation of peach stones, when a given amount of the precursor was impregnated with a constant volume of phosphoric acid solution, and the degree of impregnation (expressed as grams of phosphorous retained per gram of precursor, on a dry basis, X_p) was modified by changing both the initial concentration of the acid solution and the extent of evaporation. The analysis of the porosity of the resulting activated carbons showed that, during the impregnation stage, the phosphoric acid introduced into the material plays a double role: (i) it produces the hydrolysis of the lignocellulosic material and the subsequent partial extraction of some components, thus weakening the particle, which swells; and (ii) the acid occupies a volume which inhibits the contraction of the particle during the heat treatment, thus leaving a porosity when it is extracted by washing after carbonization. Both effects cause the micro- and mesopore volume to increase with increasing X_p. At the same time, it was found (when plotting the data for all the series of carbons prepared) that there is a good agreement between the volume of micropores and the volume occupied by the acid phase existing at the carbonization temperature ($P_2O_5.xH_2O$) up to $X_p \leq 0.4$. For larger values of X_p there is no

further increase in micropore volume, because the larger extent of hydrolysis caused by the high concentration of phosphoric acid produces a larger weakening of the cellular structure, which favours the larger development of mesoporosity.

If one takes into consideration that the yield of chemical activation by phosphoric acid is very high, of the order of 40%, and that the bulk density is also large, the possibility of further physical activation becomes very attractive for the preparation of especially activated carbons. This process of combined activation—physical after chemical—has already been tested for the same material (peach stones) previously activated with $ZnCl_2$ and later gasified with carbon dioxide [8]. Partial gasification with carbon dioxide developed surface area and porosity to a maximum for around 60–70% burn-off. In this way it was possible to prepare granular microporous activated carbons with very high apparent surface area (up to 3700 m^2/g), and reasonable bulk density.

The present work presents the results corresponding to the carbon dioxide activation of five activated carbons prepared by phosphoric acid activation (using different experimental conditions to obtain very different pore size distributions), and analyses the changes in bulk density and porosity as a function of those of the original activated carbon and the extent of carbon dioxide burn-off.

2. EXPERIMENTAL

The detailed preparation of the five activated carbons obtained by phosphoric acid activation (B1, B5, C1, C2, C3) can be found elsewhere [7]. In summary, 50 g of the starting material (peach stones, with a particle size of 2.8–3.5 mm) were impregnated with 100 ml of solutions of phosphoric acid at 85°C for

---

*To whom all correspondence should be addressed.

Reprinted from *Carbon* 34 (4), 457-462 (1996)

2 h. The mixtures were then subjected to partial evaporation, which progressively modifies the concentration of the solution and, consequently, the amount of "phosphoric acid" absorbed by the material; this amount was experimentally determined [7], and is expressed as grams of phosphorous per gram of lignocellulosic precursor on a dry basis, X_p (see Table 1). Carbons B1 and B5 were obtained using an initial 68 wt% H_3PO_4 solution whereas for the other three carbons a 85 wt% H_3PO_4 solution was used. The impregnated materials were heat treated under a flow of nitrogen for 4 h at 450°C and thoroughly washed with distilled water.

The five activated carbons were further activated by gasification with carbon dioxide at 825°C for different periods of time (Table 1). In this way five different series of activated carbons, each one covering a wide range of burn-off, were obtained. Since physical activation takes place at a much higher temperature (825°C versus 450°C) a loss of material is expected during the heating stage up to 825°C, due to the further carbonization of the sample. In order to distinguish this weight loss from the one occurring on carbon dioxide gasification, the five original activated carbons were also subjected to a heat treatment in a nitrogen flow at 825°C. Each final carbon has been designated by using the nomenclature of the initial chemically activated carbon followed by the burn-off reached in the carbon dioxide gasification (e.g. carbon C2-0 is carbon C2 heat treated in nitrogen at 825°C, and carbon C2-9 is carbon C2 gasified up to 9% burn-off)

All activated carbons have been characterized by

adsorption of $n\text{-}C_4H_{10}$ (273 K); one of the series, C2, was also characterized by adsorption of N_2 (77 K) and CO_2 (273 K).

3. RESULTS AND DISCUSSION

The yields obtained in the chemical activation process (Table 1) are high when compared with the yield of simple carbonization of the same material (unimpregnated), around 28% [8], indicating the ability of the chemical—H_3PO_4 in this case—to retain carbon material and to avoid the loss of otherwise volatile material. The differences among the yield values for the five original activated carbons are due to the impregnation method, as discussed elsewhere [7].

The weight losses due to the heat treatment at 825°C (expressed on the basis of the chemically activated product) are given in Table 2, where it is shown that they are very similar for the five original activated carbons. This means that the volatile matter content of the chemically activated carbon is a func-

Table 2. Weight loss upon heat treatment in nitrogen (825°C) of chemically activated carbons and rate of reaction for the C–CO_2 reaction at 825°C

Carbon	Weight loss (%)	$R \times 100$ (h^{-1})
B1	11	3.4
B5	9	4.0
C1	13	3.7
C2	13	4.3
C3	9	6.7

Table 1. Experimental details of phosphoric acid and carbon dioxide activation

Carbon	X_p (g/g)	Activation time (h)	Yield (%)	Burn-off in CO_2 (%)	Bulk density (g/cm³)
B1	0.21		43		0.56
B1-25		7		25	0.52
B1-47		14		47	0.40
B1-61		18		61	0.32
B5	0.36		38		0.40
B5-24		6		24	0.43
B5-49		12		49	0.33
B5-60		15		60	0.30
B5-69		18		69	0.27
B5-75		20		75	0.25
C1	0.30		36		0.45
C1-22		6		22	0.46
C1-45		12		45	0.38
C1-56		15		56	0.33
C1-64		17		64	0.30
C1-71		19		71	0.27
C2	0.34		33		0.37
C2-9		3		9	0.43
C2-30		7		30	0.36
C2-44		10		44	0.32
C2-52		12		52	0.28
C2-60		14		60	0.26
C2-68		16		68	0.24
C3	0.91		44		0.24
C3-21		3		21	0.27
C3-40		6		40	0.24
C3-54		8		54	0.22
C3-60		9		60	0.21

tion of the temperature used in the heat treatment and not of the impregnation conditions. These values, together with the weight loss measured upon carbon dioxide activation allow the calculation of the burn-off exclusively due to the C–CO₂ reaction listed in Table 1. As in the case of carbon dioxide activation of lignocellulosic chars [9], there is a linear relationship between activation time and burn-off. The slope of these lines has been used to calculate the reaction rate of each carbon with carbon dioxide (Table 2). These values are larger than those corresponding to the carbonized peach stones at the same temperature, 0.01 h^{-1}, and confirm that impregnation with phosphoric acid greatly modifies the carbonization process and the structure of the resulting carbon [4]. The increase in reactivity with the degree of impregnation, X_p, may very possibly be due to the important differences in pore size distribution, as shown below.

Figure 1 shows, as a typical example, the adsorption isotherms of $n\text{-}C_4H_{10}$ at 273 K on some carbons obtained by carbon dioxide activation of carbon C2. For the original carbon C2 the amount adsorbed at low relative pressures is low and rapidly increases up to $P/P_0 = 0.4$, remaining almost constant thereafter. Therefore, it is concluded that carbon C2 has a large adsorption capacity caused by microporosity, with a wide micropore size distribution (even extending to small mesopores). It is now important to define the effects of increasing the temperature of carbon C2 to 825°C before considering the effect of carbon dioxide activation at this temperature. As shown in Fig. 1, increasing the temperature to 825°C, carbon C2-0, produces a decrease in the amount of adsorption and a slightly sharper knee of the isotherm, indicating that the heat treatment produces not only a weight

loss (Table 2), but also an ordering of the structure leading to a narrowing of the microporosity. Similar results have been described for other materials, the phenomenon being generically termed secondary carbonization [10,11]. Activation with carbon dioxide does not change much the shape of the isotherm but considerably increases the amount of n-butane adsorbed, especially in the range of relative pressures up to 0.4, with the end result that increasing burn-off produces activated carbons with larger pore volume.

In order to better evaluate the porosity of the carbons of series C2, the Dubinin–Radushkevich equation was applied to the adsorption isotherms of N_2 (77 K), CO_2 (273 K) and $n\text{-}C_4H_{10}$ (273 K). The corresponding micropore volumes have been plotted as a function of burn-off in Fig. 2(a). There are clear differences among the values of micropore volume obtained with the three adsorptives for the original carbon C2, as it would be expected in a carbon with a heterogeneous micropore size distribution. Thus, $V_0(n\text{-}C_4H^{10}) > V_0(N_2) > V_0(CO_2)$. As established in previous work [12,13], the adsorption of CO_2 at 273 K (the high temperature of adsorption limits the relative pressure covered to a maximum of 0.03) provides the volume of narrow microporosity—up to 0.7 nm pore width—whereas the adsorption of N_2 at 77 K gives the total volume of micropores—up to 2.0 nm pore width.

The larger value of $V_0(n\text{-}C_4H_{10})$ with respect to $V_0(N_2)$ may be explained by taking into account the differences in molecular size of the adsorptives and the wide micropore size distribution of the carbons. Thus, a carbon with narrow and uniform microporosity accessible to both adsorptives should have the same value of micropore volume for nitrogen and n-butane. However, when the micropore size distribution is wide, the parameter limiting the range of micropores is not the pore width itself but the pore width/molecular diameter ratio, which is the measure of the enhancement of the adsorption potential in the micropore [14]. For these carbons, in which there is not a clear separation between micro- and mesopores, the upper limit of micropore size measured by adsorption techniques will be larger for the molecule with larger dimensions, n-butane.

When carbon C2 is activated with carbon dioxide there are two effects to be considered. First, the increase in temperature from 450°C (temperature of heat treatment used in the preparation of the chemically activated carbon) to 825°C produces a decrease in micropore volume; and secondly, activation of carbon dioxide at 825°C produces an increase in such volume. The balance of both effects for burn-offs up to 10% means a net decrease in microporosity (Fig. 2(a)). This decrease is greater for $n\text{-}C_4H_{10}$ than for N_2 and CO_2, this indicating that the range of wider microporosity is being modified to a larger extent by the secondary carbonization. When burn-off is around 30% the micropore volume and the micropore size distribution is similar to that of the

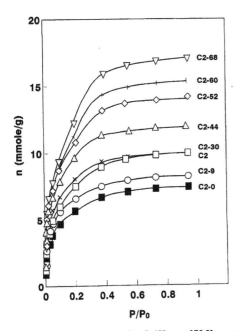

Fig. 1. Adsorption isotherms of $n\text{-}C_4 1H_{10}$ at 273 K on carbons of series C2.

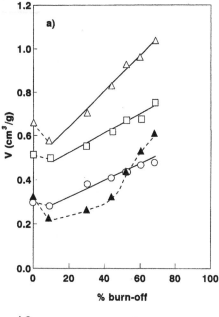

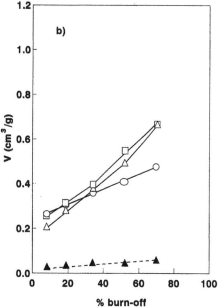

Fig. 2. Micropore volume of activated carbons as deduced from adsorption of ($\triangle$) $n\text{-}C_4H_{10}$, ($\square$) N_2, ($\bigcirc$) CO_2, and ($\blacktriangle$) mesopore volume as a function of burn-off for (a) series C2 and (b) series D.

original carbon, and only for burn-offs larger than 40% the carbons have higher micropore volume and wider micropore size distribution than the original carbon, since the increase in $V_0(CO_2)$ is less pronounced that for $V_0(N_2)$ or $V_0(n\text{-}C_4H_{10})$.

The shape of the isotherms shown in Fig. 1 indicate that the volume of $n\text{-}C_4H_{10}$ adsorbed at high relative pressures, about 0.95, is larger than $V_0(n\text{-}C_4H_{10})$. The difference between the two values for each carbon, which for the sake of comparison can be taken as

the volume of mesopores, has been included in Fig. 2(a). The tendency shown is, in general terms, similar to that followed by $V_0(n\text{-}C_4H_{10})$, although there is a more pronounced development of mesoporosity at burn-offs above 40%.

The porosity of the activated carbons obtained from carbon C2 (Fig. 2(a)) contrasts with that of activated carbons prepared by carbon dioxide activation of the char prepared from the same precursor, peach stones, or a very similar precursor, olive stones [9]. The more complete set of data for the latter (series D) will be used for the comparison (Fig. 2(b)). The evolution and volume of narrow microporosity, $V_0(CO_2)$, is very similar in carbons of series C2 and D, but in series D the volume of micropores deduced from the adsorption of N_2 and n-butane is rather lower, with a small contribution from mesoporosity. As a consequence of this, the total porosity of carbons of series D is not as large as in series C2. If one takes into account that, besides these differences, the yield for the carbonization process (the equivalent step needed in series D) is much lower than the yield in chemical activation, as a consequence of the role of the chemical (phosphoric acid in this case) in retaining volatile matter, the advantage of the combined phosphoric acid–carbon dioxide activation becomes apparent.

The evolution of porosity on carbon dioxide activation of the other series of chemically activated carbons are, in general terms, similar to those described for series C2. However, since impregnation was carried out under different experimental conditions, the porosity of the initial carbon is different in each case, as deduced from the adsorption isotherms of $n\text{-}C_4H_{10}$ in Fig. 3. Although a larger value of X_p

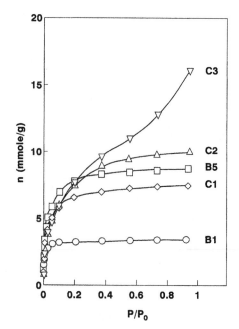

Fig. 3. Adsorption isotherms of $n\text{-}C_4H_{10}$ at 273 K on original chemically activated carbons.

means an increase in the amount adsorbed (as a consequence of the increase in the amount of phosphorous retained, X_p, after impregnation—see Table 1), the initial concentration of the acid solution also affects the development of porosity. This is shown in the isotherms of Fig. 3. Thus, carbon B5, impregnated with a 68 wt% H_3PO_4 solution until a value of X_p of 0.36 is reached, adsorbs a larger amount of n-butane at low relative pressures and a lower amount at high relative pressures than carbon C2 (85 wt% H_3PO_4 solution, $X_p = 0.34$). This behaviour has been described for other carbons prepared with different concentrations of phosphoric acid [7]. On the other hand, it is important to note that for carbons C1, C2 and C3 (85 wt% H_3PO_4 solution) the amounts adsorbed at $P/P_0 = 0.1$ are very similar and the differences caused by the increase in X_p (see Table 1) are clear at higher relative pressures, in contrast with the behaviour of carbons B1 and B5.

Figure 4 shows plots of the amount of n-C_4H_{10} adsorbed near saturation ($P/P_0 = 0.95$) and bulk density as a function of the burn-off reached during carbon dioxide activation. There is in all cases an initial decrease in porosity (due to heat treatment to 825°C), followed by a large development of porosity. Furthermore, the increase in porosity with increasing burn-off is similar in all cases and consequently the differences among the series are maintained above 20–30% burn-off. The evolution of bulk density of the carbons is parallel to that of pore volume. In fact, the decrease in porosity (caused by the narrowing of pore size) produced during the initial heat treatment from 450 to 825°C means an increase in bulk density,

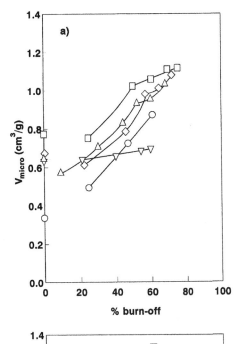

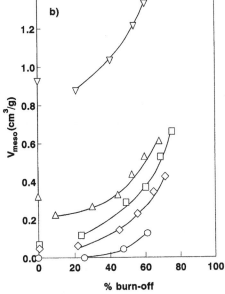

Fig. 5. Evolution of (a) micropore volume and (b) mesopore volume of activated carbons; symbols as in Fig. 4.

which decreases with increasing pore volume. The set of values of Fig. 4 indicate that the preparation process followed here permits the preparation of granular activated carbons with large adsorption capacity and reasonable bulk density; for instance a pore volume of 1.2 cm^3/g for a bulk density of 0.33 g/cm^3 (carbon C1-56). Since the pore volume increases exponentially and the bulk density decreases linearly, the amount of n-butane adsorbed by the carbon, expressed as volume per volume of carbon, increases with burn-off. If additionally one takes into account that the amount adsorbed at low relative

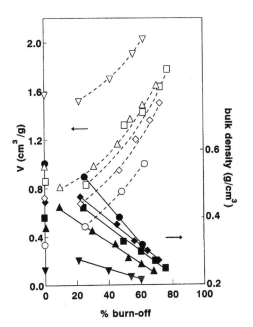

Fig. 4. Evolution of volume of n-C_4H_{10} adsorbed at $P/P_0 = 0.95$ (open symbols) and bulk density (closed symbols) upon activation with carbon dioxide of carbons of series: ○, B1; □, B5; ◇, C1; △, C2; and ▽, C3.

pressures is low, these carbons could find potential application in storage of hydrocarbons.

The fact that the volume adsorbed at $P/P_0 = 0.95$ for series B5, C1 and C2 are similar does not imply a similar pore size distribution. This is shown when the values of Fig. 4 are separated into V_0, micropore volume, and $V_{0.95} - V_0$, mesopore volume; these values have been plotted in Fig. 5. If additionally one takes into account the adsorption isotherms of Fig. 3, Fig. 5 permits the evaluation of the effects of gasification with carbon dioxide of the phosphoric acid activated carbons. Although the main effect of gasification is similar to the one described for other carbons, creation and widening of existing pores, with predominance of one or another as a function of burn-off, the porosity of the initial carbon will condition the final porosity as follows:

(1) In essentially microporous carbons, with narrow and uniform microporosity (carbon B1 or the char obtained in the carbonization of the unimpregnated precursor), there is only development of microporosity (Fig. 5(a)) with a very small contribution of mesoporosity (Fig. 5(b)).

(2) In essentially microporous carbons with wide microporosity (carbons B5, C1 and C2) there is a rapid development of microporosity up to 50% burn-off, although not as rapid as in 1. Pore widening predominates at larger burn-offs, with conversion of micropores into mesopores. The end result is a less noticeable increase in the volume of micropores and a larger increase in mesopore volume. The order followed by the micropore volume, B5 > C1 = C2, and mesopore volume, C2 > C1 > B5, is the same as in the chemically activated carbons before carbon dioxide activation.

(3) In carbons with very heterogeneous porosity (C3), for which the volume of micropores is even lower than the volume of larger size pores, the conversion of micro- to mesopores already occurs at low burn-off and consequently the volume of micropores is maintained and only that of mesopores increases.

4. CONCLUSIONS

Physical activation with carbon dioxide of previously phosphoric acid activated lignocellulosic material (peach stones) results in granular activated carbons with high adsorption capacity and a wide range of pore size distribution. In this way it is possible to prepare activated carbons with large pore volume (up to 2.0 cm^3/g) compatible with a reasonable bulk density. The porosity developed by carbon dioxide activation is a function of the porosity of the initial H_3PO_4 activated carbon, this making possible the preparation of carbons ranging from essentially microporous with a very small proportion of mesopores, to carbons in which mesoporosity constitutes the largest contribution (up to 1.3 cm^3/g).

Acknowledgement—This work was supported by DGICYT (Project No. PB91-0747).

REFERENCES

1. J. Macdowall, European Patent 0 423 967 A2 (1991).
2. R. Greinke, I. Lewis and D. Ball, European Patent 0 467 486 A1 (1992).
3. J. Laine, A. Calafat and M. Labady, *Carbon* **27**, 191 (1989).
4. M. Jagtoyen and F. Derbyshire, *Carbon* **31**, 1185 (1993).
5. M. Jagtoyen, M. Thwaites, J. Stencel, B. McEnaney and F. Derbyshire, *Carbon* **30**, 1089 (1992).
6. J. Illán-Gómez, M. J. Muñoz-Guillena, C. Salinas-Martínez de Lecea and A. Linares-Solano, *Proc. Carbon'90*, p. 68. Paris, France (1990).
7. M. Molina-Sabio, F. Rodríguez-Reinoso, F. Caturla and M. J. Sellés, *Carbon* in press.
8. F. Caturla, M. Molina-Sabio and F. Rodríguez-Reinoso, *Carbon* **29**, 999 (1991).
9. F. Rodríguez-Reinoso, *Fundamental Issues in Control of Carbon Gasification Reactivity* (Edited by J. Lahaye and P. Ehrburger), NATO ASI Series E-192, p. 533. Kluwer Academic, The Netherlands (1991).
10. V. Minkova, M. Razvigorova, K. Gergova, M. Goranova, L. Ljutzkanov and G. Angelova, *Fuel* **71**, 263 (1992).
11. P. Ehrburger, J. Lahaye and E. Wozniak, *Carbon* **20**, 433 (1982).
12. J. Garrido, A. Linares-Solano, J. M. Martín-Martínez, M. Molina-Sabio, F. Rodríguez-Reinoso and R. Torregrosa, *Langmuir* **3**, 76 (1987).
13. F. Rodríguez Reinoso, J. Garrido, J. M. Martin Martínez, M. Molina Sabio and R. Torregrosa, *Carbon* **27**, 23 (1989)
14. S. J. Gregg and K. S. W. Sing, *Adsorption Surface Area and Porosity* 2nd edn, pp. 113, 207. Academic Press, London (1982).

CARBON HONEYCOMB STRUCTURES FOR ADSORPTION APPLICATIONS

K. P. GADKAREE

Corning Inc., SP-FR-5-1, Sullivan Park, Corning, NY 14831, U.S.A.

(Received 30 June 1997; accepted in revised form 4 December 1997)

Abstract—Activated carbon honeycomb structures based on synthetic precursors are described. These strong, highly durable honeycombs are continuous interpenetrating structures of activated carbon and a ceramic with adjustable broad range of carbon percentage (5–95 wt%). Dynamic adsorption performance of a particular honeycomb structure containing 18 wt% carbon is described with respect to various adsorbates, i.e. butane, toluene, formaldehyde, isopropanol etc., and the importance of space velocity in determining the adsorption performance as well as the effect of structural parameters on the performance is shown. © 1998 Elsevier Science Ltd. All rights reserved.

Key Words—A. Activated carbon, A. carbon honeycombs, D. adsorption properties.

1. INTRODUCTION

Activated carbon is a very important material industrially, with applications in a variety of areas [1] such as adsorbers in air and water pollution control, catalysts in the chemical and petrochemical industries, electrodes in batteries and supercapacitors, and purifiers in the food and pharmaceutical industries. Typically activated carbon is made from naturally occurring materials such as wood, coal and nutshell flour [2,3] etc. via high temperature, inert atmosphere processing followed by activation to create porosity in the nanometer size range. This porosity imparts special adsorption characteristics to carbon and makes it useful in the variety of applications mentioned above. As a result of the natural raw materials used as well as the necessary processing steps associated with the manufacture, the material is obtained in a finely powdered form, and is granulated later to make it suitable for handling on a large scale [1]. In spite of the widespread use of this type of product, there are some drawbacks associated with it. First, the variations in the natural material make it difficult to control the properties of the carbon from batch to batch. Secondly, the granular material has to be used in traditional packed beds in many of the pollution control applications. Although functional, such beds have inherent drawbacks such as high pressure drop associated with the flow through the packed media, particle entrainment, channeling etc. For applications such as battery electrodes fine powders may only be used with binders and metallic current collectors, which result in poor utilization of the carbon properties. To obtain carbon with controlled and reproducible properties, synthetic starting materials may be used. The synthetic materials typically include polymeric resins. Several studies [4–6] have been carried out on synthetic raw material-based carbon. These carbons eliminate the first draw-

back of the variability in raw material source and have the added advantage of control of the carbon properties through control of the synthetic material structure. In spite of these advantages the synthetic material-based carbons have not been widely available commercially presumably due to high cost. Recently a new type of product, activated carbon fibers based on synthetic raw materials has been introduced. A number of studies of the properties of these fibers and the performance advantages the fibers demonstrate, have been published [7–9]. The fibers, however, have not found widespread commercial use. One of the drawbacks of the fibers is that these small diameter ($\sim 10\ \mu m$ diameter) fibers have to be made into a structural shape to put into a device and this requires more difficult procedures such as weaving into shapes. These procedures are expensive. Availability of activated carbon in a monolithic form with controlled adsorption properties is thus desirable.

The main objective of the various devices in industrial applications is to make the special nanoporosity in the carbon accessible to a flowing stream or more accurately the components in the stream, so that these components may be adsorbed and removed from the stream. The better the efficiency of the device in doing so, the better the carbon is utilized. It is thus important that the device has as high a geometric surface area per unit volume as possible. In the case of the packed beds, although the surface area per unit volume is high, in many cases all of it may not be accessible to the fluid stream because of the preferential flow patterns that may be established. Another problem with the packed beds is that the carbon pellets in the beds are sufficiently large in diameter to prevent full utilization of carbon because of the diffusional resistance associated with diffusion through the macropores before the components are adsorbed in the meso or the microporosity.

Reprinted from *Carbon* **36 (7-8)**, 981–989 (1998)

A natural structural shape that may eliminate the problems associated with the traditional products mentioned above is the honeycomb shape. This shape is known to have a very high geometric surface area to volume ratio, a feature which has resulted in widespread use of honeycomb structures in applications such as automotive catalyst carriers. The high surface area provides high contact efficiencies between the substrate and the flow stream. If activated carbon is made available in the honeycomb form, its adsorption capacity may be utilized better and a more efficient device design may result. The pressure drop for a honeycomb structure based system will be significantly lower than a packed bed system, with the extent of decrease in pressure drop depending on the structural parameters of the honeycombs.

Although attempts have been made to form activated carbon honeycombs and some industrial products such as ozone filters in laser printers have resulted, these honeycomb products are not widely used. The main reason for the commercial honeycombs not being successful is that these honeycombs have poor durability. The honeycombs are fabricated mainly by extruding finely powdered carbon into honeycomb shapes with a polymeric binder. The bonding between the carbon particle and the binder is typically poor which results in low strength, low durability honeycombs. In liquid streams the liquid preferentially adsorbs at the interface between the particle and the binder weakening the interface still further and severe durability problems result. These honeycombs therefore have been used only in less demanding applications such as for ozone removal in laser printers.

A new class of activated carbon honeycombs have been developed, which eliminate the binders associated with the honeycomb structures mentioned above. These highly adsorbent honeycombs have a continuous carbon structure. Since no binders are used, these inert high strength honeycombs are highly durable and provide an attractive alternative to the traditional packed bed system. In the following one of the concepts for fabricating these honeycombs is presented. The dynamic adsorption performance of any activated carbon system is the key to determining the usefulness of the system in any application. Although the major advantage of the honeycomb structure is the very open structure which results in a low pressure drop, this same open structure may result in poor adsorption performance. Preliminary data is thus presented on dynamic adsorption performance of one type of such a honeycomb structure with various adsorbates and the effect of some of the process and structural parameters on this performance is discussed.

2. CONCEPT AND EXPERIMENTAL DETAILS

The honeycombs discussed in this report were fabricated as composite structures of ceramics and carbon. In principle the process is as follows. Highly porous ceramic honeycombs are first fabricated. These honeycombs are fabricated from various ceramic compositions starting with clays. Cordierite-based compositions, which are used for fabricating low thermal expansion honeycombs for automotive catalyst support were used for this work. The honeycombs are commercially available from Corning Inc., Corning, NY. Various clays are mixed together with polymeric binders and extruded through steel dies in honeycomb shapes. The honeycombs are then fired to high temperatures ($\sim 1500°C$) to burn out binders and to react and sinter clays to form cordierite honeycombs. These honeycombs may be fabricated with a wall thickness of anywhere from 0.1 mm and higher and with cell densities as high as 95 cells per square cm. The cells may be rectangular, triangular, hexagonal or other shapes. Another important factor that can be adjusted is the wall porosity. By utilizing appropriate compositions, the wall porosity may be adjusted from 10% to 70%. Most of the work described in this report was done with cordierite honeycombs with 62 cells per square cm, a wall thickness of 0.19 and 0.29 mm and wall porosity of $\sim 50\%$. The mean pore size for these honeycombs is ten microns. Figure 1 shows the pore size distribution of the honeycombs measured by the mercury porosimetry technique.

The honeycomb is impregnated with high carbon yield polymeric resin of low viscosity. The resin is allowed to soak into the ceramic honeycomb structure. The excess resin is then drained and the resin-coated honeycomb is subjected to a drying and curing cycle to crosslink the resin. The resin forms an interpenetrating network with the ceramic. The cured honeycomb is then subjected to carbonization and activation in an inert atmosphere to form the composite carbon–ceramic honeycomb. The resin chosen for this work was a phenolic resole from Occidental Chemical Co., Niagara Falls, NY, because of two important characteristics. This resole has a viscosity of 100 cP. This low viscosity allows the impregnation and draining step to be carried out with ease. In addition the inexpensive phenolic resins have a very high carbon yield ($\sim 50\%$ of the cured weight), thus

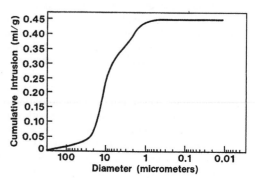

Fig. 1. Pore size distribution of ceramic honeycomb.

reducing the cost of the carbon produced. The aqueous resole contains about 65% solids. The honeycombs are simply dipped in the resin, allowed to soak for a few minutes and then drained of the excess resin by blowing air through the cells, before being subjected to a drying and curing process. The coated honeycombs are air dried and then dried at 95°C and cured at 150°C to crosslink the resin. Most of the cured resin remains in the cell wall porosity with a thin layer of resin on the surface. The coated honeycomb is then subjected to a carbonization and activation process. The carbonization is carried out at 900°C in nitrogen. The activation is carried out at the same temperature in carbon dioxide to obtain burnoff in 25–30% range. On carbonization a ceramic–carbon composite structure is formed. This structure is monolithic with carbon forming a continuous structure inseparable from the ceramic backbone. As the SEM micrographs show, it is difficult to see the boundary between the carbon and the ceramic. The SEM's of the ceramic honeycomb and the composite honeycomb are shown in Fig. 2.

Figure 2(a) shows the polished cross section of the ceramic as well as the composite carbon honeycomb. It is clear from the micrograph that the wall thickness of the honeycomb does not change significantly as a result of the carbon impregnation and the large

porosity in the honeycomb wall is not strongly affected. Figure 2(b) shows a high magnification view of the fracture surface of the honeycombs. It is clearly seen that the carbon has formed on the ceramic and also occupied the small pores in the wall. The activation step generates the desired porosity in the carbon for adsorption without affecting the strength of the structure significantly. The carbon–ceramic composite structure has 30–40% higher strength than the precursor ceramic honeycomb. Table 1 shows the comparison of the strength of the honeycombs.

The dynamic adsorption properties of the honeycombs mentioned above were measured with an apparatus, the schematic of which is shown in Fig. 3. As shown in the figure various gases such as nitrogen, butane and toluene stored in pressurized tanks are metered through flowmeters in appropriate proportions and mixed before passing through the sample

Table 1. Axial crush strength of the honeycombs (2.54 mm min^{-1} speed)

Ceramic honeycomb (0.19 mm wall)	6.3 MPa ($\pm 10\%$)
Composite carbon honeycomb (0.19 mm wall)	8.6 MPa ($\pm 6\%$)
Composite carbon honeycomb (0.29 mm wall)	13.8 MPa ($\pm 8\%$)

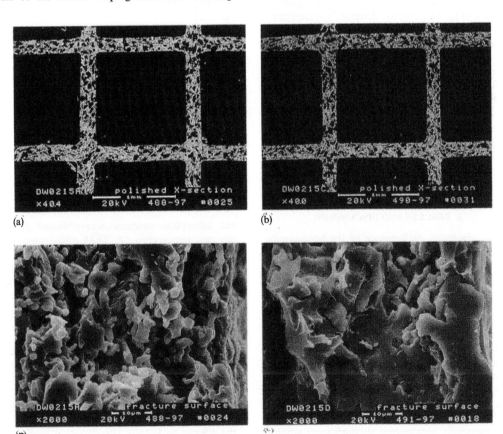

Fig. 2. SEM of ceramic and composite carbon honeycomb.

Schematic Of Dynamic Adsorption Apparatus

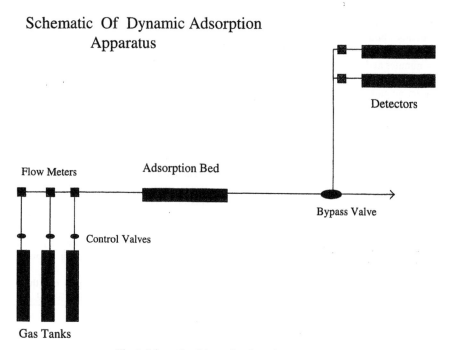

Fig. 3. Schematic of dynamic adsorption apparatus.

chamber. A small fraction of the mixed gases is diverted to the hydrocarbon detector. The calibrated detectors measure the hydrocarbon levels in the gas streams. All the equipment is computer controlled and the data is automatically stored and plotted to generate the breakthrough curves. The adsorption performance of the samples is measured as a function of time. Adsorption isotherms of the carbon are measured on an Omnisorp 100 from Coulter Inc. using standard procedures. Although methods have been developed to fabricate anywhere between 5 and 100 wt% carbon honeycombs, all the data given in this report was generated on first generation honeycombs with about 18 wt% carbon.

3. RESULTS AND DISCUSSION

The adsorption performance of the honeycombs depends on several material and process parameters. The material parameters include the carbon adsorption porosity on the honeycomb, the carbon percentage as well as the wall thickness of the honeycomb. The process parameters include the flow rate and concentration of the adsorbate, the adsorption potential which is dependant on the carbon structure as well as the adsorbate properties such as molecular weight. Although these carbon honeycombs have been made with a range of carbon structures which affect the adsorption performance, the carbon structure used for this work is described by a type I isotherm shown in Fig. 4. The carbon in this case is thus essentially microporous although some adsorption at higher relative pressure is evident. The nitro-

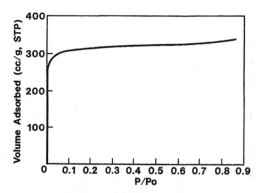

Fig. 4. Nitrogen adsorption isotherm of carbon.

gen adsorption isotherm was obtained at liquid nitrogen temperature. A TEM micrograph of the carbon structure is shown in Fig. 5. As the figure shows the carbon has a very regular structure with platelet spacing of 0.7–0.8 nm. This carbon structure is unusual and changes very significantly as a function of material and process parameters. Detailed studies of these structures will be published separately.

The adsorption performance of a device is measured in terms of the efficiency of removal of a certain constituent from a flowing stream. The performance is generally given in terms of a breakthrough curve, which shows the ratio of effluent concentration to influent concentration as a function of time [10]. The breakthrough capacity, i.e. amount adsorbed until the effluent to influent concentration reaches 0.95, depends on parameters such as flow rate, concen-

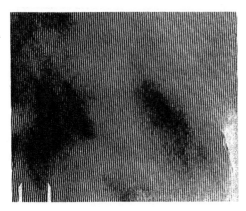

Fig. 5. TEM of carbon nanostructure.

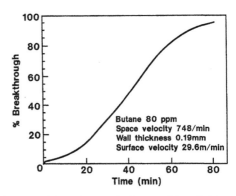

Fig. 6. Breakthrough curve for butane, thin-walled honeycomb (0.19 mm wall).

tration of adsorbate and the amount of carbon in the honeycomb. In the literature, the breakthrough curve is given as a function of surface velocity calculated from volumetric flow rate divided by the empty cross-sectional area of the bed.

Although honeycomb structures may have advantages mentioned above such as low pressure drop, an important factor may substantially affect the performance negatively. Honeycombs have straight flow paths with very small cell dimensions. Under standard conditions encountered in practice, the flow through the honeycomb cell is laminar. Laminar flow means that there is very little mixing of the fluid stream during the flow through the honeycomb. As a result the efficiency of contact between the carbon surface and the contaminant to be removed may be very low and the adsorption efficiency may be negatively affected, especially at the high flow rate low concentration conditions encountered in practice. Experiments were thus done to evaluate adsorption efficiencies for butane and toluene, an aliphatic and an aromatic hydrocarbon, at 80 ppm concentration. These two hydrocarbons at the specified concentration are used as model compounds for adsorption behavior in certain industrial applications. Formaldehyde and acetaldehyde adsorption performance was evaluated at 30 ppm concentration. These two components are present in diesel exhaust. Finally, isopropanol adsorption performance was evaluated at 80 and 300 ppm concentration. Isopropanol is a typical solvent in solvent recovery applications. The data obtained with all these compounds at the specified concentrations and flow rates is thus expected to give an idea of the performance of these new structures under the dynamic conditions encountered in practice.

Figure 6 shows the dynamic adsorption performance of a 400 cell honeycomb with 0.19 mm wall thickness for butane adsorption at 80 ppm inlet concentration. The honeycomb was 2.54 cm in diameter and 3.8 cm long. The adsorption performance was evaluated at a flow rate of 15 000 cm³ min⁻¹ of the nitrogen-containing butane at 80 ppm. As the figure

shows, the effluent concentration is 1.6 ppm or 2% of the influent at one minute. There is thus an immediate breakthrough. As the adsorption sites are filled up, the effluent concentration begins to increase and eventually reaches 95% of the influent level, i.e. 76 ppm at 75 minutes. Integrating the area above the curve gives the total adsorption capacity of the honeycomb. In this case the capacity is 127.9 mg. Since there is immediate breakthrough, it is clear that the length of the honeycomb is not sufficient to develop the mass transfer zone fully. The surface velocity in this case is 29.61 m min⁻¹. The honeycomb was then removed from the apparatus and regenerated in a 150°C oven. The experiment was then repeated with the same honeycomb at the same concentration of butane (80 ppm) but at a surface velocity of 46 m min⁻¹. The data showed that there was immediate breakthrough and the initial adsorption efficiency was 95% (4 ppm breakthrough) in this case. Increasing the surface velocity further to 100 m min⁻¹ results in a further decline in initial efficiency and the initial effluent concentration is 6 ppm. Figure 7 shows a comparison of these three curves. The nature of the curves is similar to the curves obtained from packed beds and so there does not seem to be a heavy penalty associated with using an open honeycomb structure with straight flow

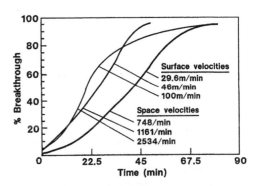

Fig. 7. Effect of surface velocity on breakthrough curve for butane.

channels and low pressure drop in terms of adsorption behavior.

All the breakthrough curves are labeled in terms of space velocity. The space velocity is defined as the flow rate of the fluid stream divided by the volume of the bed. This space velocity may be visualized as the number of times the entire fluid amount in the bed is exchanged per unit time and may also be called the turnover frequency. The concept of space velocity is routinely used in catalysis literature. In adsorption literature, however, this concept is not commonly used. Data is normally presented in terms of surface velocity. As shown later in this paper, space velocity is a more fundamental parameter and is very useful when comparing data or scaling up the system even for adsorption systems. The space velocity is defined here based on the empty volume of the bed. In that sense it is not the true turnover frequency, because the bed has the honeycomb in it. Only the void volume should be taken into account if the true turnover frequency or true space velocity is to be found. As the honeycomb structural parameters such as cell density or wall thickness or wall porosity change, this true space velocity will change. It is shown later that in spite of this limitation the empty bed space velocity is still a very effective indicator of the performance of the bed as long as similar systems are compared, in addition to being a much simpler parameter to calculate than the true space velocity.

It is desirable to obtain high adsorption efficiencies from small bed volumes. A parameter that can be changed to improve adsorption efficiency for the honeycombs is the wall thickness. Increased wall thickness at the same porosity will result in increased amount of carbon per unit volume of the honeycomb, improving its adsorption capacity. Thicker walls reduce the open area for flow through the honeycomb thus increasing the true surface and space velocity, but also increase the contact efficiency between the contaminant to be removed and the carbon. A balance between these two factors will determine whether the adsorption efficiency increases or decreases under given conditions. If the adsorbate diffuses at a high enough rate to the interior of the wall and empty adsorption sites are available for adsorption continuously, then a thicker-walled honeycomb should give better adsorption efficiencies and capacities. To test this hypotheses honeycombs with the same composition and cell density but with 0.29 mm wall thickness were fabricated, compared to 0.19 mm wall honeycombs used for earlier experiments. The thicker-walled honeycomb was treated in the same manner to form a carbon composite honeycomb as described previously and its adsorption performance tested with 80 ppm butane under identical conditions, i.e. 80 ppm butane, 15 000 cm^3 min^{-1} flow rate over a 2.54 cm diameter, 3.8 cm long honeycomb. Figure 8 shows the data. This data may be compared with the data in Fig. 6. As the figures show the adsorption performance of

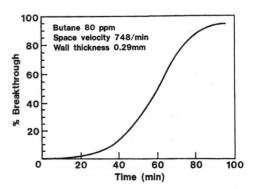

Fig. 8. Butane adsorption on thick-walled honeycomb (0.29 mm wall).

the thicker-walled honeycomb (0.29 mm) is clearly superior to the thin-walled honeycomb (0.19 mm). The initial adsorption efficiency is 100% with no butane breakthrough for about 15 minutes. The adsorption capacity, i.e. the butane adsorbed up to 95% breakthrough is 185 mg compared to 127 mg for the thin-walled honeycomb. The thicker-walled sample picked up more resin and hence had more carbon. The amount of carbon on the thick- and thin-walled samples was 2.21 and 1.45 g, respectively. This 52% increase in carbon results in a 46% increase in adsorption capacity. In spite of the increased true space velocity (the empty bed space velocity remains the same) the thicker-walled honeycombs give significantly enhanced adsorption performance, demonstrating that the honeycomb structures may be designed to obtain desired adsorption performance.

Toluene adsorption experiments were carried out on the honeycomb samples under identical conditions to those of butane experiments. Figure 9 shows the data for the thin-walled sample. Influent concentrations were 80 ppm of toluene. The toluene adsorption capacity was 427 mg, substantially higher than the butane capacity of 127 mg for the sample. Increasing the wall thickness (form 0.19 mm to 0.29 mm as described previously) increased the toluene capacity to 579 mg. To evaluate the relative effect of surface velocity and space velocity on adsorption

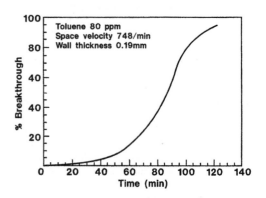

Fig. 9. Toluene adsorption–breakthrough curve.

performance, experiments were done with the 0.19 mm wall thickness samples as follows. The first set of experiments were done with samples 2.54 cm in diameter and 3.81 cm long at 15 000 $cm^3\,min^{-1}$ flow rate and at 30 000 $cm^3\,min^{-1}$ flow rate. The corresponding space velocities are 748 and 1496 min^{-1}. The toluene concentration was 80 ppm in both cases. Another experiment was done with a sample 2.54 cm in diameter but 7.62 cm in length with a flow rate of 30 000 $cm^3\,min^{-1}$. Since the cross-sectional area of the samples for both 30 000 $cm^3\,min^{-1}$ flow rate experiments is the same, the surface velocity remains the same, however, the space velocity changes because the volume is different. The surface velocities are 29.56 and 59.13 m min^{-1}, respectively, for the 15 000 and 30 000 $cm^3\,min^{-1}$ flow rates. The space velocities for the 3.81 and 7.62 cm long samples at 30 000 $cm^3\,min^{-1}$ flow rate are 1496 and 748 min^{-1}, respectively. Figure 10 shows the data obtained. As the figure demonstrates, in spite of doubling the surface velocity from 29.56 m min^{-1} to 59.13 m min^{-1} the adsorption performance remains identical if the space velocity is constant. As the space velocity changes from 748 min^{-1} to 1496 min^{-1}, however, there is significant drop in adsorption performance. It is thus clear that space velocity is a more meaningful parameter as far as adsorption performance is concerned. Surface velocity on the other hand is not an independent variable affecting the adsorption performance.

Formaldehyde adsorption experiments were conducted on the honeycomb samples to evaluate performance for the case of small molecules at very low concentration. Figure 11 shows the adsorption data at 748 min^{-1} space velocity for thin-walled samples. It is seen that formaldehyde is adsorbed at a high efficiency initially (no breakthrough), but breakthrough occurs soon after. The breakthrough is delayed with the thick-walled samples compared to the thin-walled sample as expected. The nature of the formaldehyde breakthrough curves is different from the butane and toluene curves shown earlier.

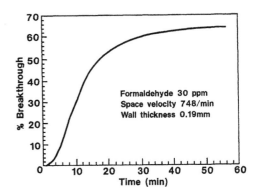

Fig. 11. Formaldehyde adsorption on thin-walled (0.19 mm) honeycomb.

After the initial breakthrough the curves rise rapidly, however, in both the cases shown these curves do not rise to complete breakthrough, i.e. equal influent and effluent concentrations. The curves rise to a certain value and then flatten out at a constant effluent to influent ratio or adsorption efficiency. For the thin-walled sample this effluent to influent concentration ratio is 68% and for the thick-walled sample the ratio is 56%. This behavior is unexpected. As in the case for butane and toluene, the curves should attain 100% breakthrough. Initially all the adsorption sites are vacant and hence there is no breakthrough. All the adsorbate is easily adsorbed. As the vacant sites fill up the breakthrough occurs and when the adsorption sites are all filled up there is 100% breakthrough. For the case of formaldehyde, however, this does not appear to be the case. After attaining a certain breakthrough value the curve flattens out and does not rise for a considerable period of time. A possible explanation for this behavior is as follows. For very small highly volatile molecules, adsorption potential is low. Particularly under the high flow rate low concentration conditions, it is very difficult to adsorb these compounds from a flowing stream. Adsorption efficiency will depend on how efficiently the molecules come in contact with the carbon, which is determined by wall thickness, flow rate and concentration as well as by the size of the pores relative to the molecules. Only the smallest size pores with pore widths close to that of the adsorbate size will be effective. Figure 12 shows the pore size distribution obtained on these samples utilizing the Micromeritics ASAP2000 equipment and DFT software. As the figure shows there is a significant distribution of pore sizes in the micropore range. The pore sizes fraction that will be effective for formaldehyde adsorption under the given conditions is difficult to determine based on the breakthrough curve data since many factors contribute to the adsorption performance. A qualitative explanation may be as follows. For formaldehyde adsorption only a small fraction of pores are effective because of its low adsorption potential. Initially all the *effective* adsorption sites are vacant

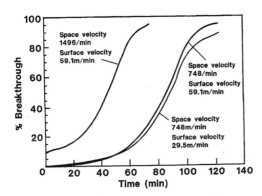

Fig. 10. Effect of the surface velocity and space velocity on toluene adsorption.

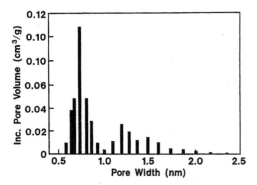

Fig. 12. Micropore size distribution of carbon.

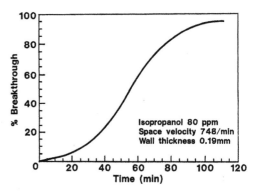

Fig. 13. Isopropanol adsorption performance at 80 ppm.

and so all the molecules are adsorbed. As the molecules are adsorbed, these molecules begin to diffuse inside the wall. The surface sites in contact with the flowing stream are filled up quicker than the molecules could diffuse inside the wall, and so breakthrough occurs and adsorption efficiency begins to drop. However, as the surface sites become available due to diffusion of the adsorbed molecules to the interior, more adsorption can take place. When a dynamic equilibrium is reached between the diffusion rate and the adsorption rate, the curve flattens out. Eventually, when all the sites are filled up the effluent concentration should rise and attain the influent concentration value. For larger molecules with higher adsorption potential, a much larger fraction of pores are effective adsorbers. There is thus sufficient time available for the molecules to diffuse into the interior, while other unfilled pores are being filled. As a result breakthrough cannot occur until a much larger proportion of pores is filled up and continues to increase in magnitude until 100% breakthrough is observed. The experiment with formaldehyde cannot be continued very long to verify whether the curve rises to 100% breakthrough because of practical considerations, i.e. the limited amount of formaldehyde stored in the tanks, and the number of tanks needed etc. Once adsorbed, the molecules could diffuse into larger pores in the sample, since high flowrate stream is not a factor any more. In the thick-walled samples the curve flattens out at a lower breakthrough level because of the higher efficiency of contact discussed earlier. Experiments were also done with acetaldehyde at 30 ppm concentration and 748 min^{-1} space velocity. The breakthrough curve flattens out in this case at 88%, a higher level than 68% obtained with formaldehyde as expected. Isopropanol experiments were done at 80 and 300 ppm levels to check the suitability of the honeycomb structures in solvent recovery type of applications. The isopropanol adsorption curve for both concentrations of 80 and 300 ppm at 748 min^{-1} space velocity is similar to the standard breakthrough curves obtained with butane and shown in Figs. 13 and 14. At 300 ppm concentration of isopropanol, an experiment was carried

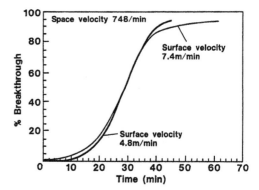

Fig. 14. Effect of space velocity on isopropanol adsorption performance.

out to evaluate the effect of variation in surface velocity at constant space velocity as described earlier for toluene. Figure 14 shows again that the performance obtained is equivalent as long as the space velocity is kept constant, even though the surface velocity doubles, again indicating that space velocity is a more fundamental parameter in dynamic adsorption measurement.

4. CONCLUSIONS

New activated carbon honeycomb structures based on synthetic raw materials have been demonstrated. These honeycombs with continuous, uninterrupted carbon structure are strong and highly durable. A large variety of such honeycombs with controlled cell density, wall thickness and cell geometries can be fabricated. In spite of the very open structure of these honeycombs, high adsorption efficiencies may be obtained even with high flow rate, low concentration flow streams. This behavior has been verified with a variety of standard adsorbents such as butane, toluene, formaldehyde, acetaldehyde and isopropanol. Adsorption efficiencies and capacities may be adjusted by adjusting wall thickness.

It has been found that in spite of the open structure, the honeycombs give breakthrough curves similar to packed beds in the case of butane, toluene, isopropanal etc. The breakthrough curve has a very different nature for flow streams containing a low concentration of small molecules such as aldehydes. The curve rises to a certain value of breakthrough level and then plateaus out. This behavior was explained in terms of dynamic equilibrium between adsorption and diffusion of the adsorbate molecules into the honeycomb carbon walls.

It has also been shown that space velocity is a fundamental parameter which controls adsorption performance. This parameter should be used in characterizing adsorption performance rather than surface velocity as is traditionally done.

Acknowledgements—The author wishes to thank J. F. Mach for fabrication of the honeycomb samples, T. P. Grandi for dynamic properties measurement, Professor M. Jaroniec of Kent State University for carbon pore size distribution measurement and D. Pickles for TEM of the carbon nanostructure.

REFERENCES

1. Jankowska, H., Swiatkowski, A. and Choma, J., *Active Carbon*. Ellis Harwood, Chichester, U.K., 1991.
2. Sun, J., Hippo, E. J., O'Brien, I. and Crelling, W. S., *Carbon*, 1997, **35**, 341.
3. Hussein, M. Z., Tarmizi, R. S., Zainal, Z. and Ibrahim, R., *Carbon*, 1996, **34**, 1447.
4. Park, S., Yasuda, E., Akatsu, T., *et al. Carbon*, 1995, **33**, 1377.
5. Hishiyama, Y., Kaburagi, Y. A. and Inagaki, M., *Carbon*, 1993, **31**, 773.
6. Hatori, H., Yamada, Y. and Shiraishi, I., *Carbon*, 1993, **31**, 1307.
7. Suzuki, M., *Carbon*, 1994, **32**, 577.
8. Stoeckli, F., Centento, T. A., Fuertes, A. B. and Muniz, J., *Carbon*, 1996, **34**, 1201.
9. Jones, C. W. and Koros, W. J., *Carbon*, 1994, **32**, 1419.
10. Tien, C., *Adsorption Calculations and Modeling*. Butterworth–Heinemann, London, 1994, p. 123.

(B) Modifications to Porosity

COATING OF ACTIVATED CARBON WITH SILICON CARBIDE BY CHEMICAL VAPOUR DEPOSITION

R. Moene,[a,*] H. Th. Boon,[a] J. Schoonman,[b] M. Makkee,[a,†] and J. A. Moulijn[a]

[a]Department of Chemical Process Technology, Section Industrial Catalysis, Delft University of Technology, Julianalaan 136, 2628 BL Delft, The Netherlands

[b]Department of Inorganic Chemistry and Thermodynamics, Section Applied Inorganic Chemistry, Delft University of Technology, Julianalaan 136, 2628 BL Delft, The Netherlands

(*Received 5 April 1995; accepted in revised form 8 November 1995*)

Abstract—Coating of activated carbon with silicon carbide by chemical vapour deposition (CVD) has been investigated to improve the oxidation resistance and the mechanical strength of activated carbon extrudates. The oxidation resistance has been analyzed by thermal gravimetric analysis in air; the temperature at the maximum rate of oxidation (T_{max}) is used to compare the modified carbons. Selective deposition of SiC by reacting $SiCl_4$ with the carbon surface cannot be achieved below 1400 K. Silicon deposition has been encountered in all cases. Coating of activated carbon using a $CH_4/SiCl_4$ mixture results in SiC deposition at 1376 K. The oxidation resistance of this modified activated carbon has been improved by 150 K ($T_{max} = 1025$ K), while the side crushing strength improved by a factor 1.7. The residual surface area was 176 m^2/g. SiC coatings have also been obtained by decomposing CH_3SiCl_3 at temperatures above 1200 K. The side crushing strength of the extrudates improved by a factor of 1.4, while the resistance against oxidation remained similar to that of the original carbon. The residual surface areas and pore volumes averaged 530 m^2/g and 0.33 ml/g, respectively. Both methods of SiC deposition result in surface areas which are high enough for catalyst support applications. Evaluation of the infiltration performance of this SiC-CVD process using CH_3SiCl_3 shows that 20–95% of the SiC has been deposited inside the extrudates. The residual porosity of the extrudates is evaluated using a general mathematically developed chemical vapour infiltration design chart, which correlates initial Thiele moduli with the porosity after deposition. Good agreement is obtained between the experimental data and the design chart. Copyright © 1996 Elsevier Science Ltd

Key Words—Activated carbon, modification, chemical vapour deposition, silicon carbide, oxidation resistance, mechanical strength.

1. NOTATION

a	Fitting constant in eqn (6)
ASA	Active surface area
C_p	Heat capacity (J mol^{-1} K^{-1})
D_{eff}	Effective diffusion coefficient (m^2 s^{-1})
D_{gas}	Diffusion coefficient in the gas phase (m^2 s^{-1})
$D_{Knudsen}$	Knudsen diffusion coefficient (m^2 g^{-1})
E_a	Activation energy (kJ mol^{-1})
$\Delta_f H$	Heat of formation (J mol^{-1})
k	Reaction rate constant (mol m^{-3} s^{-1})
k_0	Pre-exponential factor (mol m^{-3} s^{-1})
L	Characteristic diffusion length (m)
m	Weight (kg)
m_0	Initial weight (kg)
n	Reaction order (—)
P	Pressure (Pa)
R	Reaction rate (mol m^{-2} s^{-1})
S	Absolute entropy (J mol^{-1} K^{-1})
S_{BET}	Surface area determined by the BET method (m^2 g^{-1})
S_{ext}	External surface area of one extrudate (m^2)
S_v	Specific surface area per unit volume (m^2 m^{-3})
S_t	Surface area determined by the t-method (m^2 g^{-1})
SCS	Side crushing strength (N)
T_{max}	Temperature at maximum rate of oxidation during temperature programmed oxidation in air (0.167 K s^{-1})
V_{ext}	Volume of one extrudate
V_{micro}	Micro pore volume determined by the t-method (cm^3 g^{-1})
V_{pore}	Total pore volume (cm^3 g^{-1})
ϵ	Residual porosity
ζ	Carbon conversion
φ	Thiele modulus
ρ	Density (kg m^{-3})
τ	Tortuosity

2. INTRODUCTION

The use of activated carbon as a catalyst support is mainly limited by its sensitivity towards reactions with oxygen and hydrogen [1]. The reactivity with oxygen can be lowered by modifying the carbon surface in such a way that the number of sites, which are active in the oxidation, is lowered. These active sites (often referred to as the active surface area, ASA [2]) are responsible for the cyclic desorption of CO/CO_2 and adsorption of oxygen, which results in

---

*Present address: Koninklijke/Shell-Laboratorium, Amsterdam, Badhuisweg 3, 1031 CM, Amsterdam, The Netherlands.

†To whom all correspondence should be addressed.

burning the carbon support. The reactivity towards molecular hydrogen of bare activated carbon is low. Transition or noble metals, applied on this support, however, catalyze the hydro-gasification of carbon into methane. Two different mechanisms can be distinguished. The first type of hydro-gasification is metal catalyzed dissociation of hydrogen, which subsequently "spills over" to defects in the basal planes to form methane. Secondly, at temperatures higher than 100 K carbon can dissolve in the metal crystallite, diffuse to the gas–metal interface and react with hydrogen to methane. Research has primarily been focused on achieving an increased oxidation resistance by decreasing the ASA. A decrease in ASA can be achieved by heat treatment or impregnation with metal phosphates, chlorine compounds, and boron glasses [3,4], which all improve the oxidation resistance to a certain extent. Stegenga et al. [5] investigated the possibility of applying silicon carbide on the carbon surface. They impregnated activated carbon with tetraethoxysilane ($Si(C_2H_5O)_4$, TEOS), which was subsequently decomposed into SiO_2, followed by a heat treatment at 2273 K acquiring the SiC. This has resulted in an increase in oxidation resistance of 100 K, while a considerable amount of surface area was retained. Furthermore, the side crushing strength was improved by a factor two. An alternative method of modifying surfaces by SiC deposition to achieve an improvement in oxidation resistance is chemical vapour deposition (CVD). Thus, carbon/carbon composites are coated with silicon carbide which acts as a diffusion barrier for carbon and oxygen [6]. Another mode of operation exists in the surface modification and densification of porous substrates. This technique, referred to as chemical vapour infiltration (CVI), is one of the few in which ceramic composites can be manufactured consisting of undamaged fibres or whiskers embedded in a ceramic matrix. Thus, a very high toughness and strength can be combined, disclosing numerous applications of ceramic materials at demanding process conditions. Methyltrichlorosilane (CH_3SiCl_3, MTS) is frequently used to achieve either an oxygen resistant SiC coating or a ceramic composite consisting of toughening fibres and a SiC matrix. Table 1 shows examples of this and several other SiC precursors and their application. To achieve an optimal oxidation resistance, it is important that stoichiometric SiC is deposited. Although the C/Si molar ratio in methyltrichlorosilane is one, deposition of silicon is encountered at temperatures below 1200 K, whereas carbon is co-deposited with SiC above 1800 K. The mechanism of SiC deposition can be regarded as two independent sub-systems, i.e. the deposition of carbon and the deposition of Si [22,23]. Equal rates will result in stoichiometric SiC. Detailed knowledge of the kinetics in the gas phase and on the surface is, however, limited, especially for chlorine containing SiC precursors like CH_3SiCl_3. The gas-phase kinetics for the SiH_4–hydrocarbon system is relatively well understood [19,20,24], here, more knowledge of the surface chemistry is needed. Thermodynamic calculations are widely used for a first insight in the condensed phases which are stable under process conditions [22,23,25]. Furthermore, the gas-phase composition can be determined at thermodynamic equilibrium. Considering a certain amount of gaseous components (varying from 6 to 45), it has been shown that $SiCl_4$ and CH_4 are the most abundant Si and C components in the Si–Cl–C–H system below 1000 K. $SiCl_2$ and C_2H_2 are formed at temperatures higher than 1800 K. Deposition of stoichiometric SiC is predicted at certain H_2/CH_3SiCl_3 ratios which depend on temperature and pressure. On a thermodynamic basis, silicon co-deposition is found at high H/Si ratios, whereas carbon is co-deposited at low H/Si ratios.

In the present paper chemical vapour deposition of SiC is evaluated as a modification technique to achieve strong activated carbon particles with a high surface area and a high oxidation resistance. An obvious prerequisite, identifiable prior to this investigation, is the formation of SiC layers with the capability of shielding the underlying carbon substrate while retaining a major part of the porous structure. Thermodynamic calculations are carried out to establish preliminary indications for optimal deposition conditions by determining equilibrium gas-phase compositions and stable solid phases at various conditions. Reactive CVD (using the activated carbon as carbon source) and conventional CVD (using $SiCl_4/CH_4$ or CH_3SiCl_3) are performed on activated carbon extrudates and the thermodynamic calculations are evaluated. Subsequently, the oxygen reactivity, residual surface area, and side crushing strength are evaluated according to their modification technique. Finally, the infiltration process will be compared with the results obtained by mathematical modelling of CVI.

3. THERMODYNAMICS

Table 2 shows the gaseous and solid components which are incorporated in the calculations. The thermodynamic data (i.e. $\Delta_f H^{298\,K}$, $S^{298\,K}$, and $C_p^{800-2000\,K}$) are taken from JANAF Thermochemical Tables [26]. The calculations are carried out using atomic mass balances and minimization of the overall Gibbs energy using the programme Solgasmix [27]. It should be noted that only the relative amounts of atoms determine final thermodynamic equilibrium compositions, hence, no distinction can be made between carbon originating from graphite or methane. Thermodynamic calculations are carried out for atomic Si/(Si + C) ratios varying from 0 to 1 and for H/Si ratios of 40 and 400, at 10 kPa, and temperatures between 800 and 1800 K. Figure 1 displays an example of the mixture of gaseous components with equilibrium pressures exceeding 10^{-10} bar for the system Si/C = 1 and H/Si = 40. $SiCl_4$ and CH_4 are the most abundant gaseous SiC precursors at low temperatures. The equilibrium pressure of methyl-

Table 1. SiC precursors and their application

SiC precursor	Conditions of synthesis		Application	Ref.
	Temperature (K)	Pressure (kPa)		
CH$_3$SiCl$_3$	1523–1873	n.r.	coating	[7]
	1400	1.7	coating	[8]
	1173–1273	10 to 35		[9]
	1173–1373	10 to 100	composite (CVI)	[10]
	1252–1270	2 to 13.3		[11]
CH$_3$SiCl$_3$/CH$_4$	1473–1723	4.6	coating	[12]
SiCl$_4$/CH$_4$	1400–1600	100	coating	[13]
(CH$_3$)$_2$SiCl$_2$	1473–1600	n.r.	coating	[14,15]
SiCl$_4$/Carbon	1500-1700	100	coating	[16–18]
SiH$_4$/C$_x$H$_y$	1773–1923	100	electronic	[19–21]

n.r. = not reported.

Table 2. Components used in thermodynamic calculations

Ar	HCl	CH$_4$	CH$_2$Cl$_2$	CCl$_3$	SiHCl$_3$	SiCl$_2$
H	Si (s)	C$_2$H$_2$	CH$_3$Cl	CCl$_4$	SiH$_2$Cl$_2$	SiCl$_3$
H$_2$	SiC (s, ß)	C$_2$H$_4$	C$_2$HCl	C$_2$Cl$_2$	SiH$_3$Cl	SiCl$_4$
Cl	C (s, graphite)	CHCl	CCl	C$_2$Cl$_4$	SiH$_4$	CH$_3$SiCl$_3$
Cl$_2$	CH$_3$	CHCl$_3$	CCl$_2$	C$_2$Cl$_6$	SiCl	—

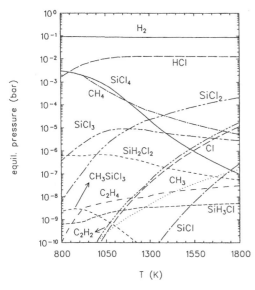

Fig. 1. Equilibrium gas phase composition (including SiC, Si or carbon formation) for H$_2$/MTS = 20 and P = 10 kPa.

trichlorosilane does not exceed 10^{-8} bar, which originates from the low bonding energy of the Si–C bond (290 kJ/mol) in the MTS molecule relative to the Si–Cl (359 kJ/mol) and C–H bonds (338 kJ/mol). Above 1300 K the equilibrium pressure of SiCl$_2$ surpasses that of SiCl$_4$. The most abundant C-precursor shifts from methane to ethyne (C$_2$H$_2$) above 1700 K. The trends found in these calculations agree with those reported in the literature for the Si–H–Cl–C system. Figures 2 and 3 show the condensed phases in the temperature range of 800 to 1800 K as a function of the ratio of silicon and carbon in the

system. The effect of H/Si ratio has been investigated as well.

Three areas are encountered in Fig. 2, i.e. regions of SiC + C, SiC, and SiC + Si deposition. Increasing the H/Si ratio from 40 (Fig. 2(a)) to 400 (Fig. 2(b)) results in a broader range of conditions in which Si is co-deposited, silicon is exclusively present at high Si/C ratios and low temperatures. This evolves from the equilibrium shift of the following reaction to the right-hand side:

$$SiCl_4(g) + 2H_2(g) \rightleftarrows Si(s) + 4HCl(g) \qquad (1)$$

Furthermore, the region in which solid carbon is present decreases in size with increasing amounts of hydrogen at temperatures below 1200 K. This is the consequence of the equilibrium shift of the following reaction to the right-hand side:

$$C(s) + 2H_2(g) \rightleftarrows CH_4(g) \qquad (2)$$

The CVD diagrams have also been determined at 100 kPa (1 bar). Qualitatively, no differences are found compared to Fig. 2, except for the low temperature SiC area. This region is somewhat extended into the SiC + C region.

From these calculations it can be generally concluded that equimolar amounts of Si and C in the system are beneficial for stoichiometric SiC deposition. Furthermore, increasing the H/Si ratio from 40 to 400 decreases the amount of solid carbon, whereas the amount of co-deposited Si is increased. This might suggest optimal conditions for SiC deposition at H$_2$/SiCl$_4$ ratios below 40. The reverse of reaction SiCl$_4$ + 2H$_2$ + C $\rightleftarrows$ 4HCl + SiC, however, becomes significant at very low H$_2$/HCl ratios and thus a minimal excess of hydrogen is indispensable for SiC formation.

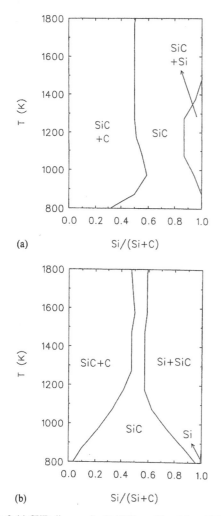

(a)

(b)

Fig. 2. (a) CVD diagram for $H_2/SiCl_4 = 20$ and $P = 10$ kPa. Condensed phases as a function of gas-phase composition and temperature. (b) CVD diagram for $H_2/SiCl_4 = 200$ and $P = 10$ kPa. Condensed phases as a function of gas-phase composition and temperature.

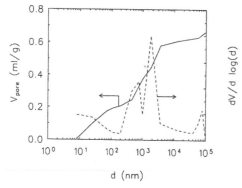

Fig. 3. Pore volume distribution of RW08.

4. EXPERIMENTAL

4.1 Materials

The modifications have been performed on steam activated, peat based, Norit activated carbon extru-

dates, type RW08. Prior to modification, the extrudates were separated according to degree of activation. The properties of the fraction used are shown in Table 3. The pore size distribution for the meso- and macro-pore region, determined by mercury porosimetry [28], is shown in Fig. 3. The typical poly-disperse porous structure [29] is evident. Two major contributions to the macro-pore structure are displayed, pores with radii of 250 and 1000 nm. The micro-pore region cannot be identified with Hg-porosimetry.

Hydrogen (99.99%), Ar (99.99%), and CH_4 (99.995%) were obtained from Hoek Loos. Purification of the gases was carried out by passing them through a bed of Pd/Al_2O_3 (hydrogen) or Cu/Al_2O_3 (Ar, CH_4) followed by a molecular sieve (5A) for water removal. Silicon tetrachloride (99%) was obtained from Aldrich Chemical Company, methyltrichlorosilane (98%) from Janssen Chimica, and were all used without further purification. Graphite plates ($10 \times 5 \times 1$ mm) were obtained from Johnson Matthey (batch 35142).

4.2 Chemical vapour deposition of silicon carbide

A schematic diagram of the CVD apparatus is shown in Fig. 4. A thin bed of extrudates (200 mg, 1 mm bed length) was positioned in a hot-wall tubular quartz reactor (internal diameter of 42 mm) to assure deposition under differential conditions. Prior to CVD, the substrates were heated in hydrogen with 0.167 K/s up to 1400 K (above the highest deposition temperature applied), followed by cooling to the deposition temperature, and adjusting to sub-atmospheric pressure. A graphite substrate, which was placed just upstream of the extrudates, was used to determine the rate of SiC deposition. The reactant gas was composed by bubbling argon through an evaporator filled with $SiCl_4$ or MTS (held at 410 or 400 K, respectively), which was subsequently mixed with hydrogen (and CH_4) before entering the reactor. The total gas flow rate equalled 5.6 ml/s (STP) and 27.8 ml/s (STP) for the $SiCl_4$ and CH_3SiCl_3 experiments, respectively.

4.3 Scanning electron microscopy

The activated carbon extrudates and graphite substrate were analyzed by scanning electron microscopy (JEOL JSM-35, Au sputtering of 4 minutes) to investigate the morphology of the deposited material. Furthermore, the growth rate has been determined from the layer thickness on the graphite substrate. From a representative area, a micrograph was taken and a mean thickness was determined by measuring the SiC layer at 11 places at equal distance from each other [30].

4.4 Surface area measurements

Nitrogen isotherms at 77 K were recorded on a Carlo Erba Sorptomatic 1800. Prior to measurement the samples were degassed at 423 K and 0.1 kPa. The BET surface area (S_{BET}), t surface area (S_t) and micro-

Table 3. Physical properties of the activated carbon extrudates

S_{BET} (m²/g)	V_{pore} (ml/g)	S_t (m²/g)	V_{micro} (ml/g)	$\rho(Hg)$ (kg/m³)	$\rho(He)$ (kg/m³)	Ash (wt%)	Length (mm)	Diameter (mm)
947	0.60	112	0.41	661	2167	5.2	3.0	0.81

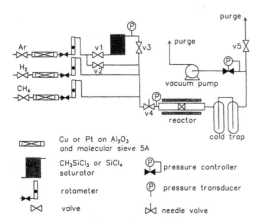

Fig. 4. Scheme of the CVD apparatus.

pore volume (V_{micro}) were determined according to the literature [31,32].

4.5 Temperature programmed desorption

Temperature programmed desorption (TPD) is carried out on home-made equipment by heating 200 mg activated carbon with 0.167 K/s to 1273 K in a helium–argon gas mixture (97/3 v/v). The gases evolved were recorded by a mass spectrometer (Varian Mat44 S) in multiple ion detection mode. Calibration was carried out by decomposing calcium oxalate monohydrate.

4.6 X-ray diffraction

Diffraction patterns were recorded on home-made equipment using a Cu-Kα. beam by measuring the Debije–Scherrer pattern.

4.7 Thermogravimetric analysis

A Setaram TAG 24 S thermobalance was used to determine the oxidation resistance. The samples (~30 mg) were heated with a heating rate of 0.167 K/s in air to 1273 K, while recording the weight change. The temperature at the point of maximum rate of weight loss (T_{max}) is generally assumed to be a suitable property for comparing the oxidation resistance of the activated carbon.

4.8 Side crushing strength

The side crushing strength was determined with a Schleuniger-2E. The average value of ten samples was used.

5. RESULTS

5.1 Pretreatment

Heating the activated carbon in hydrogen up to 1400 K (this first step of the modification procedure

will be referred to as the pretreatment), resulted in a total weight decrease of 8.4%. The difference in oxidation behaviour compared to the original carbon is depicted in Fig. 5. The hydrogen pretreatment increases the resistance against oxidation by 80 K. This is shown by the identical increase in temperature at which the oxidation starts and the temperature at the maximum rate of oxidation (T_{max}). The improvement in oxidation resistance is the result of reducing the ASA by decomposing the oxygen containing surface groups into H_2O, CO_2, and CO, and no significant change in TSA. Thus, the number of defects in the graphitic structure and, hence the oxygen reactivity, has been decreased. The decomposition of the oxygen containing surface groups of the fresh carbon during this initial step has been analyzed by temperature programmed desorption in a helium–argon mixture. The results are shown in Fig. 6. Single oxygen functional groups generally produce CO onto decomposition of carbon–oxygen complexes, whereas CO_2 is formed from decomposing carboxylic acids. The weight losses of H_2O, CO_2, and CO were 2.4, 3.0, and 1.7 wt%, respectively. The total weight loss is somewhat lower than that during the H_2 pretreatment. SEM analysis of the extrudates prior and after the pretreatment showed that the macro-structure does not change. Figure 7(a) and (b) are indicative for the structure of the carbon extrudates after the H_2 pretreatment. Conglomerated

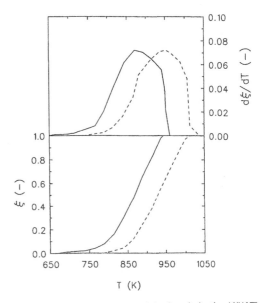

Fig. 5. Carbon conversion (ζ) and the first derivative ($d\zeta/dT$) for the original (—) and H_2- pretreated (—) carbon during TGA in air (heating rate 0.167 K/s).

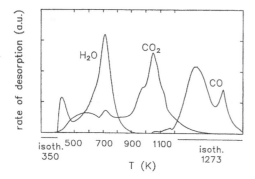

Fig. 6. Temperature programmed desorption of H_2O, CO_2, and CO from fresh RW08.

carbon particles of a grained structure can be distinguished, which are typical for a steam activated carbon. The outer surface appears to consist of granules of smaller diameter, as displayed in Fig. 7(b). The side crushing strength of the original and pretreated carbon was identical, i.e. 20.9 ± 4.7 N.

In the following section, the three types of SiC formation will be discussed separately. The activated carbon which has been heated to 1400 K in H_2 will be taken as reference material in all cases.

5.2 Reactive CVD of $SiCl_4$ on activated carbon ($C + SiCl_4 + 2H_2 \rightarrow SiC + 4HCl$)

A ratio of $H_2/SiCl_4$ equal to 20 has been used in all cases. The CVD temperature ranged from 880 to 1380 K. After modification no differences in colour are encountered. Figure 8 displays the amount of solid deposits, determined by TGA, and T_{max} as a function of deposition temperature. It is shown that the deposited amount decreases with increasing temperature and that the oxidation resistance has only slightly been improved compared to the H_2 pretreated carbon. XRD revealed only silicon as crystalline phase, no crystalline SiC has been detected.

5.3 CVD of SiC using $SiCl_4$ and CH_4 on carbon ($SiCl_4 + CH_4 \rightarrow SiC + 4HCl$)

The results of this modification are displayed in Table 4. The use of methane as an additional carbon source results in sole SiC deposition at a reaction temperature of 1375 K. Silicon deposition is encountered at lower temperatures. Activation of relatively stable methane is probably the rate limiting step in the formation of SiC. An additional improvement (compared to the H_2 pretreated carbon) in oxidation resistance of 75 K has been achieved after reaction at 1375 K. The side crushing strength of this sample was 35.9 ± 5.6 N, which is an improvement of a factor 1.7 compared to the original carbon.

It should be noted although the activated carbon contains large amounts of ash components, which can act as gasification agent (see Table 3), an experiment of only $SiCl_4$ and H_2 in the vapour phase did not result in the formation of any SiC.

Fig. 7. Morphology of (a) the external surface and (b) the internal surface of RW08 after H_2 treatment.

5.4 CVD of SiC using CH_3SiCl_3 on activated carbon ($CH_3SiCl_3 \rightarrow SiC + 3HCl$)

After modification with methyltrichlorosilane above 1220 K, the extrudates and graphite substrate were covered with a shiny metallic grey film. X-ray diffraction showed deposition of stoichiometric β-SiC. SEM analysis revealed a cauliflower-like SiC morphology, as shown in Fig. 9(a) and (b), which is the typical SiC structure resulting from using this precur-

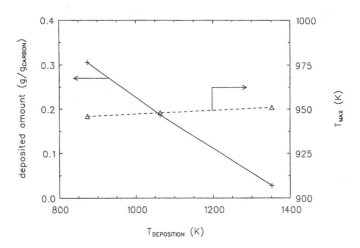

Fig. 8. Deposited amount (+) and T_{max} (Δ) as a function of deposition temperature.

Table 4. Modification of activated carbon by CVD using H_2/$SiCl_4$/CH_4 (reaction time = 6 hours)

H_2SiCl_4 (mol/mol)	CH_4SiCl_4 (mol/mol)	T (K)	P (kPa)	Deposited amount (g/g_{carbon})	T_{max} (K)	S_{BET} (m^2/g)	V_{pure} (ml/g)
170	1.8	1091	10	0.28 (Si)	945	431	0.23
170	1.8	1376	10	0.30 (SiC)	1025	176	0.12

sor under these conditions [33]. Deposition of SiC on graphite results in a more compact and dense layer than on activated carbon, which originates from the less rougher surface morphology. A crack in the SiC layer sometimes appears after prolonged deposition, which is caused by differences in expansion coefficients between graphite/activated carbon and SiC. Differences in nucleation can easily be identified from Fig. 9(b) and (c). The number of nucleation sites on activated carbon exceeds that of graphite as shown by the initially more grained structure of the first SiC deposited on activated carbon. The first phase of growth in a CVD process results in a thin microcrystalline layer of random orientation, imposed by the roughness of the substrate surface and barely limited growth of the first nuclei [34]. The influence of the substrate surface decreases at increasing layer thicknesses; a finely denticulated morphology is then formed, on top of which multi-star twin tips can be discerned. Owing to the extensive surface roughness of activated carbon, this grained SiC morphology remains visible even at substantial SiC layer thicknesses. Figure 9(d) displays a top-view of SiC deposited on activated carbon. As can be seen in this micrograph the major part of the deposited SiC is present within the outershelf of the extrudates up to 100 μm. The porous structure is still evident, while cracks probably originate from the different expansion coefficient between the SiC and the ativated carbon, as can be seen in Fig. 9(c). This difference will cause the SiC layer cracks during cooling from deposition temperature to room temperature.

5.4.1 *Oxygen reactivity* The T_{max} of the modified extrudates varies between 985 and 1120 K, sug-

gesting an additional increase in oxidation resistance of 35 to 170 K. The establishment of the exact position of the T_{max} is, however, difficult, owing to various maxima in the first derivative as displayed by Fig. 10. Four maxima in the range of about 950 to 1150 K can be distinguished for a heating rate of 0.167 K/s. The number of maxima decreases with decreasing heating rates (0.0167 and 0.00833 K/s), pointing to the limitation of oxygen diffusion into the extrudate during oxidation by inhomogeneous carbon overlayer protection than the existence of different SiC species, which results in different onsets of oxidation.

5.4.2 *Textural properties* Textural properties are of primary importance for utilizing these modified activated carbon extrudates as catalyst support. Table 5 displays the surface area and pore volume for several extrudates. For convenience, the data are also presented normalized to initial amount of carbon. This allows additionally the identification of the amount of surface area and pore volume lost solely by the weight increase due to SiC deposition. The surface areas (total and micro) and pore volumes (total and micro) after modification declined by around 50% of the initial value. Half of this decrease can be attributed to the weight increase due to SiC deposition. The remainder is probably caused by blocking parts of the porous structure.

5.4.3 *Side crushing strength* Analysis of the strength of the modified activated carbon results in the 95% confidence regions of the side crushing strength (SCS) as displayed in Fig. 11. The extrudates are strengthened by an average factor of 1.4 after SiC-CVD modification. A direct relationship between

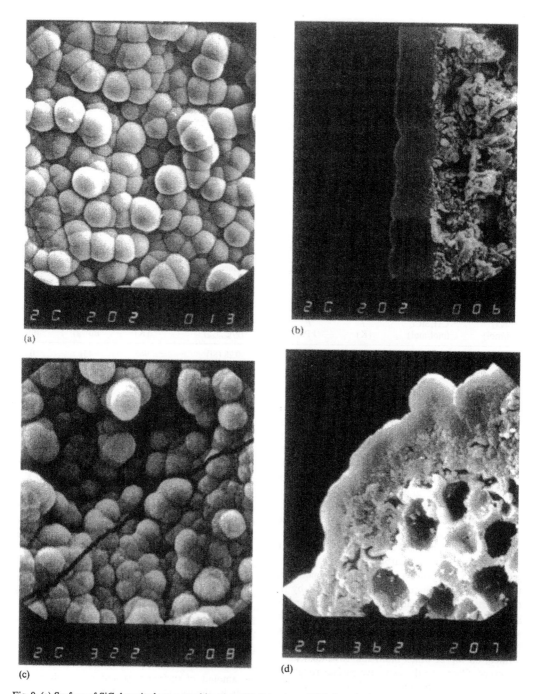

Fig. 9. (a) Surface of SiC deposited on a graphite plate. (b) Side view of SiC deposited on a graphite plate. (c) Side view of SiC deposited on activated carbon. (d) Surface of SiC deposited on activated carbon.

the synthesis procedures and corresponding SCS is, however, difficult to obtain.

5.4.4 *Kinetics of SiC deposition* The reaction rate of SiC formation has been determined by measuring the layer thickness of SiC deposited on the graphite substrates. The growth rate at 1222 and 1273 K as a function of the MTS concentration is shown in Fig. 12. Figure 13 displays the corresponding Arrhenius plot. The dependence of reaction rate on the concentration appears to be negligible in the parameter range investigated. A zero order dependence of the concentration on the reaction rate can be assumed based on Figs 12 and 13.

6. DISCUSSION

6.1 *Evaluation of the catalyst support properties*

In the modification of activated carbon by SiC-CVD a compromise has to be met. First, the SiC

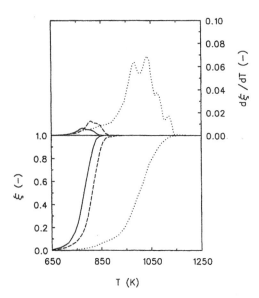

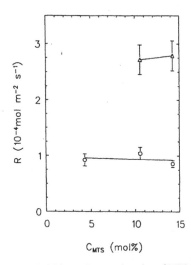

Fig. 12. Rate of SiC formation as a function of MTS concentration at 1223 K (O) and 1273 K (Δ). The lines are used to guide the eye.

Fig. 10. Conversion and first derivative of a MTS modified carbon during TGA in air, heating rate: 0.00833 K/s (—), 0.0167 K/s (– – –), and 0.167 K/s ($\cdots$).

Table 5. Textural properties of SiC-CVD modified activated carbon

Conditions		Textural properties							
T (K)	C_{MTS} (mmol/m^3)	S_{BET} (m^2/g, m^2/g$_c$)		V_{pore} (ml/g, ml/g$_c$)		S_t (m^2/g, m^2/g$_c$)		V_{micro} (ml/g, ml/g$_c$)	
1273 (2)	100	532	729	0.33	0.46	49	67	0.23	0.32
1242 (3)	102	533	712	0.33	0.44	53	71	0.23	0.31
1222 (4)	104	512	672	0.31	0.41	44	58	0.23	0.30
1223 (5)	42.6	576	712	0.37	0.45	60	74	0.25	0.31
1224 (6)	140	510	712	0.31	0.43	45	63	0.23	0.32
1223 (7)	69.8	571	714	0.35	0.43	47	59	0.26	0.33
1273 (8)	134	471	670	0.28	0.40	38	54	0.21	0.30

The numbers in brackets refer to Fig. 16 and Table 7.

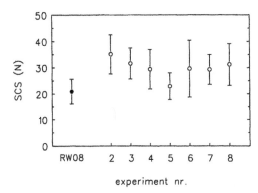

Fig. 11. Confidence regions (95%) of the side crushing strength (SCS) for various modified carbons. The experiment numbers refer to Table 5.

layer has to be thick and dense, and uniformly deposited in the pores to ensure a considerable strength improvement, and to guarantee adequate protection against oxidation. Secondly, a sufficiently large surface area has to be available after modification for catalyst support applications. This means

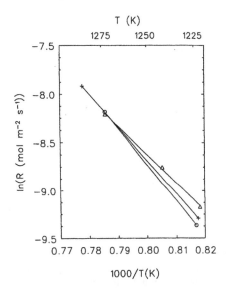

Fig. 13. Arrhenius plot for various MTS concentrations: 4.33 mol% (+); 10.5 mol% (Δ); and 14.2 mol% (O). The lines are used to guide the eye.

that the SiC layer has to be as thin as possible. This controversy should be borne in mind when evaluating the various properties of the modified activated carbon. Reactive CVD is on first sight an elegant way to achieve both objectives. Here, the activated carbon acts as the carbon source for SiC formation. Hence, deposition of SiC is expected to be limited to the surface where carbon is exposed to $SiCl_4$. Utilization of an additional gaseous carbon source (e.g. CH_4, CH_3SiCl_3) can enhance the deposition rate. A selective deposition, however, cannot be obtained. In the section below, the influence of the pretreatment and modification technique on the oxidation resistance and mechanical strength will be discussed in detail. Heating activated carbon extrudates to the temperature for SiC deposition results in changes in the surface composition of the substrates. Comparison of the pretreatment in hydrogen and an argon–helium mixture results in comparable amounts of desorbed molecules, i.e. 8.5 and 7.1 wt%, respectively. The difference can be explained by the difference in final temperature, i.e. 1400 and 1273 K, respectively. The presence of hydrogen is expected to alleviate the decomposition of the oxygen containing surface groups (e.g. by termination of the dangling bonds which remain on the surface after decomposition). However, the total amount of desorbed molecules at temperatures above the synthesis temperature of the activated carbon (around 1300 K) is expected to be roughly similar in argon and hydrogen regarding the inertness of graphite in hydrogen environments. Two conclusions can be drawn from these observations. First, the amount of oxygen containing surface groups on activated carbon decreases significantly during heating to the temperature of deposition. This will substantially decrease the ASA and, hence, the oxygen reactivity. The T_{max} is elevated by 80 K compared with the untreated carbon. Additionally, since the formation of SiC is expected to start at surface defects (i.e. oxygen containing functional groups) in the graphite lattice, the number of nucleation sites for SiC growth will diminish as well. This might result in differences in the morphology of the SiC layer, and thus the oxidation resistance. Secondly, desorption of CO starts at a temperature of 1050 K and just drops off after an isothermal period of one hour at 1273 K, which implies that the ASA is changing in the major part of the pretreatment trajectory. To achieve an identical starting material in all experiments, similar pretreatments are indispensable to assure the presence of identical amounts of oxygen containing functional groups and, hence, an identical oxygen reactivity and similar amounts of nucleation sites for SiC formation at the start of the experiment. These considerations have been the basis for utilizing the pretreatment procedure as described above.

Typical TGA profiles of all types of modification are displayed in Fig. 14. It is evident that, except for the MTS modification, T_{max} represents a good parameter for the evaluation of the oxidation resistance.

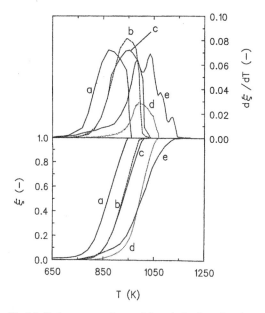

Fig. 14. Carbon conversion and first derivative of various modifications during TGA in air, heating rate 0.167 K/s: (a) RH08, (b) H_2-pretreated, (c) $SiCl_4$, (d) $CH_4/SiCl_4$, and (e) CH_3SiCl_3.

Oxidation of the MTS modified carbons results in an increased T_{max}, although the start of the oxidation is similar to that of the original carbon. This means that no real improvement in oxidation resistance has been achieved and that the increment in T_{max} originates from diffusion limitations of oxygen during oxidation.

6.1.1 *Reactive CVD using $SiCl_4$* RCVD at temperatures below 1400 K shows no significant increase in T_{max}. In all cases considerable amounts of silicon are deposited which proves that RCVD is not a suitable technique for attaining improvements in the oxidation resistance. Thermodynamically, silicon carbide deposition is found to be possible at temperatures above 800 K. Hence, it is concluded that SiC formation is kinetically controlled by the low reactivity of the activated carbon, resulting in silicon deposition.

6.1.2 *CVD of SiC using a mixture of $SiCl_4$ and CH_4* From the results presented above, it can be concluded that both objectives (improvement of oxidation resistance and mechanical strength) can simultaneously be achieved by SiC-CVD using a $CH_4/SiCl_4$ mixture at 1380 K. The oxidation resistance increases by 150 K (80 K for the pretreatment and 70 K by SiC deposition) with a concomitant side crushing strength improvement of a factor 1.7. The surface area of this material equals 176 m²/g, which suffices for catalytic purposes. The decrease in S_{BET} originates mainly from blocking the micro-pore structure which implies that infiltration of SiC has been achieved. The results support the conclusion that this type of SiC deposition indeed lowers the oxygen reactivity and that the low reactivity of the activated carbon limits the SiC formation in the RCVD set-up.

6.1.3 *CVD of SiC using CH₃SiCl₃* Deposition of stoichiometric β-SiC has been achieved at 1200 K utilizing MTS as precursor. The oxidation resistance is, however, *reduced* compared to the H₂-treated carbon. Curve e in Fig. 14 displays the early onset of oxidation, which might originate from the presence of hydrogen chloride (HCl) during deposition, which can restore a part of the ASA lost during the H₂ pretreatment. The side crushing strength improved by a factor 1.4. The deposition of SiC at the exterior of the particle has imposed diffusion problems for oxygen during oxidation, which resulted in multiple T_{max} values for one sample. The presence of this layer will undoubtedly aggravate the mass transfer towards the particle, and thus limit the use as catalyst support at high temperatures.

6.2 Evaluation of the MTS-CVI technique

MTS is a well known SiC precursor for the synthesis of SiC based composites, which allows evaluation of the modification in comparison with those reported in the literature. No influence of MTS concentration on the reaction rate could be distinguished. It should, however, be noted that the number of these experiments is not sufficient by far to determine the crucial parameters of SiC deposition from MTS. However, qualitative trends can be determined. The reaction order with respect to MTS appears to be zero order, which is in accordance with the results obtained by Loumagne *et al.* [35]. The activation energy calculated from the Arrhenius plot amounts to 279 ± 17 kJ/mol with a pre-exponential factor of 7.26 ± 0.64 × 10⁷ mol m⁻² s⁻¹. This activation energy is considerably higher than that reported by Loumagne *et al.*, i.e. 160 kJ/mol. To explain this discrepancy, it is illustrative to display several kinetic expressions which have appeared in the literature during the last decade (Table 6). The difficulty in kinetic modelling lies in the fact that SiC formation from MTS is governed by gas-phase decomposition reactions as well as by surface reactions. This means that the reactor geometry (e.g. the volume surface area ratio) and other experimental parameters like residence time of MTS in the hot zone of the reactor will influence the activation of MTS significantly. Gas-phase decomposition has been analyzed theoretically

[39] and experimentally by mass spectrometry [40]. SiCl₄, HCl, and CH₄ species are observed on decomposing MTS in hydrogen at 1428 K and 100 kPa. These results agree fairly well with the equilibrium pressures shown in Fig. 1. However, the way to extract information from these data establishing the active species in the SiC formation is ambiguous. It might be anticipated that MTS and a SiCl₄/CH₄ mixture show identical deposition characteristics. However, the SiC reaction rate and morphology from a SiCl₄/H₂/CH₄ and MTS/H₂ mixture differ significantly. Furthermore, compared to a SiCl₄/CH₄ mixture, the temperature necessary to deposit crystalline SiC is 200 K lower using MTS. This probably results from the gas-phase formation of CH₃ radicals from MTS, which increases the reactivity of the carbon precursor:

$$CH_3SiCl_3(g) \rightarrow CH_3(g) + SiCl_3(g) \qquad (3)$$

Thus, stoichiometric SiC can be deposited at lower temperatures compared with the SiCl₄/H₂/CH₄ mixture. From this it is evident that the rate of decomposition of MTS and reaction or recombination of radicals will have a major influence on the kinetics found for SiC deposition. This aspect of MTS as SiC precursor is one of the origins of disagreements between the kinetic models found by various authors, using different reactor set-ups.

The exact amount of SiC which is deposited inside the extrudates has to be estimated in order to assess the infiltration performance of the SiC-CVI procedure. The rate expression of SiC deposition based on the data given above is:

$$R = 7.26 \times 10^7 \exp\left[\frac{-279000}{RT}\right] \qquad (4)$$

Typical reaction rates are 1–4 nm/s at 1223 and 1286 K, respectively. The polydisperse porous structure of activated carbon hampers the determination of a mean pore radius and, hence, (1) the pore closure time, (2) the amount of SiC deposited inside and outside the pores, and (3) the effective diffusion coefficient of methyltrichlorosilane. As a first rough estimate, the largest pores ($r = 1000$ nm, $V_{pore} = 0.6$ cm³/g) are assumed to determine the infiltration process. The smaller pores are supposed to be closed in the early stage of the infiltration process. The time necessary for pore closure of the 1000 nm pores is determined using eqn (4). SiC is deposited at the exterior of the extrudate in the remaining process period. The amount of SiC deposited after pore closure is subtracted from the total amount of deposited SiC. This yields the amount of SiC infiltrated in the porous structure. Subsequently, the residual porosity is calculated. The results are displayed in Table 7. The amount of deposited SiC varies between 0.22 and 0.42 g_{SiC}/g_{carbon}. No correlation is found between the reaction rate and total amount of SiC deposited on the carbon. The infiltration (SiC infiltration) ranges from 21 to 94% and improves for (a) decreasing temperatures and (b) increasing concentrations of MTS. The first effect is commonly

Table 6. Reported kinetic models for SiC deposition from MTS

Experimental conditions				$k_0 \exp(-E_a/RT)C_{MTS}^n$	
T (K)	P (kPa)	H₂/MTS	n	E_0 (kJ/mol)	Ref.
1073–1373	100	0.5–1	1	120	[36]
1250–1300	3.3	5–20	0	188	[37]
1073–1373	20	0.7–4	1	255	[38]
1175–1225	5–10	1–5	0.1–1	160 ± 40	[35]
1075–1125	5–10	3–5	2.5	>300	[35]

Table 7. Infiltration of SiC in activated carbon

Exp. no.	Experimental conditions			Deposition characteristics		
	T (K)	C_{MTS} (mmol m^{-3})	P (kPa)	$R \times 10^5$ (mol m^{-2} s^{-1})^a	Depos. amount (m_{SiC}/m_{carbon})	SiC infiltr. (%)
1	1286	40.5	10	36.5	0.224	21.3
2	1273	100	10	27.2	0.371	66.3
3	1242	102	10	15.8	0.336	81.3
4	1222	104	10	10.4	0.313	89.4
5	1223	42.6	10	9.2	0.236	88.7
6	1224	140	10	8.56	0.397	94.2
7	1223	69.8	5	7.33	0.250	93.4
8	1273	134	10	27.9	0.423	69.5

^{a} Determined from SiC deposition on graphite.

encountered in CVI-processes; by reducing the reaction rate with respect to the diffusion rate, infiltration is enhanced, resulting in higher amounts of SiC in the pores. The second effect originates from the zero-order rate expression found for MTS. Increasing the MTS concentration does not affect the deposition rate of SiC. However, an increase in bulk concentration of MTS imposes a higher concentration gradient in the pore (and hence a higher diffusion rate). Moreover, depletion of MTS will occur at higher infiltration depths compared to lower bulk MTS concentrations.

To compare the influence of deposition conditions on the residual porosity, a plot of the residual porosity versus the Thiele modulus is applied. The Thiele modulus for a nth order reaction is derived similarly to the procedure described elsewhere [41]. The tortuosity is assumed to be 4 [42]. The gas-phase diffusion coefficient is taken from Sheldon and Besmann [11]. The total void fraction of the extrudates is derived from the He and Hg densities given in Table 3, and equals 0.666. This results in an initial void fraction for pores with radii larger than 1000 nm ($V_{pore, r>1000 nm}$ = 0.6 ml/g, Fig. 4) of 0.40. The calculation of the surface area of this pore system is based on the pore volume of the region of the 1000 nm pores, i.e. 0.2 ml/g (Fig. 4), and equals 0.4 m^2/g. From these numbers the effective diffusion coefficients and the corresponding Thiele moduli are determined. For a full evaluation the Thiele moduli are to be related to the residual porosity of the extrudates. The residual porosity is defined in the following equation:

$$\epsilon = \frac{\left(V_{pore, r>1000nm}\right)_{final}}{\left(V_{pore, r>1000nm}\right)_{initial}} \quad (5)$$

The results of the calculations are shown in Table 8.

A similar approach had previously been applied to

the results of the mathematical modelling of isothermal chemical vapour infiltration [41]. The CVI process had been modelled as a transient process in which the pore geometry changes in time owing to the deposition of solids. The influence of various types of surface kinetics on the deposition profile and the porosity of the material after pore closure had been evaluated. A general correlation had thus been found between the residual porosity and the Thiele modulus, which was independent of the kinetic model implemented. This correlation or design chart is fitted by eqn (6) which is based on the equation that describes the relationship between the effectiveness factor and Thiele modulus in heterogeneous catalysis.

$$\epsilon = 1 - \frac{3}{a-\varphi}\left[\frac{1}{\tanh(a-\varphi)} - \frac{1}{a-\varphi}\right] \quad (6)$$

where the constant a equals 20 ± 1 (95% confidence region) which shows that more stringent values of the Thiele modulus are needed to arrive at low residual porosities compared to reaching an effectiveness factor of one in the initial stage of the densification process. Figure 15 displays the Thiele moduli and corresponding residual porosities of the experiments reported in Table 7 relative to the design chart. For comparison, data reported by Sheldon and Besmann [11] are included. Their results are obtained from the infiltration of a carbon fibre bundle with SiC. The average pore diameter of this preform

Table 8. Thiele moduli and residual porosities for the MTS modified activated carbons

Exp. no.	1	2	3	4	5	6	7	8
φ	2.47	1.38	0.99	0.79	1.25	0.70	0.96	1.19
ϵ	0.97	0.87	0.85	0.85	0.89	0.80	0.88	0.84

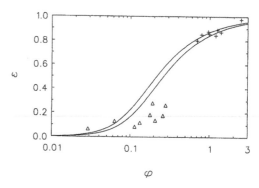

Fig. 15. The residual porosity (ϵ) versus the Thiele modulus (φ) for data of Table 7 (+) and from Ref. [11] ($\triangle$). The solid lines embody the 95% confidence region of the design chart [41].

ranges typically from 10 to 20 μm, which explains the considerably lower residual porosities achieved by these authors compared to those obtained for the modification of the activated carbon extrudates.

The theoretical design chart and experimentally obtained data display a good correlation for infiltrations in both low and high residual porosity regimes. This allows the use of Fig. 15 as a convenient pre-experimental design chart for optimization of CVI procedures. Numerical modelling and experimental validation are, of course, still indispensable for evaluating the process and the material produced.

7. CONCLUSION

Modification of activated carbon by SiC-CVD using $SiCl_4/CH_4$ results in improvements in oxidation resistance (up to 170 K) as well as in the side crushing strength (factor 1.7). The residual surface area is 176 m^2/g, which suffices for catalyst support applications. Additional advances require properties of the SiC layer to be deposited on the carbon surface which are conflicting to the characteristics necessary for catalyst support utilization, i.e. a thin SiC coating to ensure the preservation of sufficient surface area. Selective deposition by reactive CVD cannot be achieved below 1400 K, silicon deposition has been encountered in all situations. Application of CH_3SiCl_3 (MTS) embodies the use of an additional gaseous carbon source. Here, SiC deposition and infiltration have been observed. The oxidation resistance did not improve; the side crushing strength enhanced by a factor of 1.4. Utilization of these modified extrudates as catalyst support is feasible regarding their final specific surface area of 530 m^2/g. A zero-order relationship was found between the MTS concentration and the rate of SiC formation, the activation energy has been found to be 279 kJ/mol. The quantity of SiC deposited inside the extrudates ranges from 20 to 95% of the total amount deposited. Comparison of the residual porosity of the extrudates with an earlier mathematically developed chemical vapour infiltration design chart shows good agreement. [7–10,12–18,21,36–38]

Acknowledgements—This research was part of the Innovation-Oriented Research Programme on Catalysis (IOP number 90017) and was financially supported by the Ministry of Economic Affairs of The Netherlands.

REFERENCES

1. H. Jüntgen, *Fuel* **65**, 1436 (1986).
2. N. R. Laine, F. J. Vastola and P. L. Walker, *J. Phys. Chem.* **67**, 2030 (1963).
3. N. Murdie, E. J. Hippo and W. Kowbel, In *Proc. Carbon '88, Internt. Carbon Conf.*, p. 74 (1988).
4. P. Vast, G. Palavit, L. Montagne, J. L. Boulliez and J. Cordier, In *Proc. Carbon '88, Internt. Carbon Conf.*, p. 582 (1988).
5. S. Stegenga, M. van Waveren, F. Kapteijn and J. A. Moulijn, *Carbon* **30**, 577 (1992).
6. D. B. Stinton, T. M. Besmann and R. Lowden, *Ceram. Bull.* **67**, 350 (1988).
7. D. J. Cheng, W. J. Shyy, D. H. Kuo and M. H. Hon, *J. Electrochem. Soc.* **134**, 3145 (1987).
8. F. J. Buchanan and J. A. Little, *Surf. Coat. Technol.* **46**, 217 (1991).
9. E. Fitzer, W. Remmele and G. Schoch, In *Proc 7th European Conf. on CVD*, Journal de Physique, C5-209 (1989).
10. F. Christin, R. Naislain and C. Bernard, In *Proc 7th Int. Conf. on CVD* (edited by T. O. Sedwick and H. Lydtin), p. 499. Electrochemical Society, Princeton, (1979).
11. B. W. Sheldon and T. M. Besmann, *J. Am. Ceram. Soc.* **74**, 3046 (1991).
12. D. H. Kuo, D. J. Cheng and W.J. Shyy, *J. Electrochem. Soc.* **137**, 3688 (1990).
13. A. Parretta, A. Camanzi, G. Giunta and A. Mazzarano, *J. Mater. Sci.* **26**, 6057 (1991).
14. T. Kaneko, T. Okuno and H. Yumoto, *J. Crystal Growth* **91**, 599 (1988).
15. T. Kaneko, T. Okuno and I. Kabe, *J. Crystal Growth* **108**, 335 (1991).
16. H. Vincent, J. L. Ponthenier, L. Porte, C. Vincent and J. Bouix, *J. Less Comm. Met.* **157**, 1 (1990).
17. H. Vincent, J. L. Ponthenier, C. Vincent and J. Bouix, *Rev. Sci. Technol. Def.* **2**, 89 (1988).
18. J. Bouix, M. Cromer, J. Dazord, H. Mouricoux, J. L. Ponthenier, J. P. Scharff, C. Vincent and H. Vincent, *Rev. Int. Hautes Tempér. Refract. Fr.* **24**, 5 (1987).
19. C. D. Stinespring and J. C. Wormhoudt, *J. Crystal Growth* **87**, 481 (1988).
20. J. H. Koh and S. I. Woo, *J. Electrochem. Soc.* **137**, 2215 (1990).
21. J. M. Harris, H. C. Gatos and A. F. Witt, *J. Electrochem. Soc.* **118**, 335 (1971).
22. J. E. Doherty, *J. Metals* **6**, 6 (1976).
23. G. S. Fischman and W. T. Petuskey, *J. Am. Ceram. Soc.* **68**, 185 (1985).
24. M. D. Allendorf and R. J. Kee, *J. Electrochem. Soc.* **138**, 841 (1991).
25. A. I. Kingon, L. J. Lutz, P. Liaw and R. F. Davis, *J. Am. Ceram. Soc.* **66**, 558 (1983).
26. M. W. Chase Jr, C. A. Davies, J. R. Downey Jr, D. J. Frurip, R. A. McDonald and A. N. Syverud, JANAF Thermochemical Tables, *J. Phys. Chem. Ref. Data* **14** (1985).
27. G. Erikson, *Chemica Scripta* **8**, 100 (1975).
28. Internal report A8804, Norit Research Laboratory (1991).
29. H. Jankowska, A. Swiatkowski and J. Choma, *Active Carbon*, p. 79. Ellis Horwood, New York (1991).
30. *Perry's Chemical Engineers' Handbook*, 6th edn, pp. 2–85. McGraw-Hill, New York (1984).
31. S. Brunauer, P. H. Emmett and E. Teller, *J. Am. Chem. Soc.* **60**, 309 (1938).
32. J. H. de Boer, B. G. Linsen and Th. J. Osinga, *J. Catal.* **4**, 643 (1965).
33. F. J. Buchanan and J. A. Little, *Surf. Coat. Technol.* **46**, 217 (1991).
34. D. J. Cheng, W. J. Shyy, D. H. Kuo and M. H. Hon, *J. Electrochem. Soc.* **134**, 3145 (1987).
35. F. Loumagne, F. Langlais and R. Naslain, in *Proc. 9th European Conf. on CVD*, (edited by T. Mäntylä), p. 527 (1993).
36. K. Brennfleck, E. Fitzer, G. Schoch and M. Dietrich, In *9th Int. Conf. on CVD* (edited by M. D. Robinson *et al.*), p. 649. Electrochemical Society, Pennington, NJ (1984).
37. T. M. Besmann and M. L. Johnson, In *Proc. Int. Symp. on Ceramic Materials and Components for Engines* (edited by V. J. Tennery). American Ceramic Society, Westerville, OH (1989).
38. E. Fitzer, W. Fritz and G. Schoch, *High Temp. High Pres.* **24**, 343 (1992).
39. M. D. Allendorf, C. F. Melius and T. H. Osterheld, In *12th Int. Conf. on CVD*, PV 93-2 (edited by K. F. Jensen and G. W. Cullen), p. 20. Electrochemical Society, Pennington, NJ (1993).
40. Y. Yeheskel, S. Agam and M. S. Dariel, In *11th Int. Conf. on CVD*, PV 90-12 (edited by K. E. Spear), p. 696. Electrochemical Society, Pennington, NJ (1990).
41. R. Moene, J. P. Dekker, M. Makkee, J. Schoonman and J. A. Moulijn, *J. Electrochem. Soc.* **141**, 282 (1994).
42. C. N. Satterfield, *Heterogeneous Catalysis in Practice*, 2nd edn. McGraw Hill, New York (1980).

EFFECTS OF ACIDIC TREATMENTS ON THE PORE AND SURFACE PROPERTIES OF NI CATALYST SUPPORTED ON ACTIVATED CARBON

SHAOBIN WANG and G. Q. (MAX) LU*

Department of Chemical Engineering, The University of Queensland, St Lucia, Qld 4072, Australia

(*Received* 20 *May* 1996; *accepted in revised form* 22 *August* 1997)

Abstract—Activated carbon as catalyst support was treated with HCl, HNO_3, and HF and the effects of acid treatments on the properties of the activated carbon support were studied by N_2 adsorption, mass titration, temperature-programmed desorption (TPD), and X-ray photoelectron spectrometry (XPS). Ni catalysts supported on untreated and treated activated carbons were prepared, characterized and tested for the reforming reaction of methane with carbon dioxide. It is found that acid treatment significantly changed the surface chemical properties and pore structure of the activated carbon. The surface area and pore volume of the carbon supports are generally enhanced upon acid treatment due to the removal of impurities present in the carbon. The adsorption capacity of Ni^{2+} on the carbon supports is also increased, and the increase can be closely correlated with the surface acidity. The impregnation of nickel salts decreases the surface area and pore volume of carbon supports both in micropores and mesopores. Acid treatment results in a more homogeneous distribution of the nickel salt in carbon. When the impregnated carbons are heated in inert atmosphere, there exists a redox reaction between nickel oxide and the carbon. Catalytic activity tests for methane reforming with carbon dioxide show that the activity of nickel catalysts based on the acid-treated carbon supports is closely related with the surface characteristics of catalysts. © 1998 Elsevier Science Ltd. All rights reserved.

Key Words—A. Activated carbon, B. chemical treatment, D. porosity, D. surface properties.

1. INTRODUCTION

The use of carbon in catalysis has been continuously growing in the last decade. Its application as catalyst support offers several advantages [1,2] because it is a relative inexpensive and inert material. Expensive supported precious metals can be easily recovered by burning off the carbonaceous support. In addition, the surface area, porosity and functional groups of activated carbons can affect the characteristics of the catalyst. These characteristics can be modified by physical or chemical treatments of the material-activated carbon support.

Depending on the carbon sources and the activation processes, Na, K, Si, Al, Zn, P, N, Fe, Ca and other elements can be present on the surfaces of activated carbons. These inorganic compounds can catalyse undesired reactions, alter the sintering resistance of the supported metals, or act as catalyst poisons and thus adversely affect catalyst performance [3–5]. Usually additional acid treatments of the support material are essential to remove, at least partially, any surface contaminants and to reduce the ash content. It has been reported that acid treatment can improve the catalytic activity of activated carbon-supported catalysts [6,7].

It is well known that the interaction between the active phase and the support depends not only on the nature of the catalyst system, but also on the preparation method. Pre-treatment of the support or addition of promoters can significantly affect such interactions as well as the dispersion of the active species. The dispersion of a supported metal is one of the main parameters relating to its catalytic activity. When carbon is the support, metallic dispersion can be more or less controlled by means of thermal and chemical modification of the support porous structure and surface oxygenated complexes.

Ni-based catalysts are effective for carbon dioxide reforming of the methane reaction. However, these catalysts all employ metal oxides as support, few researchers have ever used carbon as support [8]. Gaerrero-Ruiz et al. [9,10] have tested Co catalysts supported on an activated carbon for this reaction and their results indicated that carbon-based catalysts could give satisfactory activity. Bradford and Vannice [11,12] studied the activity of Ni/C at 450°C for this reaction, compared its activity with other catalysts, and found that specific activity based on turnover frequency followed the order of $Ni/TiO_2 > Ni/C > Ni/SiO_2 > Ni/MgO$.

In this paper, we report our studies of the effects of acid treatments on the surface properties of activated carbon and the interaction between impregnated Ni and the support during the preparation of Ni-based catalysts and the catalytic activity for the CO_2/CH_4 reforming reaction.

*Corresponding author. Tel: +61 7 33653735;
Fax: +61 7 33654199; e-mail: maxlu@cheque.uq.edu.au

Reprinted from *Carbon* 36 (3), 283–292 (1998)

2. EXPERIMENTAL

2.1 Preparation of activated carbons and catalysts

A commercial activated carbon (AC) (Calgon, 4 mm × 6 mm) was treated with different acidic solutions, such as 2N HCl, 2N HNO$_3$ and 2N HF for about 24 hours. After the treatment, samples were washed with distilled water and dried in air at 103–105°C overnight. The ash content was determined by burning off the carbon at 750°C. The carbon supports treated by acids were then referred to as AC–HCl, AC–HNO$_3$ and AC–HF, respectively.

The catalysts were prepared by a wet impregnation method. The precursor used was Ni(NO$_3$)$_2$·6H$_2$O (BDH chemicals, AR grade). The supports were impregnated with an aqueous solution of the precursor in the appropriate concentration (0.6 g ml^{-1}) to obtain Ni loadings of about 5%. The solutions were heated and constantly stirred until total elimination of the liquid. All catalysts were dried at 103–105°C overnight. Ni loadings were determined by burning off the carbonaceous supports.

2.2 Characterisation of carbon supports and catalysts

The pH of the aqueous slurry was measured as follows: the slurries were prepared in a ratio of 10 ml of water to 1 g of carbon; this mixture was stirred and the pH was measured several times until a constant value was reached. The pH meter used was microcomputer pH-vision 6071(LAZAR).

The point of zero charge (PZC), termed as the pH value required to give zero net surface charge, is related to the two intrinsic acidity constants. Acid/base titration has been widely used to determine the PZC. However, this method is not applicable to the determination of the PZC for activated carbon. Noh and Schwarz [13] proposed a mass titration method to estimate the PZC of amphoteric solids including activated carbon and it proved to be effective.

To measure the PZC of the carbon samples, three different initial pH solutions were prepared using HNO$_3$ (0.1M) and NaOH (0.1M), such as pH = 3, 6, 11. NaNO$_3$ was used as the background electrolyte. For each initial pH, six containers were filled with 20 ml of the solution and different amounts of carbon were added (0.05, 0.1, 0.5, 1 and 10% by weight). The equilibrium pH was measured after 24 hours. The plot of pH versus mass fraction shows a plateau and the PZC is identified as the point at which the change of pH is zero. The PZC is then taken as the average of the three asymptotic pH values.

The N$_2$ adsorption/desorption isotherms at −196°C of the samples were obtained using a gas sorption analyser (Quantachrome, NOVA 1200). Samples were degassed for 3 hours at 300°C prior to the adsorption analysis. The BET surface area, total pore volume and average pore radius were obtained from the adsorption isotherms. The Dubinin–Radushkevich method was used to calculate the micropore volume and the mesopore volume was determined by subtracting the micropore volume from total pore volume. The average pore size was estimated from the pore volume assuming cylindrical pore geometry by the equation, $R_p = 2V_t/S$, where V_t is total pore volume and S is the BET surface area.

The micropore size distribution was obtained as follows. N$_2$ adsorption data on a nonporous carbon [14] were used to evaluate t values at various relative pressures, P/P_0. The comparison plots, t plots, were constructed by plotting the adsorbed amount V_a versus t at various relative pressures. A modified MP method [15] was then used to calculate the pore size distribution.

Thermogravimetric (TGA) experiments were carried out in a thermobalance (Shimadzu TGA-50). Samples were loaded into a platinum pan and heated under N$_2$ atmosphere from ambient temperature to 110° and held at this temperature for 20 minutes, and then heated to 1000°C with a controlled heating rate of 10°C min^{-1}.

The XPS measurements were conducted using a PHI-560 ESCA system (Perkin Elmer). All spectra were acquired at a basic pressure of 2×10^{-7} Torr with Mg Kα excitation at 15 kV and recorded in the ΔE = constant mode, at pass energies 50 and 100 eV.

XRD patterns were obtained with a Philips PW 1840 powder diffractometer. Co Kα radiation was employed covering two angles between 2° and 90°. The mean crystallite diameters were estimated from application of the Scherrer equation.

Temperature-programmed desorption (TPD) experiments were carried out in a vertical tube furnace. A 1 g sample was placed in a quartz tube with He as carrier gas. After heated to 110°C and maintaining this temperature for 60 minutes, the temperature was raised at 5°C min^{-1} to 800°C. The gases evolved during the TPD runs were continuously monitored using a gas chromatograph (Shimaduz GC-17A) equipped with a thermal conductivity detector and a Carbosphere column.

2.3 Catalytic testing for CH$_4$ reforming with CO$_2$

The catalytic reaction under integral reaction conditions was conducted in a vertical fixed-bed reactor made of quartz tube (10 mm inside diameter) under atmospheric pressure. A 0.2 g amount of catalyst was placed on quartz wool with a bed height of 2 mm. A thermocouple was placed in the tube with one end touching on the catalyst in order to measure the bed temperature. The stream of reactants (CH$_4$ and CO$_2$) with a ratio of 1 : 1 was fed into the reactor at the flow rate of 60 ml min^{-1} (GHSV = 18 000 cm^3 g^{-1} h^{-1}). The analysis of gases was carried out using an on-line Shimadzu-17A gas chromatograph equipped with a thermal conductivity detector. A Carbosphere (80–100 mesh) column was used to separate H_2, N_2, CO, CH_4 and CO_2. Prior to the reaction, the catalyst was reduced *in situ* at 500°C in 10% H_2/N_2 flow for 3 hours. The methane, carbon dioxide and helium gases used were of ultra high purity (>99.995%). The catalytic activity was examined over a range of temperatures (600–800°C). After reduction at 500°C for 3 hours the gas flow was switched to reactant feed gas and the temperature raised to the reaction temperature at a heating rate of 10°C min^{-1} and maintained for 5 minutes. A time on stream of 5 minutes prior to rate measurement was used. CH_4 conversions were calculated as follows

$$CH_4 \text{ conv } (\%) = ([CH_4]_{in} - [CH_4]_{out})/[CH_4]_{in} \quad (1)$$

where $[CH_4]_{in}$ and $[CH_4]_{out}$ are CH_4 concentrations in the gas flow before and after passing the reactor, respectively.

3. RESULTS AND DISCUSSION

3.1 *Surface and pore structural evolution of supports*

Treatment of activated carbon with acid solutions resulted in the removal of the inorganic constituents. The ash contents of the acid-washed activated carbons are given in Table 1. It is obvious from the table that the removal of mineral matter depends upon the type of acid solution. Treatment of activated carbon with HCl and HNO$_3$ resulted in about the same reduction in ash content whereas HF treatment decreases the impurities greatly to about 1.8% ash content.

The differences in the ash removal by the various acids can be explained as a result of the ability of the acids to remove different inorganic components. It has been reported that the content of the main impurities (Al, Si, Fe, Ti, K) in carbons could be drastically diminished by HF. Other elements such as Ca and S are principally extracted by HCl. HNO$_3$ treatment removed mineral matter such as S, Fe and Ca [6,16]. Hence, it is deduced that the most of the mineral matter in the as-received carbon samples is Al and Si compounds.

Table 1 lists the pore structure variations of the activated carbon samples treated in different acids.

It is observed that the original carbon (AC) has a high surface area and a well-developed porosity. The BET surface area and total pore volume are generally enhanced due to the acid treatment. However, the increase in surface area and pore volume is found to be mostly pronounced in the micropores. HF treatment increases both micropore and mesopore volumes. The average pore radius remains about the same for all supports. Similar observations have also been reported by Miguel *et al.* [6].

The variations of micropore and mesopore volumes with total pore volume are demonstrated in Fig. 1. As is shown, micropore volume increases linearly with total pore volume increasing. Although the mesopore volume increases as the total pore volume increases, their linear relationship is not as good as that between micropore and total pore volume. More mesopore volume enhancement is found in AC–HNO$_3$. From the weight losses of carbon supports by acid treatment it is seen that HF eliminated most of the inorganic matter in carbon. Removal of these compounds resulted in the development of both micropore and mesopore volume.

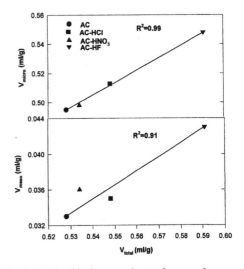

Fig. 1. Relationship between the total pore volume, and micropore and mesopore volumes.

Table 1. Effect of acid pre-treatment on activated carbon texture

	Ash (%)	Ni (%)	S_{BET} (m^2 g^{-1})	V^* (ml g^{-1})	$R^\dagger$ (Å)	V_{micro} (ml g^{-1})	V_{meso} (ml g^{-1})
AC	7.44		972	0.528	10.86	0.495	0.033
AC–HCl	5.80		1015	0.548	10.78	0.513	0.035
AC–HNO$_3$	5.56		987	0.534	10.88	0.498	0.036
AC–HF	1.77		1087	0.591	10.82	0.548	0.043
Ni/AC		4.98	888	0.478	10.76	0.448	0.030
Ni/AC–HCl		4.54	864	0.465	10.76	0.434	0.031
Ni/AC–HNO$_3$		4.48	836	0.453	10.99	0.422	0.031
Ni/AC–HF		4.38	865	0.476	10.82	0.440	0.036

*Total pore volume.
†Average pore radius.

The porosity evolution of the supports can be better illustrated by the DR plots derived from the adsorption isotherms (Fig. 2). It can be seen that the shape of the plots is similar for all samples: a linear portion at low relative pressures due to micropore filling and an upward deviation at higher relative pressure ascribed to the coverage of the surface of wider pores and an heterogeneous pore structure. However, as the mineral matter reduced with acid treatment, the treated activated carbons present DR plots of which the linear part is extended to higher relative pressures. This is characteristic of activated carbons with more narrow micropores. Because these pores had been filled with N_2 at very low relative pressures, little further adsorption occurred. This can be clearly seen from their micropore size distributions as in Fig. 3.

In Fig. 3, it is seen that there are two peaks in the pore size distributions of AC and AC–HCl. Only one peak appears in the pore size distribution of AC–HNO$_3$ and AC–HF. The height of the peak centred at about 8 Å of raw carbon support is increased due to acid treatment, which is reflected in pore volume enhancement. Acid treatments by HNO$_3$ and HF resulted in significant variations in pore size distribution. HCl treatment has a little effect on micropore size distribution. HNO$_3$ and HF increase the uniformity of the micropores of the as-received carbon.

The main effect of acid treatment on the porosity of the support is the enhancement of microporosity. More homogeneous micropore distributions are resulted from acid treatment. This could be the consequence of the selective removal of inorganic matter in carbons.

3.2 Surface acidity of the supports

The pH of the aqueous slurry and the PZC of the carbons may give a good indication about the surface oxygen complexes and the electronic surface charges

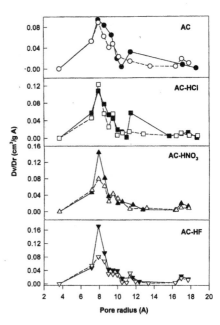

Fig. 3. Micropore size distributions of supports and catalysts.

of carbons. This surface charge arises from the interaction between the carbon surface and the aqueous solution and will determine the strength of interaction with the metal precursor in the catalyst preparation step. The complexes on the carbon surface are generally classified as acidic, basic, or neutral groups. Carboxylic, anhydride and lactone are acidic groups, while phenolic, carbonyl, quinone and ether groups are neutral or weakly acidic. Basic complexes are mainly pyrone and chromene groups. Based on the slurry pH, the nature of surface oxygen groups on the support and the dominant complexes can be deduced.

The values of pH and PZC of various carbon supports are shown in Table 2. It is seen that the untreated activated carbon shows a basic property. AC–HCl is neutral and two other carbons are found to be acidic. AC–HNO$_3$ has the highest acidity. The pH and PZC of each support are very close. This verifies the conclusion by Noh and Schwart [17] that pH$_{slurry}$ can be taken as equivalent to the PZC of the support. The pH results also indicate that acid treatment increases the acidity of the supports. The surface complexes on the carbon support have changed and more acidic groups (such as carboxylic) produced.

The increase in the acidity of the support upon acid treatment may be attributed to the removal of

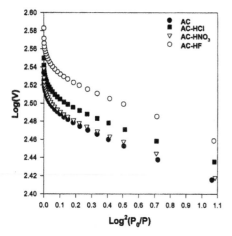

Fig. 2. DR plots from N_2 adsorption isotherms of the carbon supports.

Table 2. The pH$_{slurry}$ and PZC of various activated carbon supports

	AC	AC–HCl	AC–HNO$_3$	AC–HF
pH$_{slurry}$	8.75	6.58	2.80	3.32
PZC	8.60	6.70	2.98	3.29

inorganic compounds' leaving sites on the carbon surface which can chemisorb oxygen in air at room temperature. This would result in more oxygen surface complexes which are more acidic. HNO_3 and HF removed more inorganic constituents resulting in more acidic support. In addition, HNO_3 as an oxidant may also oxidize some complexes producing acidic groups.

3.3 Ni^{2+} adsorption capacity

Figure 4 shows the Ni^{2+} isothermal adsorption on various activated carbon supports at 25°C. It is demonstrated that the adsorption capacity of activated carbon is increased with acid treatment. The strongest acidic carbon shows the highest uptake of Ni^{2+}, while the as-received carbon has the lowest uptake. The relative order in adsorption capacity for the acid-treated carbons is not the same as that in surface area and pore volume. However, it can be closely correlated with the pH of the carbon slurry or the PZC. This indicates that metal adsorption on the support is strongly dependent on the chemical properties of the support.

Metal adsorption on catalyst supports not only depends upon the surface area and pore volume of the support but also on the acidic and basic characteristics of the support as well as the interaction between the metal precursor and support. Brunelle [18] proposed an electrostatic adsorption model that emphasised the importance of the impregnating solution pH, the PZC of the support and the charge of the adsorbing ion. According to this model, when pH is less than PZC, the surface hydroxyl groups of the support are protonated; the positively charged surface then attracts and adsorbs anions from a net negative surface charge because deprotonation and adsorption of cations occurs. In other words, if the pH of the impregnation solution is higher than the pH of the aqueous slurry of the support, then the adsorption of cations is favoured [19]. Noh and Schwarz studied Ni^{2+} ion adsorption on carbon

and found that the more oxidised carbons adsorb more nickel at a fixed condition of pH and nickel concentration [20].

3.4 Surface functional groups

Using XPS it is possible to measure directly the concentration of oxygen and other elements on or near the surface. In addition, it is possible to detect the different classes of surface functional groups in the outermost atomic layers of supports. In Table 3, C and O concentrations on different carbon surfaces are given. The values represent the ratios of the area of C and O in the topmost atomic layers. The effects of acid treatment demonstrate that HCl and HNO_3 generally increase the surface oxygen content, whereas HF decreases the surface oxygen content. Surface oxygen on carbon consists of inorganic and organic matter. Acid treatment mainly gets rid of the inorganic oxides and increases the organic oxygen-contained materials, which is reflected in the results of pH and ash content.

In Table 3 and Fig. 5, the C1s and O1s signals of carbon supports are compared. The C1s signal maxima are shifted to higher binding energy. This is because of an increase in oxidic species (alcohols, carbonates or carboxylic) and ether on the carbon surface after acid treatment. A shift of O1s signal to higher BEs may indicate an increase in groups such as hychroquinones, ether and liquid water. This conforms to the C1s signal shift variation.

Therefore, the XPS results confirm that more acidic groups such as carboxylic and ether were produced by acid treatment, resulting in the variation of the supports' pH and their adsorption on metal ions.

TPD experiments can provide interesting information about the amount, thermal stability and nature of the surface oxygen groups. During the TPD process the oxygen surface complexes desorb primarily as CO and CO_2. CO_2 proceeds from the decomposition of carboxylic, anhydride (acidic groups) and lactonic groups, whereas CO proceeds from the decomposition of phenolic, carbonyl, quinone, pyrone and anhydride (acidic) groups. The application of this technique to the study of the supports shows the great differences in surface chemistry among them.

Figure 6 presents the CO and CO_2 evolution profiles from the four AC carbons. In general, the desorption profiles are typical of those found on carbons, and the CO_2 complexes, which are responsi-

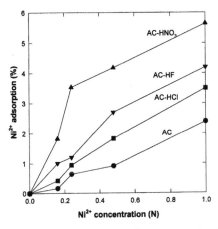

Fig. 4. Isothermal adsorption of Ni^{2+} on activated carbon supports at 25°C.

Table 3. Results of XPS analyses

Support	C (%)	O (%)	C1s signal maximum (eV)	O1s signal maximum (eV)
AC	94.6	5.4	286.0	534.1
AC–HCl	94.2	5.8	286.6	535.1
AC–HNO$_3$	91.0	9.0	286.7	535.0
AC–HF	96.6	3.4	286.5	534.1

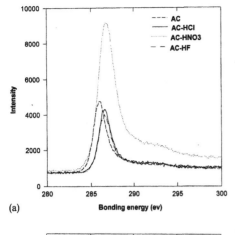

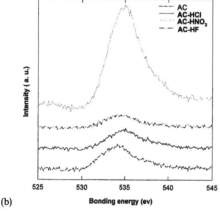

(a)

(b)

Fig. 5. XPS spectra of C1s and O1s on various carbon supports. (A) C1s, (B) O1s.

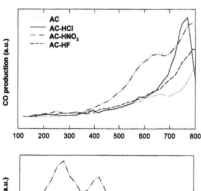

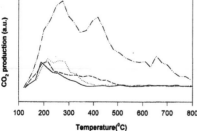

Fig. 6. TPD profiles of untreated and acid-treated carbons.

ble for the acidic sites of the carbon surface, are evolved at much lower temperature than CO complexes. AC treated with HNO_3 generates an intense oxidation that results in the evolution of large amounts of CO_2 and CO. AC–HCl, AC–HF and AC have comparable amounts of CO_2 complexes, but with different thermal stability. AC, AC–HNO_3 and AC–HF seem to have more stable CO complexes than AC–HCl because no peak is shown up to 800°C whereas a maximum at 760°C is observed for AC–HCl. In CO_2 evolution curves AC–HNO_3 shows three peaks at about 280, 400 and 650°C. AC–HF has a maximum at 200°C and a shoulder around 400°C. For AC and AC–HCl there is a broad peak occurring at 150–300°C. These results indicate that more and stable carboxyl and anhydride groups formed on AC–HNO_3 and AC–HF, which is reflected in the pH values. AC treated by HCl, however, has less thermal stability. Hence, it can be deduced that some chemical groups may be changed or removed from the AC–HCl support during the acid washing process.

3.5 Pore and surface characteristics of Ni catalysts

From Table 1, it is also seen that the impregnation process with $Ni(NO_3)_2$ solutions resulted in a decrease both in S_{BET} and V, as expected. This is due to pore blockage caused by nickel nitrate. However, the decrease is not as great as that reported by Gandia and Montes [2]. In their studies, it was found that the surface area decrease was recovered when $Ni(NO_3)_2$ was decomposed upon higher temperature treatment because the NiO particles formed during the decomposition have considerably smaller diameters than that of $Ni(NO_3)_2$ particles. In our experiments, we found the $Ni(NO_3)_2$ begins to decompose from 250°C and that the rate of decomposition reaches its maximum at around 350°C. Because degassing was conducted at 300°C in this study, during which nickel nitrate was decomposed to NiO, the S_{BET} decrease was found to be less than that reported by Gandia and Montes [2] where they used 150°C as the degassing temperature.

Listed in Table 4 are the decreases in pore volume of the catalysts due to impregnation. Impregnation generally results in larger decreases in micropores than in mesopores and acid-treated carbons show more decrease in pore volume than untreated carbon. This seems to indicate that treated carbon can have

Table 4. Decrease in porosities resulting from catalyst impregnation

Carbon	V_{micro} (%)	V_{meso} (%)
AC	9.5	9.1
AC–HCl	15.4	11.4
AC–HNO_3	15.3	13.9
AC–HF	19.7	16.3

larger adsorption capacity of nickel ions, and nickel ions can diffuse easier into the inner pores of the carbon for treated carbons.

The pore size distributions of the Ni catalysts are shown in Fig. 3. It is seen that impregnation of nickel salt on AC–HNO₃ and AC–HF greatly decreases the peak height of the pore volume at 8 Å. For samples AC and AC–HCl, the greatest decrease of volume occurs at 12 Å. This indicates that the nickel ion may deposit into a range of pores depending on the support surface properties and the interaction between the support and precursor. For acidic carbon (AC–HNO₃ and AC–HF) the interaction between the carbon and metal precursor is stronger and the nickel ion will diffuse easily into the inner pores. For basic carbons (AC and AC–HCl) the interaction between the support and precursor is not favoured and the surface basic complexes may play as centres to anchor the nickel ion.

3.6 Distribution of active catalyst species in supports

The XPS technique provides valuable information on the location and distribution of the active metal in the pore structure of the carbon support. Table 5 gives the XPS and gravimetric analysis results of Ni/C molar ratios. It is seen that the ratio $(Ni/C)_{XPS}/(Ni/C)_g$, being indicative of the ratio of surface and bulk Ni concentrations, is much higher than 1 for all catalysts except for Ni/AC–HNO₃. For the Ni/AC–HNO₃ catalyst this ratio is close to 1. This means that the larger concentration of nickel is on the surface but not in the bulk of the catalysts. It is found that the carbons treated by acids show lower $(Ni/C)_{XPS}/(Ni/C)_g$ values compared with that of the raw carbon-supported catalyst. This means that acid treatment improves the nickel distributions in porous carbon leading to more nickel ions diffused into the inner pores. This deduction can be proven by the micropore size distribution of the catalysts (Fig. 3), where pore volumes are clearly reduced because of impregnation of the metal species.

Conventional X-ray diffraction line broadening calculations have been performed in order to estimate the metallic crystallite diameter. These results are summarised in Table 5. As can be seen, the metallic crystallite size of the Ni catalysts varies depending on the acid employed for treatment and follows the order of Ni/AC–HCl > Ni/AC–HF > Ni/AC > Ni–HNO₃. HCl and HF treatment increase the metal crystallite size of the catalyst while HNO₃ treatment decreases the particle size.

Metal dispersion or metal crystallite size of the supported catalyst depends strongly on the distribution of the active phase within the support, and on the type and degree of interaction reached. During the impregnation process, the acidic or basic character of supports and of the metal precursor solution have an important influence on the way in which the precursor molecules diffuse into the support and interact with it. Some research has shown that the interaction of the metal precursor with carbon sometimes shows a positive influence on dispersion [20,21] and sometimes has a negative effect [19,22,23].

In this investigation it has been found that acid treatment increases the acidic functional groups on activated carbon which improves the metal distribution in porous carbon supports. However, AC treated by HCl showed less thermal stability. Hence, in this case more chemical functional groups will decompose during the reduction step, which made the metal ions more mobile and aggregated as larger particles resulting in its having the largest crystallite size among the four catalysts.

The results of the thermal analyses of Ni–AC catalysts are shown in Fig. 7. There are two peaks in the TGA curves of Ni catalysts. Ni(NO₃)₂ decomposed to NiO at a temperature of 200–300°C on the carbon support, which is lower than that of the unsupported salt. The peak occurring between 500–700°C corresponds to the NiO reduction temperature of Ni/C catalysts which can be proved with XRD results (Fig. 8).

The XRD pattern of the support corresponds to an amorphous material. The catalyst XRD pattern

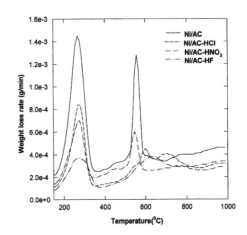

Fig. 7. Thermal analysis of Ni catalysts at 10°C min⁻¹ in N₂ atmosphere.

Table 5. Comparison between Ni/C molar ratios from XPS and gravimetric analyses

Catalyst	$(Ni/C)_{XPS}$	$(Ni/C)_g$	$(Ni/C)_{XPS}/(Ni/C)_g$	S_{Ni} (nm)
Ni/AC	0.040	0.010	4.0	10.4
Ni/AC–HCl	0.021	0.0093	2.2	21.8
Ni/AC–HNO₃	0.0089	0.0092	0.97	9.0
Ni/AC–HF	0.018	0.0090	2.0	12.7

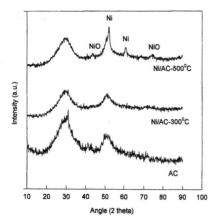

Fig. 8. XRD patterns of Ni catalysts heat-treated at different temperatures.

is very similar to that of the support, but heat treatment in N_2 at increasing temperatures progressively show peaks corresponding to nickel oxide at 300°C and metal nickel at 500°C. The lower intensities of NiO peaks occurring on the XRD pattern at 300°C are probably due to the lower amount of NiO content loading on carbon and higher dispersion. For the XRD pattern at 500°C it is seen that both Ni and NiO peaks are appearing. During the preparation for XRD measurement, contamination and exposure of the sample to air could lead to the oxidation of Ni to NiO. Hence, TGA and XRD measurements indicate that at 300°C nickel nitrate decomposes to produce nickel oxide and when heated to 500°C the interaction between nickel oxide and carbon leads to a redox reaction.

Gandia and Montes [2] studied the effect of thermal treatment on a Ni activated charcoal-supported catalyst and found that NiO can be reduced to the metallic state when the catalysts were subjected to an inert atmosphere at 500°C. The interaction between the carbon and salts for impregnated carbon upon heat treatment in inert atmosphere was also reported by other researchers [24,25]. They found that there was a redox reaction between the metal salts and the carbon, and the initial species Cr(IV) and Cu(II) were reduced to Cr(III), Cu(I) and Cu, respectively. It was suggested that the mechanism is very similar to that of metal catalysed oxidation of carbon. When the Ni/C catalysts are treated in N_2 at a sufficiently high temperature, the oxides resulting from the nitrate decomposition transfer their oxygen atoms to the support, which in turn is oxidised.

From Fig. 7 it is seen that nickel nitrate decomposes in the same temperature range. For all catalysts the redox reaction occurred at different temperatures. These results suggest that acid treatment has little effect on nickel nitrate decomposition. However, it does affect the interaction between nickel oxide and carbon. HNO_3-treated carbon is more active than the untreated carbon, while other treated

carbons show less activity, which is demonstrated from the peak shift and sharpness.

3.7 Catalytic activity of Ni-supported catalysts

Catalytic reforming of methane with CO_2 was conducted on the four catalysts at different temperatures and the CH_4 conversions and the specific activities are shown in Table 6. The metallic surface area (m^2 Ni/g Ni) was calculated from the equation proposed by Coenen [26] for spherical particles as may be expected for catalysts prepared by the impregnation method assuming a value of 6.33 $Å^2$ for the cross-sectional area per surface nickel atom and the full reduction of the supported nickel:

$$S_{Ni} = 543/D_{Ni} \qquad (2)$$

where D_{Ni} is the mean nickel particle size. It is seen that the Ni/AC catalyst demonstrates higher CH_4 conversion at 600°C and rapid deactivation between 600–700°C. After that the CH_4 conversion increases again. For other catalysts CH_4 conversion increases with increasing temperature, but the values are much lower than those on the Ni/AC catalyst. It is also found that the order in catalytic conversion among the acid-treated catalysts is as follows: Ni/AC–HNO_3 > Ni/AC–HF ~ Ni/AC–HCl, which is related to the variation in the metal crystallite size. A comparison between CH_4 conversion over catalysts and the thermodynamic equilibrium conversion shows that CH_4 conversions over all catalysts are much lower than those of thermodynamic equilibrium values. These results are quite different from those reported by Guerrero-Ruiz et al. [9,10] and Bradford and Vannice [11,12]. Guerrero-Ruiz et al. [9,10] found that CH_4 conversion over a Co/C catalyst could reach 12.1% and 33.2% at 500°C and 600°C, respectively. Bradford and Vannice [11,12] also reported that the catalytic activity of Ni/C in terms of turnover frequency (TOF) was larger than that of Ni/SiO_2 at 450°C. The different results can be ascribed to the nature of different supports employed. It has been found that the catalytic reaction of methane and carbon dioxide over catalysts was influenced by the nature of the catalyst supports. An Ni/SiO_2 catalyst based on silica gel showed much higher catalytic conversions of CH_4 and CO_2 whereas the one based on precipitated silica showed no catalytic activity for this reaction [27].

The nature of the support greatly affects catalyst activity due to the varying active surface area and acid–base property. Carbon dioxide reforming involves the adsorption and dissociation of CO_2 on catalysts. Since CO_2 is a well known acid gas, adsorption and dissociation of CO_2 may be improved with a basic catalyst.

The catalytic activity results indicate that the size of the active phase Ni and the surface basic property of the support influence the reactant conversion for this reaction. Due to the basic surface of the Ni/AC catalyst it shows the highest conversion among the

Table 6. CH_4 conversion (%) and specific activities, A_s (μmol m^{-2} s^{-1}), at different temperatures on various catalysts

Catalyst	Temperature (°C)									
	600		650		700		750		800	
	Conv	A_s	Conv	A_s	Conv	A_s	Conv	A_s	Conv	A_s
Ni/AC	15.1	1.01	9.2	0.58	5.1	0.30	10.8	0.61	12.3	0.67
Ni/AC–HCl	3.0	0.46	2.8	0.40	3.6	0.49	4.1	0.53	6.5	0.81
Ni/AC–HNO$_3$	4.0	0.25	4.1	0.25	5.0	0.28	6.6	0.36	9.3	0.48
Ni/AC–HF	2.6	0.24	2.7	0.24	3.1	0.26	4.7	0.37	8.0	0.60
TEC*	48.4		61.2		78.9		87.4		92.6	

*Thermodynamic equilibrium conversion.

four catalysts. AC–HF and AC–HNO$_3$ have similar pH but the nickel crystallite size of Ni/AC–HNO$_3$ is smaller than that of Ni/AC–HF, making Ni/AC–HNO$_3$ show higher CH_4 conversion. In terms of specific activity Ni/AC–HNO$_3$ and Ni/AC–HF show much similar activities and the order of activity follows   Ni/AC > Ni/AC–HCl > Ni/AC–HF ∼ Ni/AC–HNO$_3$. This seems to suggest that the pH of the support has greater influence on reaction.

Guerrero-Ruiz et al. [9,10] reported that the Co/C catalyst for CO_2 reforming of methane showed gradual deactivation and suddenly became inactive at 650°C. While for the Ni/MgO–C system it showed high conversion and lower deactivation rate. The deactivation seems to be due to carbon deposition over the metal as the Ni/C catalyst is efficient for the Boudouard reaction. The stability of Ni/MgO–C is due to the basic function of MgO which changes the nature of the CO_2 and CO chemisorbed species on this working catalyst. Bradford and Vannice [11,12] studied the behaviour of Ni/C catalysts for this reaction at 450°C under differential conditions and found that the Ni/C catalyst also showed deactivation and this was presumably due to carbon deposition and/or nickel sintering. However, the carbon deposition could not determined by a conventional TPO technique because, in the absence of isotopically labelled carbon, it was not possible to differentiate between oxidation of the support and oxidation of carbon deposits formed during reaction.

In general, the much lower conversions of Ni/C catalysts in CO_2 reforming of methane are due to the nature of the carbon supports. Acidity of the support can reduce the CO_2 adsorption thus resulting in lower conversion.

4. CONCLUSION

1. Acid treatment of activated carbon with different acids generally results in the removal of inorganic constituents and an increase in surface area and pore volume of the carbon. The increases in surface area and pore volume are generally prevailing in micropores. HF treatment enhances the surface area and pore volume to the greatest extent.

2. The acidity of carbon is increased by acid treatment, which results in the enhancement of adsorption capacity on the nickel ion. The surface functional carbon–oxygen groups are affected by the type of acids, which generates different decomposition profiles of the supports in inert atmosphere.

3. Impregnation of activated carbon with nickel salt produces larger decreases in the volume and surface area in micropores than that in mesopores. When nickel catalysts are heated in inert atmosphere, reduction occurs between carbon and nickel oxide. The nickel distribution on acid-treated carbons is found to be improved by acid treatment thus resulting in more metal ions diffusing into the inner pores.

4. Acid treatment influences the surface properties of nickel catalysts and consequently the catalytic activity for CO_2 reforming of methane. The low catalytic activity is closely related to the acidic surface of the acid-treated supports.

Acknowledgements—The authors would like to thank Dr B. J. Wood of the Department of Chemistry for his help in performing the XPS measurements. We also thank Mr F. Audsley of the Department of Earth Science for XRD measurements. Dr H. Y. Zhu's help on pore size distribution calculations is greatly appreciated. Partial financial support from the Australian Research Council (ARC) is also gratefully acknowledged.

REFERENCES

1. Albers, P., Deller, B. M., Despeyroux, B. M., Schafer, A. and Seibold, K., J. Catal., 1992, **133**, 467.
2. Gandia, L. M. and Montes, M., J. Catal., 1994, **145**, 276.
3. Ehrburger, P., Mahajan, O. P., Walker, P. L., Jr, J. Catal., 1976, **43**, 61.
4. Bartholomew, C. H., Agrawal, P. K. and Katzer, J. R., Adv. Catal., 1982, **31**, 135.
5. Rodrigue-Reinoso, F., Rodriguez-Ramos, I., Moreno-Castill, C., Guerrero-Ruiz, A. and Lopez-Gonzalez, J. D., J. Catal., 1986, **99**, 171.
6. de Miguel, S. R., Heinen, J. C., Castro, A. A. and Scelza, O. A., React. Kinet. Catal. Lett., 1989, **40**, 331.
7. Nakamura, T., Yamada, M. and Yamaguchi, T., Appl. Catal., 1992, **82**, 69.
8. Wang, S., Lu, G. Q. and Millar, G. J., Energy and Fuels, 1996, **10**, 896.
9. Guerrero-Ruiz, A., Rodriguez-Ramos, I. and Sepulveda-Escribano, A., J. Chem. Soc. Chem. Commun. 1993, 487.

10. Guerrero-Ruiz, A., Sepulveda-Escribano, A. and Rodriguez-Ramos, I., *Catal. Today*, 1994, **21**, 545.
11. Bradford, M. C. J. and Vannice, M. A., *Appl. Catal.*, 1996, **142**, 73.
12. Bradford, M. C. J. and Vannice, M. A., *Appl. Catal.*, 1996, **142**, 97.
13. Noh, J. S. and Schwarz, J. A., *Carbon*, 1990, **28**, 675.
14. Kaneko, K., Ishii, C., Ruike, M. and Kuwabara, H., *Carbon*, 1992, **30**, 1075.
15. Zhu, H. Y., Lu, G. Q., Maes, N. and Vansant, E. F., *J. Chem. Soc. Faraday Trans.*, 1997, **93**, 1417.
16. Calahorro, C. V., Cano, T. C. and Serrano, V. G., *Fuels*, 1987, **66**, 479.
17. Noh, J. S. and Schwartz, J. A., *J. Colloid Interface Sci.*, 1989, **130**, 137.
18. Brunedle, J. P., *Pure Appl. Chem.*, 1978, **5**, 1211.
19. Roman-Martinez, M. C., Cazorla-Amoros, D., Linares-Solano, A., Salinas-Martinez, C., Yamashita, H. and Anpo, M., *Carbon*, 1995, **33**, 3.
20. Noh, J. S. and Schwarz, J. A., *J. Catal.*, 1991, **127**, 22.
21. Prado-Burguete, C., Linares-Solano, A., Rodriguez-Reinoso, F. and de Lacea Salinas-Martinez, C., *J. Catal.*, 1989, **115**, 98.
22. Coloma-Pascual, F., Sepulveda-Escribano, A., Fierro, J. L. G. and Rodriguez-Reinoso, F., *Langmuir*, 1994, **10**, 750.
23. van Dam, H. E. and van Bekkun, H., *J. Catal.*, 1991, **131**, 335.
24. Ehrburger, P., Lahaye, J., Dziedzinl, P. and Fangeat, R., *Carbon*, 1991, **29**, 297.
25. Molina-Sabio, M., Perez, V. and Rodriguez-Reinoso, F., *Carbon*, 1994, **32**, 1259.
26. Coenen, J. W. E., *Appl. Catal.*, 1991, **75**, 193.
27. Wang, S. and Lu, G. Q., *Appl. Catal*, submitted.

Letter to the Editor

Change of porous structure and surface state of charcoal as a result of coating with thin carbon layer

A. Oya[a,*], R. Horigome[a], D. Lozano-Castello[b], A. Linares-Solano[b]

[a]Faculty of Technology, Gunma University, Kiryu, Gunma 376-8515, Japan
[b]Department of Inorganic Chemistry, University of Alicante, Apartado 99, E-03079-Alicante, Spain

Received 14 May 1999; accepted 21 May 1999

Keywords: A. Charcoal; B. Coating; C. Adsorption; D. porosity

Charcoal has been used widely for a long time as a cheap adsorbent material. At present, much work is in progress to improve the properties or to open new application fields [1-3]. We consider, however, that the following two disadvantages must be overcome in order to use this material more extensively in the future. The first is to increase the small specific surface area derived from a characteristic porous structure containing a large number of macropores. The second is to suppress the formation and release of fine carbon particles. There are many fine carbon particles on a commercially available charcoal surface and they are also formed as charcoal rods rub against each other. For example, when charcoal is used as an adsorbent material for water purification the fine particles are released to contaminate the water. This letter reports the changes of a porous structure and the surface state of charcoal through coating with a small amount of pitch and subsequent carbonizing.

Rods of Hinoki (Japanese cypress) wood were carbonized at 800°C in an industrial charcoal kiln. However, the properties and structure of the charcoal rods were considerably different from one another so that they were carbonized again at 1000°C for 1hr in a nitrogen atmosphere in the laboratory. In order to avoid data scatter, the charcoal rods were ground and grains of 2.0 - 5.6mm, after sieving and mixing, were used in the present work. The pitch used for coating was Kureha petroleum pitch GPF3 with a softening point of 247°C, a toluene insoluble fraction of 59wt% and no quinoline insoluble fraction.

Tel:+81-277-30-1350, Fax:+81-277-30-1353
e-mail:oya@chem.gunma-u.ac.jp

One gram of the pitch was dissolved in 200ml of quinoline into which 15g of the charcoal grains were then added and soaked for 20 min without stirring. The grains were taken out, dried at 200°C for one week, cured at 270°C for 4 days in air and then carbonized at 900°C for 30min in a nitrogen atmosphere. The weight increases were 3.4% and 2.5% after curing and carbonizing, respectively. A portion of the carbonized charcoal grains was finally activated under steam at 900°C for 5min, resulting in a burn-off of 10.8wt%.

Fig. 1 shows an SEM photograph of the pristine charcoal grain surface. Without treatment many fine particles are seen on the charcoal with a characteristic skeleton texture derived from the precursor wood. After

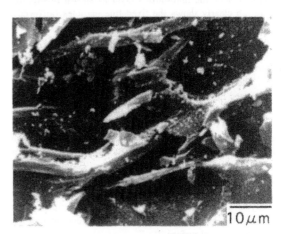

Fig. 1. SEM photograph of the pristine charcoal.

Reprinted from *Carbon* **37** (9), 1499-1502 (1999)

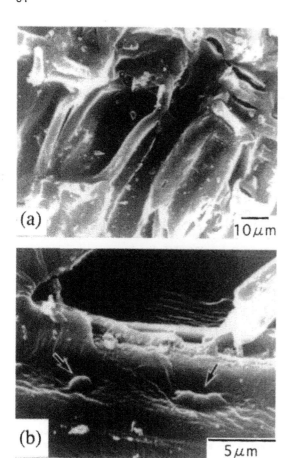

Fig. 2. SEM photographs of the charcoal coated with the pitch, after curing.

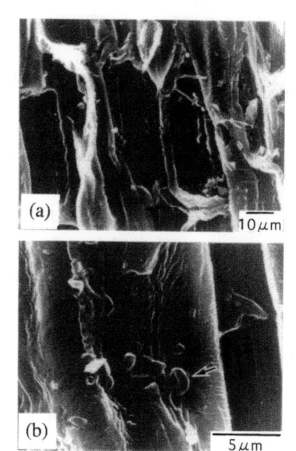

Fig. 3. SEM photographs of the charcoal coated with the pitch, after carbonizing.

coating and curing with the pitch the surface becomes smooth and the fine particles seem to be fixed on the surface with the thin carbon layer as shown in Fig. 2a. Fig. 2b clearly shows the fixed fine particles (arrowed). The surface state seen in Fig. 2 was substantially unchanged after carbonizing as seen in Fig. 3a. The firmly fixed particles were also observed in Fig. 3b as arrowed.

The smooth surface observed in Fig. 3 was seriously damaged through the activation as shown in Fig. 4. The surface became rough and many fine particles appeared as arrowed in Fig. 4b. It is clear that the activation treatment removed the thin carbon layer and, in addition, resulted in the formation of fine particles possibly through further oxidation to the charcoal grain surface. From the results it became apparent that coating with a small amount of pitch material and subsequent carbonizing is an effective way to fix the fine particles on the charcoal surface. At present we have no data on the abrasion resistance of the coated charcoal surface.

Table 1 shows data on the porous structures. It should be noted that the pristine charcoal has an important narrow microporosity (0.21 cm^3/g) which is not accessible to N_2 adsorption because it is very narrow [4,5]. The small surface area can be increased with an activation treatment, but at the same time an abundance of fine carbon particles is formed. The activation is not a useful way to complete our requirements described initially.

From Table 1 it can also be seen that, with the coating process, the narrow microporosity increases up to 0.23 cm^3/g and, interestingly, this narrow microporosity becomes accessible to N_2 adsorption, reaching a BET surface area value of 428 m^2/g. It must be strongly emphasized that the present treatment, using a small amount of pitch material, resulted in a nearly equal, large increase of specific surface area compared to that by an activation treatment of the pristine charcoal. The weight of charcoal was increased by 2.5wt% after carbonizing, so if the increase of specific surface area from 13 m^{2}g^{-1} to

Table 1 Porous structure of the samples

	BET surface area (m²/g)	Pore volume (cm³/g)				
		Macro (>50 nm)	Meso (7.5-50 nm)	Meso (2-7.5 nm)	V_{N2} (cm³/g)*	V_{CO2} (cm³/g)*
Pristine charcoal						
before activation	13	0.34	0.10	0.00	0.01	0.21
after activation	438	0.32	0.08	0.02	0.22	0.25
Carbon-coated charcoal						
before activation	428	0.27	0.18	0.02	0.24	0.23
after activation	780	0.48	0.07	0.04	0.39	0.33

*V_{N2} Micropore volume (pore size smaller than 2 nm) calculated from the application of Dubinin-Radushkevich (DR) equation to N_2 adsorption at 77 K

*V_{CO2} Narrowest micropore volume (pore size smaller than 0.7 nm) calculated from the application of Dubinin-Radushkevich (DR) equation to CO_2 adsorption at 273 K

428 m^2g^{-1} were only dependent on the coating carbon layer, calculations would show that it must have a specific surface area of more than 15000 m^2g^{-1}, which is apparently incorrect. These increases in volumes of small sized pores can be due to two possible effects: (i) interaction between the porous structure of the charcoal and the thin carbon layer, which seems to indicate that the impregnation of pitch, using quinoline, into the porous charcoal structure plays an important role, and (ii) swelling during the preparation process [6], which has opened the narrow microporosity rending it accessible to N_2 adsorption. Both points require further investigation.

Moreover, the coating process has also advantages because it allows for a better activation process; the resulting specific surface area of the coated sample is further increased after an activation treatment (from 428 m^2/g to 780 m^2/g). However, this treatment damages the smooth surface (Fig. 4) which is not in accordance with the aims of this paper, but could be useful for other applications.

Table 1 also shows that the coating treatment resulted in a decrease of the macropore and an increase of mesopore (7.5 - 50nm) volumes. The decrease in macropore volume is certainly caused by coating the charcoal surface with a thin carbon layer. The change in macroporous structure of the coated charcoal can be observed in Fig.5 which shows the pore size distribution curves obtained by mercury porosimetry. The pore volume of the coated charcoal was increased considerably after the activation process, but no change was observed between the uncoated charcoals before and after activation. Therefore, again, activation after the coating process can be interesting. This phenomenon will be examined in more detail in future.

It can be concluded from the present results that the small specific surface area of charcoal is largely increased

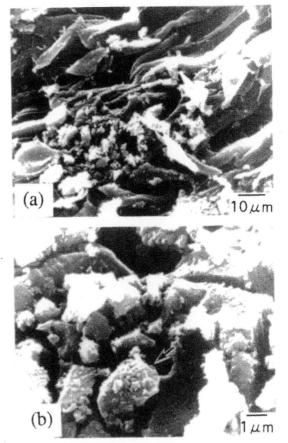

Fig. 4. SEM photographs of the charcoal coated with the thin carbon layer, after activating.

86

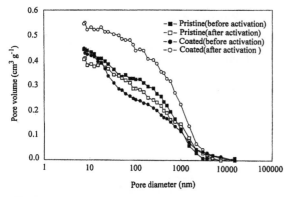

Fig. 5. Pore size distribution curves of the samples by mercury porosimetry.

by coating with a small amount of pitch and subsequent carbonization. Simultaneously this treatment is also effective in fixing the fine carbon particles on the charcoal surface. The technique is simple and practical.

Acknowledgement

This study was financially supported by a Grant-in-Aid from the Ministry of Education, Science, Sports and Culture of Japan (Project No.09243101) and CICYT (Project AMB-96-0799). Thanks are also to Kureha Chemicals Co. Ltd. for supplying petroleum pitch.

References

[1]   Bezzon G, Luengo CA, Capobianco G. Extended Abstracts Eurocarbon 98. Strasbourg, France, 1998. p.349

[2]   Yumine T, Kominami H, Abe I, Kera Y. Extended Abstracts of International Symposium on Carbon. Tokyo, Japan, 1998. p.190.

[3]   Arafune T, Ishii Y, Ogihara K, Ogura N. J. Water and Waste,1991;33:993.

[4]   Rodriguez-Reinoso F, Linares-Solano A. In: Thrower, PA, editor. Chemistry and physics of carbon, vol 21, New York: Dekker, 1988:1.

[5]   Linares-Solano, A.; Salinas-Martinez de Lecea, C.; Alcaniz-Monge,J.; Cazorla-Amoros, D.; Tanso, 1998;[No.185]: 316.

[6]   Van Krevelen DW. Coal, Typology-Physics-Chemistry-Constitution, 3rd edition, Elsevier.1993.

(C) Properties of Activated Carbons

Characterizing the ignition process of activated carbon

Y. Suzin, L.C. Buettner *, C.A. LeDuc

ERDEC, Aberdeen Proving Ground, MD 21010, USA

Received 21 November 1996; accepted 20 June 1997

Abstract

Properties of activated carbons at elevated temperature, specifically the spontaneous ignition temperature (*SIT*) and the point of initial oxidation (*PIO*), are of great significance in a number of unit operations. Although the *SIT* and *PIO* should be evaluated for each individual use, a standard method is required to compare products, insure quality, and optimize processes. Different methods for determining the *SIT* and *PIO* are compared using base (BPL, Coconut Shell, and MaxSorb) and impregnated carbons (ASC, ASZM, and ASZM-T). The *SIT* was determined by the ASTM and thermal analysis (TG and DSC) methods. The *PIO* was determined by effluent CO_2 concentration analysis, thermal analysis (TG and DSC) and temperature profiling. These comparisons show that not all of these methods have general utility, some are not easily correlated, and that there is evidence of the validity of two reaction regimes of interest. The simplest method that provides the most consistent conservative estimate of reaction commencement is measuring the *PIO* with the temperature profiling technique. © 1999 Published by Elsevier Science Ltd. All rights reserved.

Keywords: A. Activated carbon; B. impregnation; B. oxidation; C. thermal analysis

1. Introduction

Activated carbons are widely used in heterogeneous catalysis, filtration of hazardous industrial effluents, and personal protection within contaminated atmospheres. External heating, exothermic chemical reactions, and adsorption may raise the carbon temperature well above ambient, thus making the understanding of its properties at elevated temperatures of great operational and safety significance. Two such regions of interest are the temperature that causes the carbon to start significantly oxidizing (the *PIO* [1]) and the temperature that causes the bed to combust in a self-sustaining manner (the *SIT* [2,3]). The start of significant surface reaction is an important parameter in several testing procedures (i.e. insuring the surface is not altered by the process of drying a sample), in medium-temperature processes (i.e. insuring that potential chemical loading does not decrease with completed surface reactions), and in direct effluent breathing environments (dangerous CO_2 or CO

levels possibly generated). The temperature that causes spontaneous ignition is of special concern in processes of high operating temperature (i.e. incinerators or nuclear reactors), in adsorption units with high loading of organics (i.e. solvent recovery), or in operations involving impregnated carbons with low ignition points (i.e. filtration of low molecular weight toxic gases).

The *SIT* and *PIO* are not exclusively intrinsic properties of the carbon. The *SIT* and *PIO* change with the system operating conditions (i.e. oxygen availability, air flow rate, flow regime, relative humidity, bed and air temperatures, heating rate, bed dimensions, and thermal insulation), physical carbon conditions (i.e. particle size and bulk density) as well as intrinsic properties of the carbon (i.e. surface area and impregnants). Therefore, it is necessary to uniquely determine the *SIT* and *PIO* under application-specific conditions for each adsorbent of interest. Despite this caveat, a standard test method is required for analysis and optimization of processes, carbon comparison, and quality control purposes. The current methods available for *SIT* evaluation are:

1) ASTM [2]. A sample of carbon is exposed to a

* Corresponding author. E-mail: lcbuettn@apgea.army.mil

Table 1
Activated carbon properties

Type	Precursor	Ash(%)	Shape	Size	Bulk density (g cm^{-3})	Surface area (m^2 g^{-1})	Pore volume (cm^3 g^{-1})
BPL	Bituminous coal	8	Granular	12×30	0.48	>1100	0.7
MaxSorb	Petroleum coke	30[a]	Cylinder	1.5 mm, 3 mm	0.37	2060	1.1	–
Coconut Shell	Coconut shells	4	Granular	8×16	0.48	1200	–

[a]MaxSorb is chemically activated with KOH causing the high ash content.

Table 2
Composition of impregnation in activated carbons

Type	Impregnating ingredients (wt%)					
	Copper	Chromium oxide	Silver	Zinc	Molybdenum	TEDA
ASC	8	3	<0.1	–	–	–
ASZM	4–6	–	<0.1	4–6	1–3	–
ASZM-T	4–6	–	<0.1	4–6	1–3	2–4

Table 3
Experimental parameters used to simulate the ASTM method and military filters operation

	ASTM	Military filters
Bed diameter (mm)	25	31
Bed length (mm)	25	52
Superficial velocity (cm s^{-1})	50	22.9
Heating rate (°C min^{-1})	2	2

stream of heated air. Sample and air temperatures are increased at a constant rate of 2 or 3°C min^{-1} until the carbon ignites. The slopes of the sample bed or the outlet temperature profiles from pre- and post-ignition are linearly extrapolated and the intersection of the two curves is defined as the *SIT*.

2) Thermal analysis methods [4,5]. The carbon *SIT* can be determined by thermal gravimetry (TG – weight change as a function of temperature) or differential scanning calorimetry (DSC – heat flux emitted with increasing temperature). The *SIT* for each of these methods is usually defined as the intersection of the baseline and the slope at the inflection point of the sample mass or power density as a function of temperature curves, respectively.

The *SIT* is the temperature at which the oxidation reaction has gotten to the point that is self-sustaining. In contrast, the *PIO* is the temperature at which the surface properties have started to change due to the oxidation reactions reaching an arbitrary level of significance. We propose four approaches to defining the *PIO*:

1) Temperature comparison (TC) [1]. A bed and the air flowing into it are heated at a constant rate and the inlet and outlet bed temperatures are monitored. For an inert bed, the outlet will lag the inlet due to the heat capacity of the system. Once the carbon begins to oxidize, the outlet temperature will increase because of the heat released by the exothermic reaction. The *PIO* is defined as the point where the two temperature profiles intersect.

2) CO_2 emissions (CO_2P). The effluent CO_2 concentration is monitored and modeled with an Arrhenius equation. The *PIO* is defined as the lowest temperature where the natural log of CO_2 concentration can be linearly correlated with the inverse temperature using the Arrhenius equation.

3) and 4) Thermal analysis methods. The carbon *PIO* can be determined by TG and DSC by defining it as the point of deviation from the baseline values due to the exothermic processes.

The intent of this study is to compare and correlate the *SIT* and the *PIO* and the methods that generate this information. In addition, we investigate the behavior of the carbons in both the ASTM bed and in a bed configuration that more closely mimics the intended filter application (in this case using military filter dimensions and flow rates). In this study we test three unimpregnated (BPL, MaxSorb, and Coconut Shell) and three commercially impregnated activated carbons (ASC, ASZM, and ASZM-T).

2. Methods and materials

2.1. Activated carbons and impregnation

The unimpregnated carbons investigated were BPL (Calgon, Pittsburgh, PA), MaxSorb (Tokyo Zairyo,

White Plains, NY), and Coconut Shell (Sorb-Tech, Woodlands, TX). The physical properties of these carbons, according to the suppliers' literature, are listed in Table 1. All impregnated carbons were prepared [1] by the manufacturer (Calgon) from the base BPL activated carbon. The composition of these impregnants is summarized in Table 2.

2.2. Measurement of temperature and concentration profiles

The apparatus used in the current work is based on the ASTM method # D3466-76 and is extended to include CO and CO_2 concentration analysis of the effluent air stream. A carbon bed held inside a glass tube is exposed to a gas mixture of ultra high purity N_2 (80%) and O_2 (20%) (both from Matheson) flowing at a constant superficial velocity. Using a temperature-controlled oven and gas heat exchanger, the temperature of both the sample and the inlet air is raised at a rate

of $5°$ min^{-1} to at least 40°C below the expected *PIO*. For determination of the *PIO*, after equilibration at the initial temperature, the temperature is increased at a constant rate of 2°C min^{-1} to the point at which the exit temperature of the carbon bed exceeds the inlet temperature by 5–10°C. The CO and CO_2 concentrations in the bed effluent are recorded in one-minute intervals using a MIRAN-80 Infrared Analyzer. Each sample analysis was repeated 4–6 consecutive times for each sample, quenching the oxidation reaction between each iteration by cooling the bed to the initial temperature. For *SIT* determination, the samples were heated at 2°C min^{-1} until the bed ignited and, after ignition, the reaction was quenched by stopping air flow over the bed and turning off the oven.

Two different bed configurations were used in this study: one is defined by the ASTM specifications, and the other is sized to mimic a typical military filter. The *SIT* and *PIO* were measured for each carbon in both bed configurations (ASTM, CO_2P, and TC methods

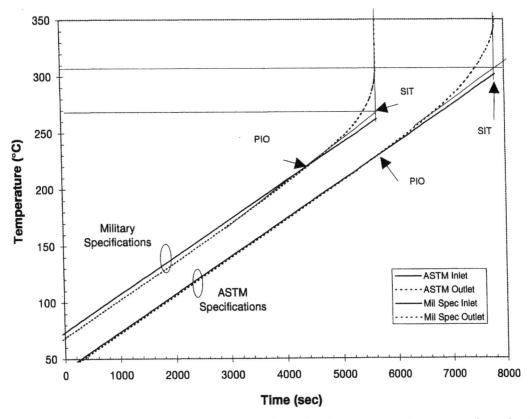

Fig. 1. Inlet and outlet temperatures are shown as a function of time online. Two separate experiments are superimposed on this graph. Two tubes were packed with ASZM carbon to either the ASTM or the end user (military filter) specifications. The method of *PIO* determination is described in the text. The application specific bed temperature profiles are shown on the same figure by shifting the graph 1000 seconds forward in time. Note that, although the ignition temperature is a sharp peak, considerable reaction occurs before the onset of ignition.

Table 4
Ignition temperatures (°C) evaluated by the various methods

Carbon	ASZM-T	ASC	ASZM	Coconut Shell	MaxSorb	BPL
SIT						
Standard method (ASTM bed specs.)	300	265	300	365	>370	>370
Standard method (military filter specs.)	275	230	260	330	340	>370
DSC	365	330	365	475	410	525
TG	370	335	370	480	455	565
PIO						
Temp. profiles (TC) (ASTM bed specs.)	150	195	225	210	265	360
Temp. profiles (TC) (military filter specs.)	155	190	225	180	230	315
CO_2 profiles (CO_2P) (ASTM bed specs.)	165	195	--	215	–	–
CO_2 profiles (CO_2P) (military filter specs.)	150	200	225	185	255	340
DSC	165	220	260	215	295	470
TG	215	290	315	325	385	500

with either the ASTM specified bed and superficial velocity or the alternative military sizing). Table 3 is a side-by-side comparison of the two bed configurations highlighting the differences in operating parameters.

2.3. Thermal analysis

DSC and TG measurements for each sample were simultaneously performed on a SETARAM instrument

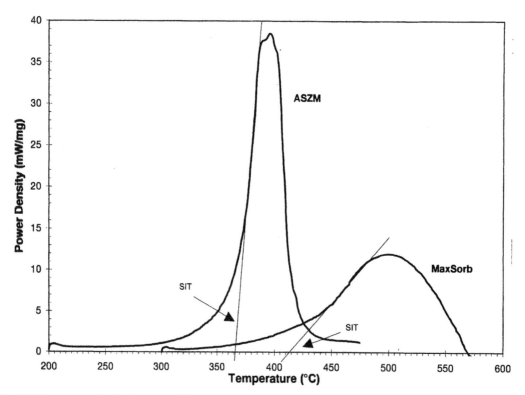

Fig. 2. DSC thermogram for MaxSorb and ASZM showing power density emitted as a function of sample temperature. Samples (3–5 mg) are tested under dry air flow in cylindrical platinum crucibles with the reference side of the TG balanced with an identical weight test sample in flowing N_2. The samples are held at 150°C to dry off residual water and then heated from 150 to 600°C at a heating rate of 5°C min^{-1} with a gas flow rate of 16 cm^3 min^{-1}. The *SIT* is defined as the temperature where the tangent to the point of inflection intersects the baseline DSC value. The *SIT* determination is shown on the figure.

(model TG–DSC 111). The samples (3–5 mg obtained from a 50 mg sample that was crushed and uniformly mixed), were tested under dry air flow in cylindrical platinum crucibles. A reference of identical mass to the test sample was balanced on the reference side of the TG. The temperature range was 150–600°C, with a heating rate of 5°C min^{-1}. The gas flow rate through the sample and reference cell was 16 cm^3 min^{-1}. Prior to each experiment, carbon samples were kept for 30 minutes at the initial temperature of 150°C to evaporate any water vapor collected during handling. The DSC and TG thermograms were computed by subtracting the reference curves obtained with the same type of carbon at identical conditions under nitrogen.

3. Results

3.1. Determination of the SIT

3.1.1. Standard method mode (SM).

In the standard method as applied in this study, the SIT is found experimentally by the intersection of the linear extrapolations of the outlet temperature profile before and after ignition. The bed conditions (sizing and flow parameters) greatly affect the SIT that is measured. This study looked at two bed conditions, those defined by the ASTM, and bed properties that mimicked an end use of interest to the authors (a military bed configuration). To show the effect of bed operating parameters, Fig. 1 indicates the determination of the SIT for ASZM carbon using both bed configurations. The SIT is defined as the intersection of the two linear extrapolations that are shown. Because the exiting temperature is not truly linear prior to ignition, there are possible deviations inherent in the determination of this line. In addition, the inlet temperature that corresponds to the ignition can be significantly overestimated. Because of the potentially life-threatening nature of a failure of this type, a more conservative estimate of the ignition temperature could be necessary. Table 4 compares the SIT values obtained with both bed configurations for each of the carbons. The effect of using the application-specific bed parameters as opposed to the ASTM bed parameters was, in every case, to lower the SIT by as much as 40°C.

3.1.2. DSC analysis.

Fig. 2 shows DSC thermograms obtained with MaxSorb and ASZM carbons. Each thermogram describes a continuous exothermic carbon oxidation

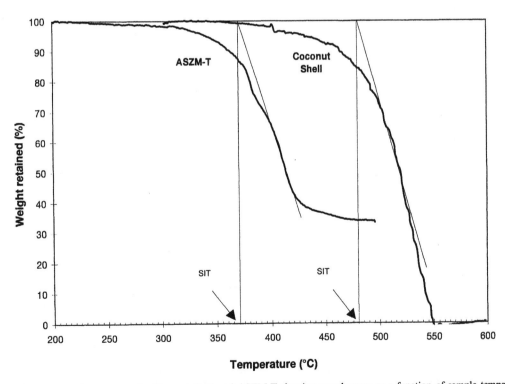

Fig. 3. Shown are TG thermograms for Coconut Shell and ASZM-T showing sample mass as a function of sample temperature. Experimental apparatus is the same as for the DSC determination. The SIT is defined as the point where the tangent to the point of inflection intersects the baseline (100% weight retained). The SIT determination is indicated on the figure.

94

terminated by complete combustion of the carbon. The *SIT* of an individual carbon was extracted from its thermogram by determining the temperature that corresponds to the intersection of the baseline with the tangent of the slope at the point of inflection for each exotherm. The *SIT*, determined by this method, is indicated on the figure for each carbon. The results for all carbons studied are summarized in Table 4.

3.1.3. TG analysis.

Similar to the previous method, the *SIT* was defined as the temperature where the intersection of the baseline (100% relative weight) and the tangent to the slope at the point of inflection for the TG thermogram occurs. Fig. 3 shows the TG results and determination of the *SIT* by this method for Coconut Shell and ASZM-T carbons. The *SIT*'s for each carbon determined by this method are summarized in Table 4.

3.2. Determination of the PIO

3.2.1. Temperature profiles (TC).

The *PIO* is determined by heating the bed and the air inlet at a constant rate; the *PIO* is defined as the temperature where the inlet and outlet temperature profiles intersect. As with the *SIT*, this temperature is a function of bed parameters and *PIO*'s were determined under bed conditions consistent with the ASTM and the end use application. Fig. 1 indicates the determination of the *PIO* under the two bed operating parameters studied.

The air inlet and bed temperature were raised to 5 to 10°C above the *PIO* and then cooled to the starting temperature. The bed inlet and outlet temperatures and effluent CO and CO_2 concentrations were monitored. Fig. 4 indicates, for 4–6 consecutive cycles, both the inlet and outlet temperatures (right scale) and the effluent

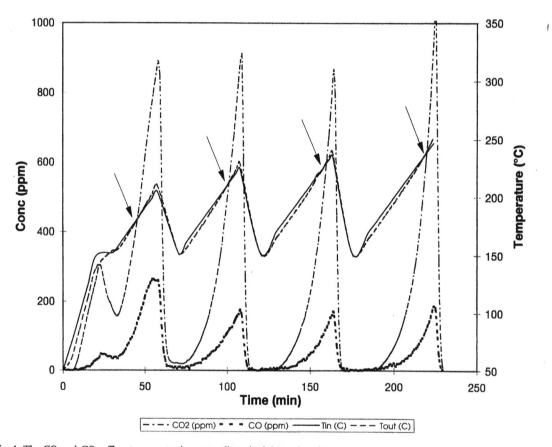

Fig. 4. The CO and CO_2 effluent concentrations as well as the inlet and outlet temperatures are shown as functions of time. The temperature of the inlet and the bed is increased by 5°C min^{-1} until the *PIO* is exceeded by 10°C, the inlet air and oven are then rapidly cooled to the equilibration temperature. This process is repeated for 4 to 6 consecutive iterations. The *PIO* is defined as the temperature where the inlet and outlet temperatures are equal. Coconut Shell carbon data is shown.

CO and CO_2 concentrations (left scale) for Coconut Shell carbon. The *PIO*, indicated by an arrow for each cycle, is the intersection of the inlet and outlet temperature profiles. To illustrate the precision of this method, Fig. 5 shows an expanded view of the time interval from 40–55 minutes in the first cycle of ASZM-T. Figs. 4 and 5 show that, (i) during the temperature ramp to the oxidation point, the outlet temperature is lower than that at the inlet (due to the heat capacity of the carbon bed); (ii) at the *PIO*, the two profiles intersect; and (iii) after the *PIO*, the outlet is greater than the inlet (caused by heat generated by carbon combustion).

It is of interest to note the continuous rise of the *PIO* with consecutive iterations found in all experiments (illustrated in Fig. 4).

3.2.2. CO_2 concentration profiles (CO₂P).

CO_2 profiles in Fig. 4 show several similarities common to all samples tested:
(1) a concentration increase of up to 350 ppm during sample heating to the pre-ignition initial temperature;
(2) a sharp increase (oxidation) to 1000 ppm or more, followed by a steep decrease (quenching) to negligi-

ble values, occurs for each of the 4–6 consecutive repetitions; and
(3) an initial large CO_2 emission, indicated by a wider concentration peak for the first repetition compared to the peaks that follow. After the initial temperature cycle, CO_2 effluent concentrations are fairly consistent from cycle to cycle.

CO_2 formation at temperatures below the *PIO* has been discussed in previous studies [4,6]. It is postulated that this pre-ignition CO_2 liberation is caused by thermal decomposition of oxygen-containing complexes. Although these complexes are largely decomposed by the initial heating stage, they still contribute to the CO_2 emission during the first temperature increase. Since the concentration profiles thereafter are reproducible, the second run was used for *PIO* evaluations based on CO_2 emission.

The location of the *PIO* based on CO_2 emission assumes that at a point of significant change in the kinetic profile of carbon oxidation takes place. The (natural) log of CO_2 concentration is shown as a function of the inverse absolute temperature in Figs. 6 and 7. In all of the carbon samples tested there was a region of linearity at the high CO_2 emission region. These lines

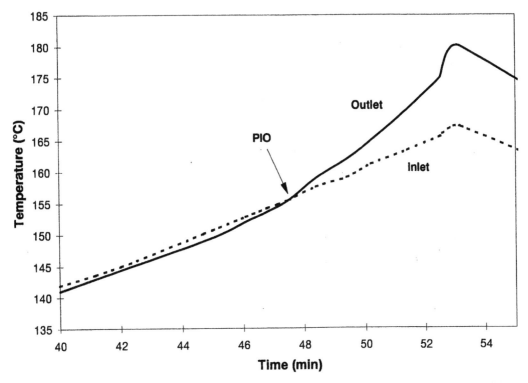

Fig. 5. Expanded scale plot of inlet and outlet air temperatures as functions of time during the first iteration in the heating cycle for ASZM-T is shown. The *PIO* is defined as the intersection of the inlet and outlet temperature profiles, shown as a sharp point. This figure demonstrates the precision of the proposed method.

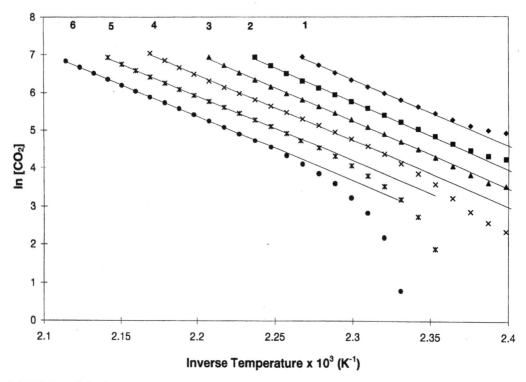

Fig. 6. ASZM-T results by the CO_2P method. An Arrhenius plot of the natural log of CO_2 concentration as a function of the inverse absolute temperature is shown for six successive temperature cycles. Each successive iteration emits a lower concentration of CO_2 at a given temperature and has a slightly higher *PIO*, but each iteration has the same activation energy (slope). The numbers above the curves correspond to the iteration number. The activation energies of the curves are as follows (in cycle order): ◆ $E_a = -34.7$ kcal mol^{-1}; ■ $E_a = -35.7$ kcal mol^{-1}; ▲ $E_a = -35.2$ kcal mol^{-1}; × $E_a = -34.4$ kcal mol^{-1}; ★ $E_a = -33.8$ kcal mol^{-1}; ● $E_a = -33.5$ kcal mol^{-1}. The *PIO* was defined as the lowest temperature that fits the Arrhenius plot to within 1% of the expected value.

were regressed, and a best fit was found. The *PIO* was defined as the lowest temperature that was within 1% of the regressed line. The *PIO*'s for each carbon, as determined by this method, were summarized in Table 4. Figs. 6 and 7 present Arrhenius plots of the ASZM-T and Coconut Shell carbons, respectively. In every iteration after the first, the activation energies (slopes of the lines) were almost constant for each carbon sample tested. The activation energies of the impregnated carbons were all approximately 40 kcal mol^{-1} and the activation energies of the non-impregnated carbons were all approximately 25 kcal mol^{-1} as shown in Table 5.

3.2.3. DSC analysis.

Expanded scale DSC thermograms of MaxSorb and BPL carbons are shown in Fig. 8. Each thermogram represents a continuous exothermic carbon oxidation terminated by consumption of the sample. The *PIO* that is determined by this method is intrinsic to the carbon and represents the point at which the exothermic reaction is significant. For comparison with the other methods

used in this work, the *PIO* of an individual carbon was extracted from its thermogram by locating the initial point of monotonic deviation from the baseline (the *PIO* was defined as the first temperature that deviated by more than 2% from the running average of the prior 5 points). These points of deviation are indicated in Fig. 2 by arrows. This definition is quite arbitrary and the *PIO* can change significantly with slight changes in the deviation required or the number of prior points that create the running average. The primary advantage of this approach is that it makes the definition consistently applied from one carbon to the next. Table 4 summarizes the *PIO*'s that were determined by the DSC method.

3.2.4. TG analysis.

Similar to the previous method, the *PIO* was defined as the point where the TG thermogram deviates continuously from the baseline (here, *PIO* is defined as the first temperature that is 0.1% smaller than the running average of the prior 5 readings). Expanded scale TG thermograms for ASC and ASZM carbons are shown

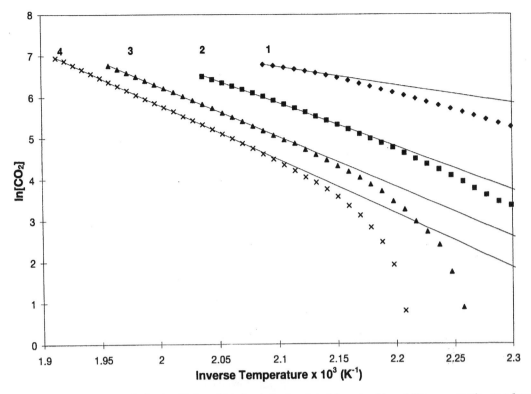

Fig. 7. Coconut Shell carbon results by the CO_2P method. An Arrhenius plot of the natural log of CO_2 concentration as a function of inverse absolute temperature is shown for six successive temperature cycles. Each successive iteration emits a lower concentration of CO_2 at a given temperature and has a slightly higher *PIO*, but each iteration, after the first, has the same activation energy (within experimental error). During the first heating cycle it is postulated that surface adsorbents are being decomposed. The numbers above the curves correspond to the iteration number. The activation energies of the curves are as follows (in cycle order): ♦ $E_a = -8.8$ kcal mol^{-1}; ■ $E_a = -21.1$ kcal mol^{-1}; ▲ $E_a = -24.2$ kcal mol^{-1}; × $E_a = -26.1$ kcal mol^{-1}. The *PIO* was defined as the lowest temperature that fits the Arrhenius plot to within 1% of the expected value.

Table 5
Activation energy from the CO_2P method (kcal mol^{-1})

Carbon	1st cycle	2nd cycle	3rd cycle	4th cycle	5th cycle	6th cycle
ASZM-T	35	36	35	34	34	34
ASC	21	38	40	42	43	43
ASZM	19	37	40	40	–	–
MaxSorb	11	22	26	–	–	–
Coconut shell	9	21	24	26	–	–
BPL	12	24	26	–	–	–

in Fig. 9. The *PIO*'s for all of the carbons obtained, using the above criteria, are summarized in Table 4.

4. Discussion

The *SIT* and *PIO* values tested in this paper by the various methods indicate that they represent two different phenomena affecting the carbons. Some carbons (especially those impregnated with TEDA in this study) can have surprisingly low *PIO*'s when compared to their *SIT* values. The standard method to define the *SIT* can lead to significant overestimations of the temperatures at which reactions commence, and caution needs to be exercised in all cases in which the exiting stream will be breathed.

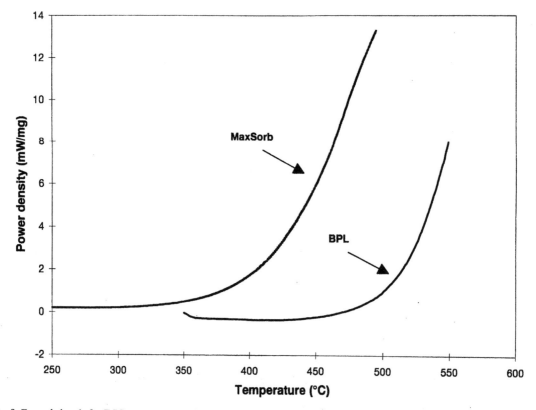

Fig. 8. Expanded scale for DSC measurements of power density as a function of sample temperature for MaxSorb and BPL carbons to show the placement of the *PIO*. The *PIO* is defined as the temperature at which the DSC reading is 2% greater than the running average of the five prior data points.

Another surprising finding is that the order of the *SIT*'s and the *PIO*'s is not consistent. A carbon can have a relatively high *SIT* (ASZM-T or Coconut Shell carbon) and have a corresponding relatively low *PIO*. In all instances the application must be considered when looking at carbon ignition properties.

The TC and CO_2P methods give *PIO* values that are comparable within experimental error. These are also the two methods that consistently yield lower values. The highest *PIO* values, on the other hand, are obtained using the TG method. The TG method gives significantly higher values for both the *SIT* and the *PIO* (by 5–110°C) compared to the DSC analysis which is performed simultaneously.

In all cases, the *SIT* was lowered significantly when the application-specific filter bed dimensions and flow rates, as opposed to the ASTM parameters, were used. This reinforces the fact that each filter application must be tested with a specific carbon as opposed to exclusively relying on the ASTM results. The deviation in *SIT* values caused by changing the bed parameters is as large as 40°C. The *PIO* was not uniformly affected by the change in the bed operating parameters, possibly indicating that this parameter has less dependence on bed operating conditions.

One of the more interesting results is that the CO_2P method yielded good consistency with the TC method, even though there appear to be two unique reactions occurring, one on the impregnated and the other on the unimpregnated carbons.

4.1. Applicability of the specific methods

4.1.1. PIO measurement.

The CO_2P method appears to be of significant use. However, due to uncertainty into the exact location of the *PIO* using the deviation from the Arrhenius plot and due to the relative difficulty in completing this measurement, it may have limited practical applicability.

Thermal analysis methods, despite yielding good reproducibility in thermogram measurement, are also hardly useful as a standard technique due to the uncertainty in locating the point on the exotherm where

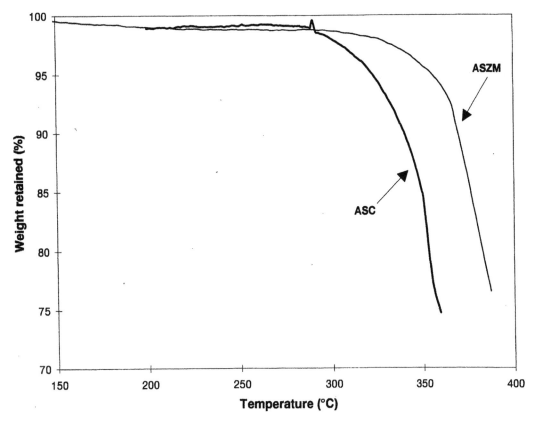

Fig. 9. TG thermograms of weight retained as a function of temperature for ASC and ASZM carbons. An expanded scale is shown to indicate the location of the *PIO*. The *PIO* is defined as the temperature at which the TG reading is 0.1% smaller than the running average of the five prior data points.

baseline deviation commences and difficulty in making measurements.

The TC method is simple to set up, gives reproducible results, and involves simple calculations. This method appears to be "forgiving" in that the bed operating parameters do not appear to have a radical effect on the *PIO* determination.

4.1.2. SIT measurement.

The ASTM test results in *SIT*'s that are consistent for one filter sizing, but are greatly dependent on the physical construction of the filter and the flow rates over the bed. In addition, in some cases, it greatly overestimates the temperature where oxidation commences. Caution must be exercised when using this method to get a bed sizing and linear flow velocity that are consistent with the intended application-specific filter.

Thermal analysis methods have the distinct advantage that they are intensive properties of the carbon; however, they generally give values of the *SIT*'s that are too high for practical use.

We strongly recommend that all carbon used in potential high temperature filtration be tested in filter dimensions consistent with the specific application using both the TC method to determine the *PIO* and the ASTM technique to determine the *SIT*. To ensure that the surface characteristics are not altered we recommend that all filter applications be operated below this *PIO* temperature. If the filter effluent is directly breathable, it is absolutely essential that *PIO* determination be made and that the filter operating temperature is maintained at a safe level below this value.

5. Conclusions

This paper serves as a warning to those that use or condition activated carbon filters in applications that involve insulated beds and/or elevated temperatures. The ASTM method for determining the spontaneous ignition temperature does not necessarily reflect the minimum temperature at which filter ignition can occur.

This ignition temperature is dependent on the physical properties of the filter, flow rates of the filtrate, and on the impregnants on the carbon.

Four methods were examined (temperature profiling, effluent CO_2 monitoring, DSC, and TG) that demonstrate the existence of a temperature where oxidation reactions commence (the so-called *PIO* [1]). Each of these independent methods indicate that reactions occur on the carbon at temperatures significantly below those temperatures that would be deemed safe by the ASTM defined *SIT*. The effect of these reactions is, at this point, unknown on the adsorptive properties of the carbon. We suggest that when conditioning beds, the *PIO* is not exceeded.

In addition, the *SIT* was determined using DSC, TG, and the ASTM method (standard and modified for larger bed size). These measurements indicate that ignition is not exclusively an intensive property of the carbon; the method that produced the lowest temperature of ignition was the method that mimicked the current application of the filter in the military setting. To ensure a safe operatoration, the ignition temperature should be determined in the application settings that it is intended to be used.

Acknowledgements

This work was performed while one of the authors (Y. S.) held a National Research Council – ERDEC Research Associateship.

References

[1] Suzin Y, Buettner LC, LeDuc CA. Behavior of impregnated activated carbons heated to the point of oxidation. Carbon (submitted).

[2] Standard test method for ignition temperature of granular activated carbon. ASTM D 3466-76 (reapproved 1993).

[3] Hardman JS, Lawn CJ, Street PJ, Further studies of the spontaneous ignition behavior of activated carbon. Fuel 1983;62:632.

[4] Akubuiro EC, Wagner NJ, Assessment of activated carbon stability towards adsorbed organics. Ind Eng Chem Res 31339–346.1992

[5] Liang SHC, Cameron LE. Differential scanning calorimetry (DSC) for the analysis of activated carbon (U). Report no. 1098, Defense Research Establishment, Ottawa, 1991.

[6] The reaction of oxygen–nitrogen mixtures with granular activated carbons below the spontaneous ignition temperature. Prepared for the US Naval Research Laboratory under Contract No. N00014-83-C-2184, GC-TR-84-385, 1984.

Activated Carbon Compendium
H. Marsh (Editor)

Carbon structure and porosity of carbonaceous adsorbents in relation to their adsorption properties

F. Haghseresht[a], G.Q. Lu[a,*], A.K. Whittaker[b]

[a]*Department of Chemical Engineering, University of Queensland, Brisbane 4072, Australia*
[b]*Centre for Magnetic Resonance, University of Queensland, Brisbane 4072, Australia*

Received 12 June 1998; accepted 30 December 1998

Abstract

A number of carbonaceous adsorbents were prepared by carbonisation at 600°C following acidic oxidation under various conditions. Effects of the chemical nature of the precursor, such as the ratio of aromatic to aliphatic carbons and oxygen content, on the chemical and structural characteristics of the resultant chars were investigated using ^{13}C NMR and Raman spectroscopy, respectively. The ^{13}C NMR spectral parameters of the coal samples show that as the severity of oxidation conditions increased, the ratio of aromatic to aliphatic carbons increased. Furthermore, it was also found that the amount of disorganised carbon affects both the pore structure and the adsorption properties of carbonaceous adsorbents. It is demonstrated that higher amount of the disorganised carbon indicates smaller micropore size. © 1999 Elsevier Science Ltd. All rights reserved.

Keywords: A. Char; B. Oxidation; C. NMR; Raman Spectroscopy; D. Microporosity; Functional groups

1. Introduction

Carbonaceous materials of high microporosity have a wide range of commercial applications such as the removal of contaminant impurities from gas and water streams. Other applications include use as catalyst supports and as molecular sieves, for separation of gaseous mixtures. The structure of a carbonaceous material is understood to be closely linked with the porosity of the material. The building block in carbons is primarily graphitic sheets, which are characterised by strong covalent bonding in the basal plane to three adjacent carbons, and weak Van der Waals bonding between the basal planes [1]. However, the structure of activated carbon is thought to be much less ordered than that of graphite. It is generally believed that two generic structural types exist. The first type consists of elementary crystallites, which are two-dimensional and analogous to graphite, with parallel layers of hexagonally-ordered carbon atoms. The second type is described as a disordered, cross-linked space lattice of the carbon hexa-gons, which results from their deflection from the planes of the graphitic lattice [2].

The extent of disorder in activated carbons and chars is known to depend on the preparation conditions and the chemical nature of the precursor. According to Oberlin et al. [3] basic structural units (BSU) of carbons are made up of small planar aromatic rings, formed during the first stage of the carbonisation. In this early stage, they are randomly oriented. However, by increasing the heat treatment temperature, they orient themselves in parallel. This process, known as local molecular orientation (LMO), depends on the heat treatment temperature and the chemical composition of the precursor. For example, it has also been shown that with an increase in the ratio of oxygen to carbon, the size of the elemental domains of bulk mesoph-ase, LMO, decreases.

Raman spectroscopy is a very useful tool in investigating structural variations of carbonaceous materials. It is established that there are five structure-sensitive lines in the Raman spectra of various carbonaceous materials. The graphite line (G band) is found at 1580 cm^{-1}. The disorder line (D band) as the result of condensed aromatic hydrocarbons [4] is found at 1360 cm^{-1}. At 1620 cm^{-1}, the line responsible for the broadening and shifting of the graphite band is visible; and the two other bands are found at 2700

*Corresponding author.
E-mail address:* maxlu@cheque.uq.edu.au (G.Q. Lu)

and 2735 cm^{-1} [5]. Other authors have shown that there is another band located at 1500–1550 cm^{-1}, arising from the imperfections in the graphite structure [6].

Raman spectral parameters have been shown to be related to various structural parameters by a number of authors. In investigating the variation of carbon structures, Tuinstra and Koenig [7] showed that the amount of crystal boundary can also be addressed as the reciprocal of the crystallite diameter in the graphite plane (L_a) [4], increased with an increase in the ratio of the intensity of the D band to the G band (I_{1360}/I_{1590}). Even though there are some disagreement, with this finding [4,8], Tuinstra's approach has received some support [9].

In analysing Raman spectra, depending on the sample type, there seems to be various opinions among various researchers in curve fitting methods for improving the accuracy in determination of spectroscopic parameters, such as peak position, band width, band shape (Gaussian, Lorentzian, or a mixture of both) and band intensity. For example, Jawahari et al [6], in their investigation of the structural variation of several commercially available carbon blacks, found that the best fitting was obtained with two Lorentzian lines at 1360 and 1600 cm^{-1} and a Gaussian band at about 1500–1550cm^{-1}. On the other hand, Dines et al. [10], using Raman spectroscopy to investigate the structural properties and bonding in a number of carbon films, found that fitting two Gaussian curves into the spectra of the developed carbon films gave the best fit.

In this work a number of microporous carbonaceous adsorbents were prepared by oxidation of coal reject in various conditions prior to heat treatment at 600°C. Their pore structures were characterised by their N_2 adsorption isotherms. Modifications in the carbon structure of the chars were investigated by examination of their Raman spectra. ^{13}C NMR spectra of oxidised coal rejects were found to give significant insights into the structural variations of the resultant chars.

2. Materials and methods

2.1. Materials

The chemicals used in this study are as follows: Phenol (99.9% Purity) supplied by Ajax Chemicals; Nitric Acid, and para-chlorophenol (PCP) from BDH Chemicals; and para-nitrophenol (PNP) by Riedel de Haen Chemicals. The coal reject (a coal waste material containing over 40% of ash) was obtained from New Hope Colleries, Queensland, Australia. It is a high volatile, bituminous coal of medium rank.

2.2. Preparation of adsorbents

A known amount of coal reject was refluxed in a fixed volume of nitric acid solution of desired concentration. The solid was then washed until the filtrate became neutral (pH=7). The solid was then dried at 100°C for 2 h prior to heat treatment at 600°C. Acid concentration, residence time and the reaction temperature were the three parameters that were varied during the pre-treatment.

The samples were labelled as follows: The number inside the bracket represents the acid concentration; the number on the left indicates the pre-treatment temperature; and the number on the right shows the treatment time length. For example, 80(6.22)6 means the char pre-treated in 6.22M nitric acid for 6 h at 80°C and 0(0)0 is the char, prepared from untreated coal reject.

2.3. Nitrogen adsorption

N_2 adsorption/desorption isotherms at 77.4K of the adsorbents were obtained using a gas sorption analyser (NOVA 1200, Quantochrome, USA). Samples were degassed overnight at 300°C in vacuum, prior to the adsorption analysis.

2.4. Phenol and para-nitrophenol adsorption

A small amount (0.05–0.15 g) of each adsorbent was weighed accurately and placed in 50 ml of 80 ppm phenol or PNP solution. The solutions were then left in a shaking water bath for three days at 25°C. After this time, the solutions were filtered and the amount of phenol or PNP in each solution was measured, using a Varian DMS90 UV Spectrophotometer at 269 nm or 317 nm, respectively. The adsorption capacities of the adsorbents were calculated from the difference between the final concentration of the adsorbate, after three days, and a phenol or PNP solution of the same concentration without any adsorbent, left in the water bath after the same period of time.

2.5. Benzene Adsorption

Benzene adsorption capacities of all adsorbents were determined thermo-gravimetrically using a micro-balance (Shimadzu TGA-50). In each experiment, 1–3 mg of sample was used, after degassing at 200°C for 2 h in the flow of nitrogen. Adsorption was carried out by allowing benzene to flow at a rate of 150 ml/min, after cooling the sample to 55°C. Our preliminary experiments showed that it was required to carry out the adsorption process at 55°C, so that equilibrium state can be achieved within 16 to 17 h.

2.6. Raman studies

Raman microscopy was employed to characterise several locations on each of the carbon materials of interest in order to ensure the reproducibility of the spectra obtained. Raman spectra were recorded using a Renishaw Raman microprobe equipped with a He–Ne laser operating at a power of 1 mW, whereas the scattered light was collected by a CCD detector. Typically, 16 scans were acquired at a resolution of 4 cm^{-1} and spectral manipulation was achieved by use of GramsResearch software (Galactic Industries).

2.7. ^{13}C NMR

All spectra were collected on a Bruker MSL300 spectrometer using cross-polarisation (CP) and magic-angle spinning at magic-angle speeds of 8–10 kHz. A standard Bruker 4 mm MAS probe was used. The speed was sufficiently high to ensure only minor contributions to the spectrum from spinning side bands. ^{1}H and ^{13}C $\pi/2$ pulse times of 3.5 μs and a contact time of 2 ms was used in all cases. The recycle delay was 3 s. These conditions are the best possible to obtain quantitative intensities under cross polarisation. The variation of intensity with cross-polarisation times indicated that the spectra were close to quantitative.

3. Results and discussion

3.1. Spectral analysis

To gain a more quantitative understanding of the Raman spectroscopic parameters of the chars, a typical set of spectral peaks were calculated by the conventional curve fitting technique. Fig. 1 shows the calculated peaks, together with the original Raman spectrum for sample 80(3.11)6. Comparison of the Raman spectra of carbonaceous materials in the available literature reveals a variety of line shapes [6]. Comparing the line shape of the spectra of other workers, such as Jawahari et al. or Dine et al. [6] with ours shows a number of differences. One of the differences is that in both of the cited cases, only one band at 1330–1360 cm^{-1} is fitted. However, the spectra of our samples showed the band stretching from 800 to 1400 cm^{-1} is made up of the disorder line at 1318–1360 cm^{-1} and two other bands. The other two bands are located at 800–1200 cm^{-1}, assigned to SP^{3} bonding carbon by Nakamizo et al. [4]. Such variation in the line shapes signifies the high degree of sensitivity of Raman Spectroscopy to structural differences in carbons. In our work, however, we were not able to relate the change in the intensity of the bands located at 800–1200 cm^{-1} to any physical features.

It can be seen that curve fitting technique varies greatly from one author to another. After carrying out a number curve fitting exercises, while keeping the positions of the bands fixed within the limits discussed so far, and varying the line shape, the best fit was found when a line shape with a mixture of 30% Lorentzian/70% Gaussian lineshapes.

In order to obtain quantitative information from the ^{13}C NMR spectra, they were all fitted with a number of Gaussian curves, whose positions were given by Trewhella et al. [11]. Curve fitting was iterated to obtain a reasonable fit. In Fig. 2 an original ^{13}C NMR spectrum with its corresponding fit is shown.

For clarity, all important parameters and ratios, as determined from the ^{13}C NMR and Raman spectra analysis, are tabulated in Table 1.

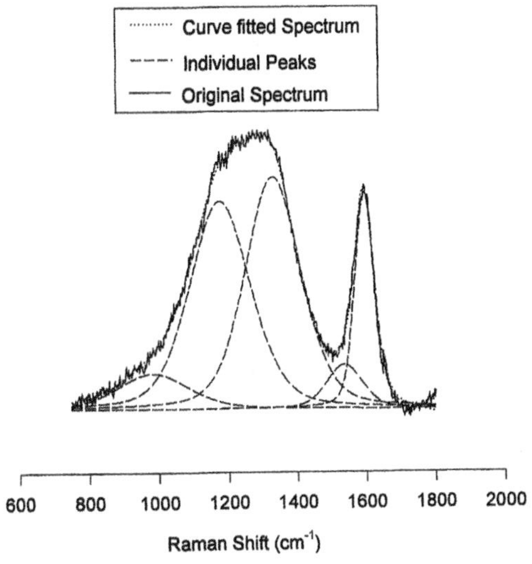

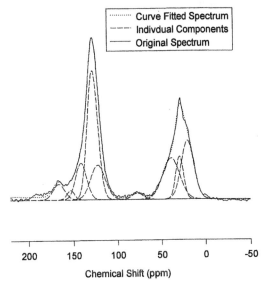

Fig. 1. Raman spectrum for sample 80(3.11)6.

Fig. 2. ^{13}C NMR Spectrum of 80(3.11)6 Prior to Heat Treatment.

Table 1
A list of the meanings of the important parameters and ratios used in this study

Symbol	Meaning	Reference
E_0	DR energy of adsorption, obtained from the N_2 adsorption	Table 2
I_{1550}/I_{1590}	Ratio of the defects in the graphene layers to the graphitic carbons, obtained from the Raman spectra	Table 2
A_{arom}/A_{aliph}	Ratio of aromatic to aliphatic carbon	Table 2
$A_{C-O}/A_{aliph+arom.}$	Ratio of total C–O groups to the sum of aliphatic and aromatic carbons	Table 2
I_{1360}/I_{1590}	Ratio of the D to the G band	Table 2
S_{BET}	Total BET surface area of the chars	Table 2

3.2. Variation of the carbon structure with different pre-treatments

Table 2 shows the ratio of the ^{13}C NMR intensity of the aromatic carbon to aliphatic carbon (A_{arom}/A_{aliph}) and the ratio of total C–O groups to the sum of aliphatic and aromatic carbon ($A_{C-O}/A_{arom+aliph}$), for most precursors. It also shows the ratio of the D to G bands (I_{1360}/I_{1590}), together with the ratio of the band due to the defects in the graphite planes, located at 1500–1550 cm^{-1} [6] (D') to G band (I_{1550}/I_{1590}), for all char samples. It is observed that A_{arom}/A_{aliph} ratios increase with an increase in the severity of pre-oxidation reactions, for all coal samples. This increase is expected, as aliphatic carbon would be oxidised more easily than aromatic carbons.

A plot of the variation in the ratio of the D band to the G band (I_{1360}/I_{1590}) of the chars, with the A_{arom}/A_{aliph} ratio of the precursors, for all samples, is shown in Fig. 3. It shows that as the amount of aromatic carbon in the precursor increases, the contribution of the disorganised carbon increases until a maximum point. However, beyond the maximum I_{1360}/I_{1590}, any increase in the amount of aromatic carbon does not cause an increase in the I_{1360}/I_{1590} ratio. This indicates that when precursors were heat treated at 600°C, where BSU of carbons, made up of fused aromatic rings, are formed [3], as the extent of aromatic carbon in the precursor increased, the contribution of

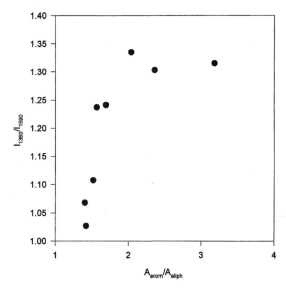

Fig. 3. Variation of the I_{1360}/I_{1590} ratio with A_{arom}/A_{aliph}.

amorphous carbon in the corresponding char also increased. Table 2 shows that the ratio of defects in the graphite structure of the chars (1550 cm^{-1}) to the G band (I_{1550}/I_{1590}) increases with an increase in the acid con-

Table 2
Summary of the spectroscopy and physical data of the chars and the oxidised coal rejects

Sample	S_{BET} (m^2/g)	E_0 (kJ/mol)	A_{arom}/A_{aliph}	$A_{totC-O}/A_{aliph+arom}$	I_{1360}/I_{1590}	I_{1550}/I_{1590}	V_t (cc/g)	V_{mic} (cc/g)	V_{meso} (cc/g)
0(0)0	107	5.65	1.386	–	0.820	0.226	0.076	0.056	0.020
80(3.11)6	354	6.24	1.405	0.0898	1.068	0.191	0.194	0.159	0.035
80(6.22)6	396	7.52	1.422	0.812	1.027	0.444	0.206	0.164	0.042
80(9.33)6	330	9.25	1.567	0.129	1.237	0.432	0.180	0.136	0.044
80(12.44)6	304	10.8	1.691	0.209	1.241	0.558	0.156	0.126	0.030
80(15.5)6	297	13.0	2.036	0.212	1.335	0.493	0.157	0.132	0.025
80(6.22)2	388	7.28	1.520	0.0708	1.107	0.379	0.221	0.172	0.049
80(6.22)23	390	15.4	2.357	0.107	1.302	0.439	0.223	0.159	0.063
60(6.22)6	404	8.48	–	–	–	–	0.218	0.171	0.047
100(6.22)6	396	8.30	–	–	1.154	0.325	0.220	0.162	0.058

centration. However, by increasing the time length and temperature of the reaction, no clear trend can be observed.

From the above discussion it can be seen that nitric acid pre-treatment led an increase in the contribution of disorganised carbon, In other words, at a given heat treatment temperature, more clusters of aromatic compounds were observed in the chars, since more aliphatic chains were oxidised and more C–O groups were formed as the results of the pre-treatment. On the other hand, no significant trends on the effects of the pre-treatment on the extent of imperfections in the graphite structure of the chars were observed.

The N_2 adsorption/desorption isotherms of a few pre-treated chars are shown in Fig. 4. The isotherms are of Type I, according to the BDDT classification [12]. In Table 2 BET surface areas (S_{BET}), total pore volumes (V_t), micropore volumes (V_{mic}), mesopore volumes (V_{meso}) of all chars, calculated on dry, ash-free basis, together with their DR adsorption energy are shown. Total pore and micropore volumes of the chars were determined from the amount of nitrogen desorbed at the relative pressures of 0.96 and 0.2, respectively; mesopore pore volume (V_{meso}) was determined from the difference of V_t and V_{mic}; and the DR adsorption energy of the chars were calculated from the linear portion of the DR plots of the chars (Fig. 5).

As a result of micropore constrictions on N_2 adsorption in carbonised materials, a higher value for micropore volume is expected from its CO_2 adsorption isotherm (at 273K) than its corresponding N_2 isotherm (at 77K) [13,14]. In our work, however, due to the lack of CO_2 data, no information about the narrower micropores could be obtained. In addition, because of pore constrictions apparent from the large hysteresis, micropore volumes

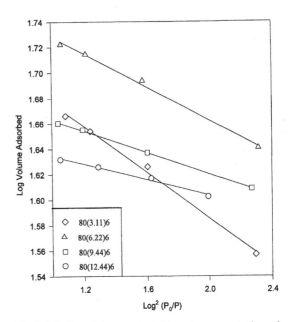

Fig. 5. DR plots of chars pre-treated in various concentrations of HNO_3.

determined from the N_2 isotherms do not represent the total pore micropore volumes. However, these pore volumes (representing larger micropores) could be used to compare pore structural changes occurred in the chars pre-treated at various conditions.

S_{BET} values together with the pore volumes of the chars in Table 2 show that increasing the severity of the oxidation treatment, generally led to a significant increase in the porosity of the resultant chars. Further insight into the pore structure evolution of the chars can be obtained by examining the variation in E_0 values of the chars, with I_{1360}/I_{1590} (Fig. 6). Fig. 6 shows that as I_{1360}/I_{1590} increases E_0 also increases. This means that the average size of micropores measured by nitrogen (larger micropores) decreases (E_0 increases) as the extent of disorganised carbon in the chars increases. As in the early stage of heat treatment, clusters of aromatic rings or BSU are randomly oriented; some void space is expected to be between them. This is easily understood. As more BSUs are formed, the space between them becomes smaller, hence smaller micropore sizes and higher E_0 values are observed.

3.3. Effect of the carbon structure on adsorption properties

By rearranging the adsorption data in terms of the number of equivalent monolayers one can obtain more insights into the adsorption behaviour of the solutes. Equivalent surface monolayer is defined as

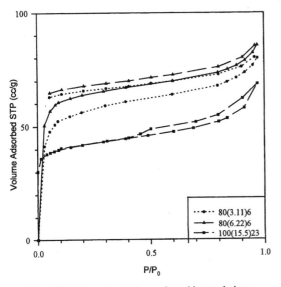

Fig. 4. N_2 Adsorption isotherms for acid-treated chars.

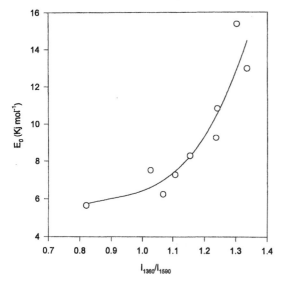

Fig. 6. Variation of E_0 with I_{1360}/I_{1590} ratio.

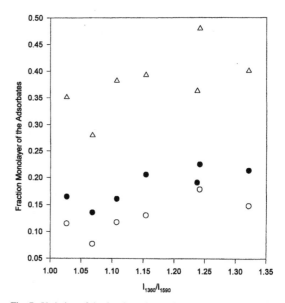

Fig. 7. Variation of the fraction of monolayers of phenol, PNP and benzene with (a) I_{1360}/I_{1590} and (b) V_{meso}/V_{mic} ($\bullet$, phenol; $\bigcirc$, PNP; $\triangle$, benzene).

$$\theta = \frac{0.001Q}{MW} N_A \frac{\sigma}{A_s} \qquad (1)$$

where N_A is the Avogadro's number, σ is the surface area of an adsorbed molecule, MW is its molar mass, Q is the adsorbed amount (mg/g), and A_s is the total surface area of the adsorbent. The surface area occupied by a molecule of an adsorbate, can be determined by the relationship given by McClellan and Harnsberger [15]:

$$\sigma = 1.091 \left(\left[\frac{MW}{\rho N_A} \right] \right) cm^2/molecule \qquad (2)$$

where ρ is the density of the solute. Using this relationship, the value of σ for benzene was calculated to be 31.3 Å^2. Density of benzene at 55°C was calculated using the Racket equation as modified by Spencer and Danner [16]. Values of σ for phenol and PNP, obtained from [17], are 43.7 and 51.9 Å^2, respectively.

Variation of the fraction of monolayers of the adsorbates with the ratio of I_{1360}/I_{1590} for phenol and PNP are shown in Fig. 7. Although the data are slightly scattered, Fig. 7 shows that with an increase in the I_{1360}/I_{1590} the fraction of monolayer increases for all solutes. This shows that as the contribution of disorganised carbon increases, which is an indication of the narrowing of the micropores (as shown in Fig. 7), the amount of adsorption of the organic compounds increases, until the micropore size decreases to a critical value (normally about the molecular size of the adsorbate). Below such a critical micropore size, the adsorption capacity would expect to decrease because the narrower micropores become inaccessible to the solute molecules. This trend can be seen in Fig. 7, though there

are not enough data points at higher I_{1360}/I_{1590} ratios to clearly back up this observation. However, we can confirm this trend from the previously reported adsorption data of phenol [18]. We showed a very clear picture that the adsorption capacity of phenol increased with increasing DR slope (E_0) and dropped drastically beyond a critical E_0 value (or below a critical micropore size).

Another possible explanation for the increase in the adsorption of the organic compounds could be due to the increase in adsorption sites, as the result of increase in the amount of disorganised carbon. Van Doorn et al. [19], using Raman spectroscopy to measure the contribution of the disorganised carbon, have shown that the reactivity of the carbons towards O_2 increased with increasing amount of disorganised carbon. It is believed that the amorphous carbons contribute more edge sites in carbons and the more edge sites more active sites for adsorption and reaction.

4. Conclusion

In this work, we have shown that by using ^{13}CNMR spectroscopy, it is possible to investigate the extent of oxidation in the precursors. As the severity of oxidation conditions increased, the ratio of aromatic to aliphatic carbons increased. The amount of disorganised carbons, as characterised by the ratio of D band to G band, affects both the pore structure and the adsorption properties of the resulting adsorbents. It was demonstrated that as the ratio of D band to G band increased, the adsorption capacity of the resulting chars for organic compounds increased. Furthermore, the increase in the extent of the disorganised carbons also indicates less space between the clusters of aromatic compounds hence smaller micropores.

References

[1] Walker Jr. PL. Carbon 1986;24:379.

[2] Smisek M. In: Smisek M, Cerny S, editors, Active carbon: manufacture, properties and application, New York: Elsevier, 1970.

[3] Oberlin A, Bonnamy S, Monthioux M, Rouzaud JN. ACS Syposium Series No. 303, Petroleum Derived Carbons, 85, (1986).

[4] Nakamizo M, Kammereck R, Walker Jr. PL. Carbon 1974;12:259.

[5] Vidano RP, Fischbach F. J Am Ceram Soc 1978;61:13.

[6] Jawhari T, Roid A, Casado J. Carbon 1995;33:1561.

[7] Tuinstra F, Koeng JL. J Chem Phys 1970;53:1126.

[8] Tsu R, Gonzalez JH, Hernandez IC. Solid State Commun 1978;27:507.

[9] Nikiel L, Jagodinski W. Carbon 1993;31:1313.

[10] Dines TJ, Tither D, Dehbi A, Mathews A. Carbon 1991;29:225.

[11] Trewhella MJ, Poplett IF, Grint A. Fuel 1986;65:541.

[12] Brunauer S, Deming LD, Deming W, Telle WE. J Am Chem Soc 1940;62:1723.

[13] Garrido J, Linares-Solano A, Martín-Martíneż JM, Molina-Sabio M, Rodriguez-Reinoso F, Torregrosa R. Langmuir 1987;3:76.

[14] Rodriguez-Reinoso F, Garrido J, Martín-Matínez JM, Molina-Sabio M, Torregrosa R. Carbon 1989;27:23.

[15] Mclelan AL, Harnsberger HF. J Colloid Interface Sci 1967;23:577.

[16] Spencer CF, Danner RP. J Chem Eng Data 1973;18:230.

[17] Caturla F, Mratín-Martínez JM, Molina-Sabio M, Rodrigues-Reinoso F, Torregrosa R. J Colloid Interface Sci 1987;124:528.

[18] Haghseresht F, Lu GQ, Zhu HY, Keller J. Pore structure and phenol adsorption characterisation of carbonaceous adsorbents prepared from coal reject. In: Characterisation of porous solids IV, Bath, England: McEnanney B, Mays TJ, Rouguerol J, Rodriguez-Reinoso F, Sing KSW, Unger KK, (Eds.), Royal Society of Chemistry, 1996.

[19] Van Doorn J, Vuurman MA, Tromp PJJ, Stufkens DJ, Moulijn JA. Fuel Process Technol 1990;24:407.

Activated Carbon Compendium
H. Marsh (Editor)

On surface of micropores and fractal dimension of activated carbon determined on the basis of adsorption and SAXS investigations

R. Diduszko[a], A. Swiatkowski[b], B.J. Trznadel[c],*

[a]*Institute of Vacuum Technology, ul. Duga 44/50, 00-241 Warsaw, Poland*
[b]*Military Technical Academy, ul. S. Kaliskiego 2, 00-908 Warsaw, Poland*
[c]*Military Institute of Chemistry and Radiometry, al. gen. A. Chruściela 105, 00-910 Warsaw, Poland*

Received 13 September 1999; accepted 6 October 1999

Abstract

Granulated commercial activated carbon obtained from hardcoal was separated for fractions, by elutriation, in an air stream of constant flow-rate. Three samples of different fall down distances from the outlet tube were demineralized. Benzene adsorption isotherms were determined at 293 K and SAXS measurements were performed for each one. The Dubinin–Izotova and Dubinin–Stoeckli equations were used for evaluation of micropore structural parameters of examined carbon samples. On the basis of these equations different methods of calculation of geometric surface area were discussed. Obtained results were compared with values evaluated by using the method proposed here. This method comprises results of SAXS and adsorption measurements. Changes of geometric surface area were confirmed by analysis of surface roughness measured by fractal dimension. The values of fractal dimension were obtained by method of calculation on the basis of adsorption data proposed previously by Ehrburger–Dolle and were compared with modification proposed here. This modification is based on well known correlation given by McEnaney. Obtained values of fractal dimension were compared with values determined on the basis of SAXS data. © 2000 Elsevier Science Ltd. All rights reserved.

Keywords: A. Activated carbon; C. Adsorption, X-ray scattering; D. Microporosity

1. Introduction

Microporous activated carbons are used in a variety of gas separation, purification and catalytic processes. Effective use of these materials requires knowledge of the pore geometry, pore size distribution and surface irregularity of the porous solids. The adsorbent activity is dependent on the magnitude of the internal surface, the pore size distribution and shape of the pores. The quantitative evaluation of geometric surface area plays an important role in the characterization of porous solids. On the other hand the surface roughness and irregularities are additional characteristics of such materials. Using the fractal analysis approach these features may be estimated on the basis of adsorption or SAXS measurements. In both cases, surface area and roughness, an experimentally convenient and

satisfactorily accurate technique of evaluation structural parameters of activated carbons is not available. The need for such technique has become increasingly urgent while one can observe a rapid development of new microporous materials for different applications.

In this paper two experimental methods for determination of geometric surface area of micropores of activated carbons are used. The first one is the benzene vapor adsorption isotherm analysis based on different adsorption models. The second one is the small angle X-ray scattering (SAXS). These two methods are based on the different models of pore shape. The slit-like pores are assumed in the case of adsorption and spherical or cylindrical shape of pores is assumed in the case of SAXS method. So, it is interesting to compare results obtained in these two ways. The interpretation of adsorption data of microporous activated carbons to obtain geometric surface area or fractal dimension is less theoretically rigorous than X-ray diffraction, although the data are significantly less expen-

---

*Corresponding author.

sive to obtain. Moreover, the adsorption method is better suited for studying the media with polydisperse pore size.

The present paper has two objectives. The first one is to establish the appropriate methodology for evaluation of the geometric surface area on the basis of adsorption and SAXS measurements. The second objective is to apply these two different experimental techniques to quantify the surface roughness of activated carbons with different levels of activation.

2. Theoretical

2.1. Adsorption method for determination of a geometric surface area of micropore

In some cases an adsorption isotherm on activated carbon with nonhomogeneous microporous structure can be approximated by the linear combination of two adsorption isotherms for adsorbents with homogeneous microporous structure [1]. If so, the two term equation of the volume filling of micropores (called the Dubinin–Izotova equation (DI)) ought to be used for a description of physical adsorption (for benzene vapor as an adsorbate):

$$W = W_{01} \exp\left[-\left(\frac{A}{E_{01}}\right)^2 \right] + W_{02} \exp\left[-\left(\frac{A}{E_{02}}\right)^2 \right] \quad (1)$$

where: W is the volume of micropores filled at temperature T and the relative pressure p/p_0, W_{01}, W_{02} are the volumes of micropores and supermicropores filled respectively [cm^3 g^{-1}], A is the differential molar work ($A = RT\ln(p_0/p)$), and E_{01} and E_{02} are the characteristic adsorption energies connected with micropore and supermicropore filling respectively [kJ mol^{-1}].

Analysis of experimental data revealed an inverse proportionality between the slit-pore half-width x_{0D} and the characteristic adsorption energy E_0 (for benzene at 293 K as the standard vapor) [2]:

$$x_{0D} = \frac{12.0}{E_0} \quad (2)$$

The mean value of slit-like half-width of these two porous structures can be expressed by [3]:

$$x_{0DI} = \frac{x_{01}W_{01} + x_{02}W_{02}}{W_{01} + W_{02}} \quad (3)$$

where: x_{01} and x_{02} are evaluated by Eq. (2) using values of characteristic adsorption energies E_{01} and E_{02} respectively. An average characteristic adsorption energy connected with the DI equation cannot be evaluated as a weighted mean of values for each kind of pores. So, it is necessary to link the left side of Eq. (3) with other formula linking the mean value of slit-like width of micropore and characteristic adsorption energy to calculate this parameter.

Such possibility gives the correlation proposed by McEnaney [4]. Applying the linear regression method to comparative analysis of the results obtained from both the SAXS and heats of immersion measurements [4] he derived the equation describing the relation between half-width of micropore and characteristic adsorption energy:

$$x_{0E} = 2.3455 * \exp(-0.0666 * E_0). \quad (4)$$

One can obtain the characteristic adsorption energy connected with the DI equation (Eq. (1)) by solving equation where left side is from Eq. (3) and right side is from Eq. (4). The solution of this equation is the value of characteristic adsorption energy for the two term equation (assigned as E_{0DI}).

The total volume of micropores is expressed by the sum of limiting volume of micropores and supermicropores:

$$W_{0DI} = W_{01} + W_{02} \quad (5)$$

The value of micropore volume evaluated by Eq. (5) and value of the mean half-width of micropore evaluated by Eq. (3) lead to value of the geometric surface area of micropores. Here, for this purpose we use the formula proposed by Dubinin [5], originally for the Dubinin–Radushkevich equation (DR) [6], and described by:

$$S_{gDI} = 1000 * W_{0DI} / x_{0DI} \quad (6)$$

Assuming the normal (Gaussian) half-width distribution of micropores and the Dubinin–Radushkevich as a local adsorption isotherm one can obtain the Dubinin–Stoeckli (DS) equation [6]. In such a case the geometric surface area of micropores is expressed by [7]:

$$S_{gDS} = \frac{1000 * W_0^0}{\delta\sqrt{2\pi}} \int_{0.25}^{x} \frac{1}{x} \exp\left[-\frac{(x - x_0)^2}{2\delta^2} \right] dx \quad (7)$$

where: x_0 is the half-width of slit-like micropore for the maximum of pore size distribution function, W_0^0 is the total volume of micropores, δ is the variance representing the distribution range. When $x = \infty$ one can obtain the total geometric surface area [3]. Moreover, it is possible to evaluate the geometric surface of micropores when $x = x_{max}$ i.e. the maximum size of micropore half-width.

However, when the activated carbon has a sufficiently developed mesopore surface (over 50 m^2 g^{-1}), it is necessary to correct each point of the experimental adsorption isotherm for the adsorption on mesopore surface using the formula [3,6,7] to obtain real geometric surface area of micropores for all mentioned methods:

$$W_{mi} = W - \gamma S_{me} \quad (8)$$

where: S_{me} is the surface area of mesopores and γ is the correction factor expressed for benzene vapor by [6]:

$$\gamma = 9.16 * 10^{-3} \exp(-A/6.35) \quad (9)$$

The surface area of mesopores may be obtained from [6]:

$$S_{me} = \frac{1}{\sigma} \int_{a_0}^{a_s} A \, da \tag{10}$$

where: σ is the surface tension and a_0 is the amount of adsorbate at the initial point of the hysteresis loop and a_s is the limiting value of adsorption. The integral given by Eq. (10) is calculated on the basis of desorption branch of hysteresis loop of adsorption–desorption isotherm.

2.2. X-ray methods for characterization of microporosity

The X-ray small angle scattering method (SAXS) is the useful diffraction experimental technique for investigations of a porous system of carbon materials. By this way in standard conditions, one can obtain structural information about scattering objects characterized by a range of dimensions approximately from 1.0 nm to 100 nm. However, it is possible to broaden that range for specific scattering structures by use of special measurement conditions [8].

In the case of activated carbons, observable scattered intensity consists of several sources: micro- and mesoporous structures, microcrystallites or clusters of mineral matters and inhomogeneity of electron density in the carbon matrix. One can neglect the influence of the last one source on scattering intensity of investigated activated carbons, because it is few orders lower then other influences. Representative experimental intensity curves are usually presented in log–log scale.

The multiple scattering was observed at small angles and in this range there is influence of a broad size distribution of the mesoporous system. Here, it is assumed that the greatest influence on intensity is derived from the microporosity system for larger angle values. So all parameters were calculated only for these ranges of intensity.

The most important parameter for micropores-reduced chord length l_r (the range of inhomogeneity also called mean intersect) was obtained directly from SAXS experiments by formula [9–11]:

$$l_r = \int_0^\infty q I(q) \, dq \left\{ \lim_{q \to \infty} q^3 I(q) \right\} \tag{11}$$

where: $I(q)$-smeared (measured) intensity and q is wave vector $q = 4\pi \sin(\theta)/\lambda$.

One can calculate the integral in the above formula only in a short range of q values (for microporous systems). For reduction termination errors we extrapolated the intensity curve from 0 to the first experimental value by an exponents formula using so called Guinier's extrapolation

and from the last experimental value to infinity by Porod's law [8].

The volume fraction of the pores in the sample of porous solid is usually calculated on the basis of real and apparent sample densities. In this work, it is proposed to calculate the volume fraction of micropores P_{mi} by taking into account only the volume of micropores and volume of solid matter for microporous system, according to:

$$P_{mi} = W_i / \left(W_i + \frac{1}{d_{He}} \right) \tag{12}$$

where: W_i is the limiting volume of micropores obtained from various adsorption isotherm equations (DI-Eq. (1) or DS [6]) and d_{He} is the real density of a porous solid.

Assuming a two phase system (pores and carbon matter) one can obtain: average chord length of pores l_p (reduced inhomogeneity) defined by

$$l_p = l_r / (1 - P_{mi}) \tag{13}$$

average chord of carbon matter l_m defined by

$$l_m = l_r / P_{mi} \tag{14}$$

The specific surface area for micropores can be calculated from [9]:

$$S_{mi} = 4*10^4 P_{mi} (1 - P_{mi}) / (l_r d_{He}) \tag{15}$$

2.3. The evaluation of fractal dimension on the basis of adsorption and SAXS measurements

The concept of fractal geometry [12] has recently been used for the description of the structural heterogeneity of solids. The new possibility in this area is offered by a recently developed equation which allows the calculation of the structural parameters of the adsorbent, i. e. micropore volume and the fractal dimension [13].

According to the kind of fractal analysis or technique used, irregular objects can be characterized by three different fractal dimensions [14]: the perimeter fractal, the surface fractal and the mass fractal. The mass fractal describes how the mass of an object varies with its size. Bezot and Hesse-Bezot [15] analyzed fractal structure of carbon black agglomerates.

In this work irregular objects of other character are analyzed. Activated carbons are obtained by activation process, which consists of a stepwise procedure of elimination of a carbon matter by an oxidizing agent.

As was proposed by Pfeifer and Avnir [16], in the case of adsorption on fractals, the PSD is described by the power law:

$$f(x) = \frac{3 - D}{x_{max}^{3-D} - x_{min}^{3-D}} x^{2-D} \tag{16}$$

where: D is the fractal dimension, x_{max} is the upper limit of micropores and x_{min} is the finest resolution at which fractality prevails. So, according to assumed limits the character of PSD may be changed. The work [17] contains discussion of this problem. Błażewicz et al. [18] assumed the upper limit equal to 1.342 nm and lower limit equal to 0.185 nm. They used the Horvath–Kawazoe equation (HK) [19] to calculate the maximum half-width of micropores (for related equilibrium pressure of benzene vapor equals to 0.175 for the beginning of hysteresis loop) and the minimum value of micropore size taken from the lower limit of the domain of the HK equation for benzene vapor as an adsorbate. The question arising now concerns the validity of concept of fractal dimension when the range of sizes covers less than one decade. As was mentioned by Ehrburger–Dolle [20] the roughness of the surface has to be considered in terms of adsorption pressure rather than in terms of geometry. So, it seems more realistic to consider the D (exiting in Eq. (16)) as an adsorption fractal dimension rather than a geometrical one (the range of relative equilibrium pressure covers over five decades). Integration of Eq. (16) defines average micropore size connected with the power law as a function of fractal dimension [21]:

$$x_f = x_{min}\left(\frac{3-D}{4-D}\right)*\left(\frac{\left(\frac{x_{max}}{x_{min}}\right)^{4-D}-1}{\left(\frac{x_{max}}{x_{min}}\right)^{3-D}-1}\right) \quad (17)$$

In the case of analysis of adsorption data, the fractal dimension D may be calculated on the basis of an equation proposed by Ehrburger–Dolle [20]:

$$D = 3 - \left[24/E_{0FR}\left(1-\frac{E_{0FR}}{E_{0DR}}\right)\right] \quad (18)$$

where: E_{0DR} and E_{0FR} are characteristic adsorption energies obtained from the DR equation and the Freundlich equation (FR) respectively. Originally [20] calculation of fractal dimension is based on the characteristic adsorption energy obtained from the DR equation. The FR equation after modification for benzene vapor takes form:

$$W = k(p/p_0)\frac{RT}{E_{0FR}} \quad (19)$$

where: k is characteristic constant.

But for $E_{0DR} < 18$ kJ mol^{-1} and/or $E_{0FR} < 12$ kJ mol^{-1} the Eq. (18) becomes [20]:

$$D = 3 - \left(\frac{8}{E_{0FR}}\right) \quad (20)$$

Determination of this parameter is difficult since the determination of domain of linearity of the DR plots is questionable. Błażewicz et al. [18] show that in such case one can use the characteristic adsorption energy obtained on the basis of the Dubinin–Astakhov equation. Here, since an adsorption isotherm is well fitted by the two term

equation (Eq. (1)), it is proposed to use for evaluation adsorption fractal dimension D the characteristic adsorption energy E_{0DI} calculated on the basis of the DI equation using McEnaney's formula (Eq. (4)).

Since the pioneering work by Bale and Schmidt [22] on the determination of the surface fractal dimensions the SAXS method has been widely used to study the structure of irregular objects. Briefly in the theory of small-angle-scattering, it is well known that the intensity of radiation scattered on a fractal surface is often proportional to a negative power of the wave vector q:

$$I \propto q^{-\alpha} \quad (21)$$

Usually this dependence is observed only when the value of q satisfies the inequality $q\xi \gg 1$, where ξ is the characteristic length of the structure producing scattering. From the value of α one can determine the fractal dimension D from relation [22]:

$$\alpha = 6 - D \quad (22)$$

when $q\xi \gg 1$, the fractal dimension is easily obtained from the slope of measured intensity as a function of wave vector in log–log coordinates. But the fractal analysis on the basis of SAXS measurements leads to two different values of fractal dimension connected with the range of values of analyzed logarithms of wave vector [23]. The mass fractal dimension may be obtained for the range of logarithm values between -2.29 and -1.80 and the perimeter fractal dimension for the range -1.2 and -0.9 [23].

3. Experimental

Granulated activated carbon obtained from hardcoal by steam activation produced by HPSDD Hajnówka (Poland) was partially ground to make uniform granules shape and dimensions. Next granules of diameter of about 1 mm was separated by sieving. Such prepared granules were fractionated in an air stream of constant flow-rate using the apparatus presented in Fig. 1. However, previously Smišek et al. [24] proposed classification of carbon samples by using elutriation in air stream of various flow-rates. This stepwise procedure was used by Kadlec and Daneš [25] to separate to in eight fractions of various specific weights. The method of classification, proposed in this work, gives the possibility to differentiate carbon samples in only one step. It is based on oblique throw, so the distance from the outlet tube is the differentiation factor. Moreover, this method is very useful even in the case of a low level of porosity differentiation in the production batch. This method may be applied to activated carbons before and after the ash removal. But, in both cases final effect of differentiation is not identical. Three samples (designated I, II, III) of different distances from the exhaust tube

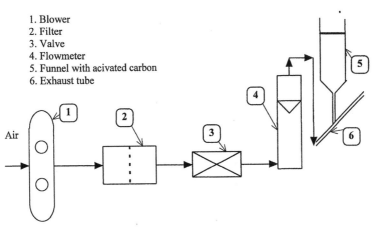

1. Blower
2. Filter
3. Valve
4. Flowmeter
5. Funnel with acivated carbon
6. Exhaust tube

Air

Fig. 1. Scheme of the apparatus for granules separation by elutriation at constant air flow.

(decreasing degree of activation) were chosen for investigations. The carbon samples were demineralized with concentrated HF and HCl acids by Korver's method [26]. The benzene vapor adsorption–desorption isotherms at 293 K were determined by gravimetric method using the McBain–Bakr vacuum balance. Fig. 2 depicts the obtained isotherms (on the left side-whole range of relative equilibrium pressures and on the right side-for pressures up to the beginning of hysteresis loop as the DR plots). Helium densities were as follows: for sample I is 2.016, for sample II is 2.015 and for sample III is 2.013 g cm^{-3}.

In addition, small angle X-ray scattering measurements (SAXS) were performed for all the studied carbon sam-

ples. A vacuum small-angle X-ray camera Kratky–Kompakt type (A. Paar, Austria) was used in these experiments [27]. Experimental parameters: scintillation counter, filtered Cu K_α radiation, sample holder 2 mm thickness with Mylar foil 6 μm window, entrance slit 30 μm, detector slit 60 μm.

Experimental results (measured intensity $I(q)$ versus wave vector q in log–log scales) are presented in Fig. 3. The indirect transformation method (ITP) [27] was used for obtaining volume distribution of micropores versus its radius. The assumed model of microporosity was simplified. In this model the microporous system consists of polydisperse distribution of spherical particles with differ-

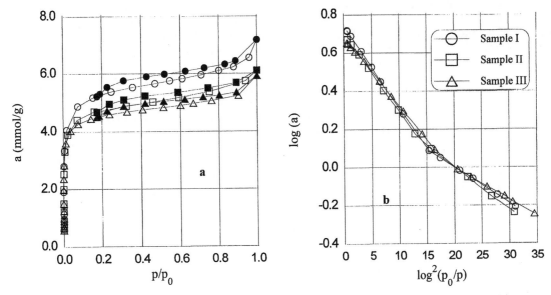

Fig. 2. Benzene vapor adsorption–desorption isotherms at 293 K on activated carbon samples: whole range of relative equilibrium pressure (a) and for pressure up to beginning of hysteresis loop as the DR plots (b).

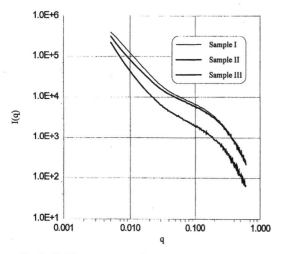

Fig. 3. SAXS measurements for examined carbon samples.

ent radius. The range of micropore radius is up to 2.0 nm and the maximum of volume distribution is for a radius lower than 0.5 nm.

4. Results and discussion

4.1. Analysis of benzene adsorption data

One can observe that the domain of linearity of DR plots of analyzed activated carbons is questionable (Fig. 2(b)). The DR plots indicate the heterogeneous micropore structure of examined carbons. So, such carbons may be analyzed by the DI or DS adsorption models. The DR equation is not valid in such a case. On the basis of obtained experimental results micropore structure parameters were evaluated from the DI equation (Eq. (1)) taking into account the correction factor expressed by Eq. (9). The values of surface area of mesopores (Eq. (10)) used in further calculations are as follows 86, 78 and 61 $m^2 \ g^{-1}$ relatively for samples I, II and III. The values of appropriate structural parameters are presented in Table 1. Based on these values it was possible to calculate mean value of micropore size x_{0DI} (from Eq. (3)) and total volume of micropores (from Eq. (5)) for determining micropore

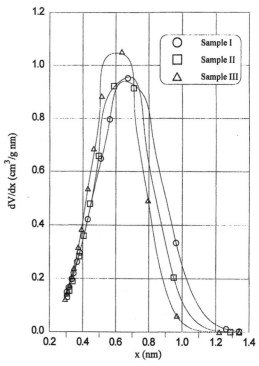

Fig. 4. The PSD curves obtained on the basis of the DS equation.

surface area S_{gDI} connected with the DI equation (from Eq. (6)).

The DR equation assumes homogeneous micropore structure, the DI equation takes into account two kinds of micropores and the DS equation was derived under assumption of normal distribution of micropore sizes. The pore size distribution (PSD) was evaluated on the basis of the DS equation [6] using Rosenbrock's optimization procedure [28]. It gives the possibility to compare the micropore surface area obtained on the basis of the simple structural model (DI) with the micropore surface area obtained assuming the heterogeneous microporous structure of activated carbon. The PSD curves obtained on the basis of the DS equation are presented in Fig. 4 and structural parameters are listed in Table 2.

The both adsorption models (DI and DS) give similar values of related structural parameters. The values of x_{0DI} and x_{0DS} differ less than 3%. The differences between

Table 1
The micropore structure parameters obtained on the basis DI equation

Sample	W_{01} (cm^3 g^{-1})	E_{01} (kJ mol^{-1})	W_{02} (cm^3 g^{-1})	E_{02} (kJ mol^{-1})	W_{0DI} (cm^3 g^{-1})	x_{0DI} (nm)	S_{gDI} (m^2 g^{-1})
I	0.21	27.16	0.26	14.16	0.47	0.67	700
II	0.24	24.89	0.22	14.17	0.46	0.66	710
III	0.21	27.53	0.21	14.45	0.42	0.63	660

Table 2
The micropore structure parameters obtained on the basis DS equation

Sample	W_0^0 (cm^3 g^{-1})	x_{0DS} (nm)	δ (nm)	S_{gDS} (m^2 g^{-1})
I	0.47	0.68	0.20	755
II	0.42	0.65	0.17	700
III	0.40	0.61	0.15	650

values of W_{0DI} the total and the values of W_0^0 the limiting volume of micropores do not exceed 10%.

4.2. Analysis of small angle X-ray scattering results

The ITP calculation results in real space as a volume distribution versus its radius $D_v(R)$ are presented in Fig. 5. The range of the micropore radius is up to about 1.5 nm and the maximum of volume distributions for radius lower than 0.5 nm. So, such low range of micropore sizes leads to disputable results especially in fractal analysis.

The evaluated values of microporosity volume fraction P_{mi} using Eq. (12), reduced chord lengths l_r (Eq. (11) with extrapolations to 0 and infinity q values), average chord lengths for micropores l_p and carbon matter l_m (Eqs. (13)

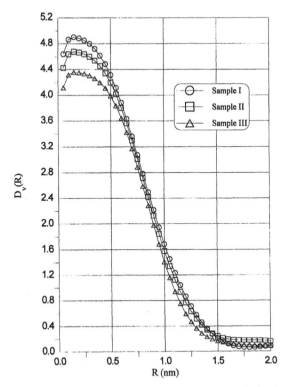

Fig. 5. The ITP calculations results obtained on the basis of SAXS measurements for examined activated carbon samples.

Table 3
The micropore structural parameters obtained from SAXS measurements

Sample	l_r (nm)	P_{mi} (cm^3 g^{-1})	l_p (nm)	l_m (nm)	S_{mi} (m^2 g^{-1})
Parameters based on the DI Eq. (1)					
I	0.659	0.49	1.28	1.35	750
II	0.662	0.48	1.28	1.37	750
III	0.680	0.46	1.25	1.49	725
Parameters based on the DS equation [6]					
I	0.659	0.49	1.28	1.36	750
II	0.662	0.46	1.22	1.45	745
III	0.680	0.45	1.23	1.52	725

and (14)) and specific microporous surface areas S_{mi} (Eq. (15)) are presented in Table 3. Different total volumes of microporous structure W_i obtained from different adsorption calculation methods were used for evaluation of the microporous volume fraction P_{mi}.

The average lengths of micropores l_p are quite similar for all samples but average chords of carbon matter l_m increases. One can conclude therefore that average size and shape of micropores are similar, but the average number of micropores decrease in order I–II–III. Because of that, the specific area of micropores decreases in similar order.

Values of specific areas of micropore, obtained on the basis of SAXS method, are higher than values obtained on the basis of adsorption methods. Results obtained on the basis of SAXS measurements are higher by about 8% using the DI equation and about 6% using the DS equation (if we compare the average values for three samples). The higher values obtained by SAXS method may be explained by accessibility for X-ray scattering of the finest or closed micropores and inaccessibility of these pores for an adsorbate (in a case of benzene vapor as an adsorbate lower limit of half-width of slit-like pore is equal 0.185 nm [18], while Dubinin assumed 0.25 nm [7]).

Analysis using SAXS method should be, in our opinion, limited only to micropores as proposed in this paper because the mesopores structure is only partially accessible for X-ray scattering and it is a source of significant errors (connected with multiple diffraction).

The comparison of results obtained by the SAXS method with results obtained by the adsorption methods (absolute values and differences between the values) leads to the conclusion that better agreement was found for analysis using the DS equation.

4.3. Fractal analysis

The use of the noninteger fractal dimension D as an operative measure of the surface irregularity and as a measure of porosity was carried out on the basis of two relations connected with adsorption measurements and on the basis of SAXS investigations. The results of calculations based on proposed here modification of Ehrburger–Dolle's method [20] of evaluation of fractal dimension D_{DI} are presented in Table 4. It seems that the values obtained here are more realistic than values calculated according to original method of Ehrburger–Dolle. The values of fractal dimension obtained on the basis of adsorption data increase due to decreasing of porosity development of analyzed activated carbon samples. One can observe a similar trend in the case of fractal dimension obtained on the basis of SAXS investigation (values of D_{SAXS} are listed in Table 4). But values of D_{SAXS} are significantly higher than those obtained on the basis of the adsorption method. The values of the perimeter fractal dimension D obtained for the range of values of logarithms of wave vector between -1.2 and -0.9 are lower than 2 and so can not be compared with that obtained on the basis of adsorption method. Previously Sahouli et al. [29] used SAXS and nitrogen adsorption data to compare values of fractal dimension for different kinds of carbon blacks. They used the fractal version of the FHH (Frenkel–Halsey–Hill) theory [30] for analysis of adsorption measurements results. The comparison between values obtained here on the basis of benzene vapor adsorption data and SAXS method and the values presented in work [29] leads to the conclusion, that the analysis based on the SAXS method rather overestimates the value of fractal dimension D. As was mentioned before the adsorption fractal dimension is based on measurement cover at least five decades of relative equilibrium pressures, while the SAXS analysis is based only on a narrow range of micropore sizes.

Jaroniec et al. [21] basing on adsorption measurements found linear dependence between fractal dimension D and average micropore size connected with the power law x_f (Eq. (17)). In this work two different methods of fractal analysis were used. So, it is interesting to check out the validity of such dependence in the case of results obtained on the basis of adsorption and SAXS investigations for the activated carbon samples. The values of x_f were calculated using Eq. (17) and lower and upper limits ($x_{min} = 0.185$, $x_{max} = 1.342$ nm) these same for both methods of fractal analysis and placed with the related values of D on a graph (Fig. 6). The linear regression carried out for the set of these values leads to relation:

$$D = 6.21 - 5.50\,x_f \tag{23}$$

The correlation coefficient for the points shown in Fig. 6 is

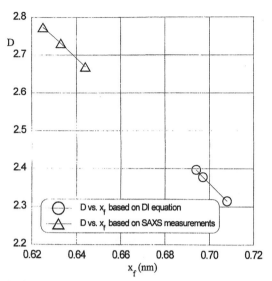

Fig. 6. Relation between fractal dimension and average micropore size associated with the power law.

equal 0.999. This value confirms the linear form of the dependence between the value of D and the value of x_f. The slope of the obtained relationship is different than that obtained by Jaroniec et al. [21], which was equal to 6.17. This difference is connected with the values of x_{min} and x_{max} of fractality prevailing range. Here, similarly like in work [18] as it was mentioned previously, the value of lower limit of the HK equation [19] domain was taken as the finest resolution $x_{min} = 0.185$ nm. On the other side, the solution of the HK equation for the relative equilibrium pressure equal 0.175 (for beginning of hysteresis loop) was taken as the upper limit $x_{max} = 1.342$ nm [18]. So, these values are based on the statistical thermodynamics. It seems that these values are better suited for such calculations than the arbitrary assumptions in work [21]. Recalculation taking into account values $x_{min} = 0.1$ nm and $x_{max} = 1.0$ nm (as in the paper [21]) leads to the value of slope of linear dependence equal to 6.17. The lower limit of fractality is strictly connected with a size of adsorbate molecule. So it ought to be chosen according to used adsorbate. On the other hand upper limit is related to the end of micropore filling process. This upper limit may be evaluated using the Horvath and Kawazoe adsorption model for relative equilibrium pressure for beginning of

Table 4
The fractal dimensions obtained on the basis of adsorption and SAXS measurements

Sample	k (mmol g^{-1})	E_{0FR} (kJ mol^{-1})	E_{0DI} (kJ mol^{-1})	D_{DI}	D_{SAXS}
I	7.76	12.39	19.17	2.315	2.669
II	7.01	12.88	19.34	2.378	2.731
III	6.84	13.45	20.30	2.397	2.773

hysteresis loop. It means that the slope of linear dependence is a function of adsorbate–adsorbent interactions. These interactions are the base of the HK equation [19].

In both calculation procedures similar linear dependencies are observed for points obtained on the basis of adsorption and SAXS data. It seems that the above linear relations are independent of the experimental method used as a base of determination of the fractal dimension.

The relationship Eq. (23) is valid for values of x_f from the range between the lower limit equal to 0.584 nm for the value of fractal dimension $D = 3$ and the upper one equal to 0.766 nm for the value of fractal dimension $D = 2$. Discussion of such narrow range of the value of x_f was carried out by Ehrburger–Dolle [20] on the basis of results obtained by Jaroniec et al. [21]. She used relation Eq. (2) to calculate the lower and upper limits of characteristic adsorption energy E_0. Similarly, linking these limits with the McEnaney's formula (Eq. (4)) one can calculate values of limits of characteristic adsorption energy range for minimum and maximum micropore size connected with the power low x_f. The deduced values are 16.81 kJ mol^{-1} (for $D = 2$) and 20.88 kJ mol^{-1} (for $D = 3$). Obtained very narrow range of characteristic adsorption energy values is not adequate to values possible for benzene vapor adsorption on activated carbons. So, it seems that it is not the relation between the values of x_f and the values of E_0.

The question arising now concerns the connection of average micropore size x_f based on power law with other surface characteristics. Analysis of data presented in Tables 1–3 leads to the conclusion that, for analyzed samples, there is a valid dependence between value of x_f and the geometric surface areas obtained on the basis of formula described by Dubinin's relationship Eq. (6). The best agreement between values of geometric surface area of micropores calculated using Eq. (6) with the value of x_f was found for data obtained on the basis of SAXS measurements (maximum deviation up to 5%). This conclusion will be examined in the near future. Further studies are in progress in order to confirm the relation between average micropore size x_f connected with the power law and geometric surface area of micropores and to determine a relation between the value of fractal dimension D and the value of x_f on the basis of independently confirmed method of evaluation of fractal dimension.

5. Conclusions

1. Analysis of adsorption data in a case of structurally heterogeneous activated carbons, as was examined here, leads to similar values of structural parameters obtained on the basis of the DI or the DS adsorption models.
2. Proposed here method of analyzing of SAXS data with taking into account adsorption results leads to more realistic values of structural parameters of activated

carbons porosity. Obtained values of specific area of micropores are confirmed by these obtained on the basis of the DI or the DS equations.
3. To take advantage of McEnaney's formula, linking mean micropore size with characteristic adsorption energy, gives possibility to extend the fractal analysis performed, according to Ehrburger–Dolle's method, on the basis of adsorption data on structurally heterogeneous activated carbons.
4. The average micropore size connected with power law is strictly dependent on assumed lower and upper limits of fractality range. But analysis of relationship between this parameter and fractal dimension shows that it is rather independent of characteristic adsorption energy.

Acknowledgements

This work was partially supported by Research Project No 0 T00A 032 12 sponsored by Committee of Scientific Research (Poland).

References

[1] Izotova TI, Dubinin MM. Zh Fiz Khim 1965;39:2796.
[2] Dubinin MM. Izvest Akad Nauk SSSR, Ser Khim 1979:1691.
[3] Dubinin MM. Carbon 1989;27:457.
[4] McEnaney B. Carbon 1987;27:69.
[5] Dubinin MM. Carbon 1980;18:355.
[6] Dubinin MM. Carbon 1985;23:373.
[7] Dubinin MM. Carbon 1987;25:593.
[8] Guinier A, Fournet G. Small angle X-ray scattering, New York: Academic Press, 1982.
[9] Janosi A, Stoeckli HF. Carbon 1979;17:465.
[10] Bota A. J Appl Cryst 1991;24:635.
[11] Cohaut N, Guet JM, Diduszko R. Carbon 1996;34:674.
[12] Avnir D, editor, The fractal approach to heterogeneous chemistry, Chichester: Wiley, 1989.
[13] Pfeifer P. Preparative chemistry using supported reagents, New York: Academic Press, 1987.
[14] Mandelbrot BB. In: Freeman WH, editor, Fractal: form, chance and dimension, New York, 1997.
[15] Bezot P, Hesse-Bezot C. Carbon 1998;36:467.
[16] Pfeifer P, Avnir D. J Chem Phys 1983;79:3558.
[17] Trznadel BJ, Światkowski A. Adsorp Sci Technol 1999;17:303.
[18] Błazewicz S, Światkowski A, Trznadel BJ. Carbon 1999;37:693.
[19] Horvath G, Kawazoe K. J Chem Eng Japan 1983;16:470.
[20] Ehrburger-Dolle F. Langmuir 1994;10:2052.
[21] Jaroniec M, Gilpin RK, Choma J. Carbon 1993;31:325.
[22] Bale HD, Schmidt PW. Phys Rev Lett 1984;53:596.
[23] Schmidt PW. In: Brumberger H, editor, Modern aspects of small angle scattering, Amsterdam: Kluwer Academic, 1995.
[24] Smišek M. Černy S, Minařova J. Chemicki prumysl 1962;12:237.

[25] Kadlec O, Daneš V. Collection Czechoslov Chem Commun 1967;32:693.

[26] Korver JA. Chemisch Weekblad 1950;46:301.

[27] Glatter O, Kratky O, editors, Small angle X-ray scattering, New York: Academic Press, 1982.

[28] Rosenbrock HM. Comp J 1960;3:175.

[29] Sahouli B, Blacher S, Brouers F, Sobry R, Van den Bossche G, Diez B, Darmstadt H, Roy C, Kaliaguine S. Carbon 1996;34:633.

[30] Zerda TW, Yang H, Gerspacher M. Rubber Chem Techn 1992;65:130.

(D) Applications

ACTIVATED CARBON FOR GAS SEPARATION AND STORAGE

S. Sircar,* T. C. Golden and M. B. Rao

Air Products and Chemicals Inc., 7201 Hamilton Boulevard, Allentown, PA 18195-1501, U.S.A.

(*Received 6 February 1995; accepted in revised form 14 July 1995*)

Abstract—Activated carbons offer a large spectrum of pore structures and surface chemistry for adsorption of gases, which are being used to design practical pressure swing and thermal swing adsorption processes for separation and purification of gas mixtures. The activated carbons are often preferred over the zeolitic adsorbents in a gas separation process because of their relatively moderate strengths of adsorption for gases, which facilitate the desorption process. Three commercial applications of activated carbons, (a) trace impurity removal from a contaminated gas, (b) production of hydrogen from a steam–methane reformer off gas, and (c) production of nitrogen from air, are reviewed. Four novel applications of activated carbons for gas separation and purification are also described. They include, (a) separation of hydrogen–hydrocarbon mixtures by selective surface flow of larger hydrocarbon molecules through a nanoporous carbon membrane produced by carbonization of a polymer matrix, (b) gas drying by pressure swing adsorption using a water selective microporous carbon adsorbent produced by surface oxidation of a hydrophobic carbon, (c) removal by selective adsorption and in-situ oxidation of trace volatile organic compounds from air by using a carbon adsorbent–catalyst composite, and (d) storage of compressed natural gas on high surface area carbons.

Key Words—Activated carbon, adsorption, gas separation, natural gas.

1. INTRODUCTION

Separation and purification of gas mixtures by adsorption has become a major unit operation in the chemical and petrochemical industries. The phenomenal growth in this area is demonstrated by Fig. 1, which plots the number of worldwide patents on gas separation by adsorption issued every year during the last twenty years [1]. The two main reasons for this development are (a) the commercial availability of a large spectrum of microporous adsorbents (zeolites, activated carbons, aluminas, silica gels, polymeric adsorbents), with varying pore structures and surface properties, which can be used to selectively adsorb specific components of a fluid mixture, and (b) the possibility of designing many different process schemes under the generic category of pressure swing adsorption (PSA) and thermal swing adsorption (TSA), for a given separation need, using the available adsorbents. The technology has evolved to be extremely versatile and flexible, owing to the large choice of adsorbent materials and their use in the design of innovative separation processes.

Activated carbons produced from coal, petroleum, vegetable and polymeric precursors have played a major role in this development. Table 1 lists some of the key commercial applications of gas separation and purification by adsorption using activated carbons.

Different precursors, methods of carbonization and activation procedures have been used by many commercial manufacturers to produce a large variety of activated carbons for gas separation applications.

A wide range of pore volumes, pore structures, pore size distribution, density, ash content and hardness, as well as surface chemistry (degree of polarity, types of oxygen and hydroxyl groups on surface, active surface area) has been incorporated into the final products. Table 2 lists some of these physical properties and Fig. 2 shows the cumulative pore size distributions for a selected number of commercially produced activated carbons. These were obtained from manufacturers' data sheets. The mean pore size of these carbons can be of the order of the molecular diameter of the gaseous adsorbates (3–5 Å for molecular sieve carbons, MSC) or several times bigger than that (20–50 Å for others). The pore size distribution may vary over a large range (RB and XE 340), a medium range (BPL and Witcarb) or a very narrow range (MSC). The BET surface areas and total pore volumes can vary between 400 and 3000 m^2/g and 0.34 and 1.8 cc/g, respectively. These large differences in the physico-chemical properties of different activated carbons generate a large difference (and thus choice) in the adsorptive characteristics of the components of a gas mixture to be separated. For example, let us consider the adsorption of CO_2 and H_2 on these carbons. Figure 3 shows the adsorption isotherms (amount adsorbed, n against equilibrium pressure, P) of pure CO_2 on various activated carbons of Table 2 at a temperature of 303 K and in the pressure range 0.1–60 atm. These isotherms were measured in our laboratory. All isotherms are Type I (Brunauer classification) because of the predominately microporous nature of these carbons, but they have significantly different shapes and offer a large range of adsorption capacities for CO_2 at a given partial

---

*To whom all correspondence should be addressed.

Reprinted from *Carbon* 34 (**1**), 1-12 (1996)

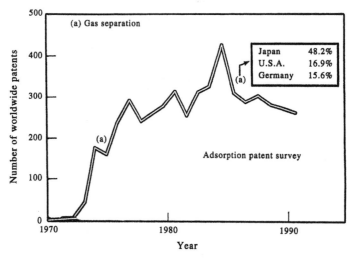

Fig. 1. Adsorption patent survey.

Table 1. Key commercial applications of activated carbons in the gas separation and purification industry

Goal	Process	Reference
Trace impurity removal	TSA	[20,21]
Solvent vapor removal and recovery	TSA,PSA	[20,22]
Air separation	PSA	[4–10]
Carbon dioxide–methane separation from landfill and biogases	PSA	[19]
Removal of CO_2 from flue gas	PSA	[22]
Hydrogen and carbon dioxide recovery from steam–methane (SMR) reformer off gas, coke oven gas, ethylene off gas	PSA	[3,18]

pressure. The Henry's Law constants (K) at 303 K and the isosteric heats of adsorption of CO_2 at the limit of zero coverage ($q°$) for these carbons are given in Table 3. They indicate weak (RB, PCB) to very strong (MSCV, Witcarb) sorption of CO_2. The saturation capacities for CO_2 sorption on these carbons also vary significantly in accordance with their total pore volumes. Table 3 also shows the Henry's Law constants at 303 K for adsorption of pure H_2 on these carbons. Although H_2 is very weakly adsorbed by these carbons, the Henry's Law selectivity of adsorption between CO_2 and H_2 at 303 K varies between 50 (PX21) and 275 (MSCV) for these carbons. Thus, the co-adsorption of H_2 from a CO_2–H_2 mixture can differ substantially on these carbons.

The desorption characteristics of CO_2 from these carbons are also strikingly different. Figure 4 shows the isothermal–isobaric desorption of CO_2 from an activated carbon column, which is initially saturated with pure CO_2 at 303 K and atmospheric pressure, by purging the column with pure H_2 at the same temperature and pressure. These data were measured in our laboratory. The figure shows the amount of H_2 purge needed to remove certain fractions of initial CO_2 present in the column. It demonstrates that the quantity of H_2 needed to remove CO_2 from the

column depends heavily on the strength of adsorption of CO_2 on the carbon (low for RB, high for MSCV).

The separation efficiency (adsorbent inventory and energy) of a gas mixture by TSA or PSA process is governed by the capacity and selectivity of adsorption of the more strongly adsorbed component of the mixture as well as by the ease of desorption of the adsorbed species. Thus, the above example shows that different activated carbons will offer substantially different practical separation efficiencies for CO_2–H_2 gas mixture. The design of the optimum separation system will require a judicious combination of the adsorption characteristics of the activated carbon and the adsorptive process scheme so that the adsorbent inventory and energy of separation are minimized for a given product specification. The spectrum of available activated carbons, however, provides a very large choice for optimum design.

The separation of CO_2–H_2 mixture can also be carried out by selectively (thermodynamic) adsorbing CO_2 on zeolite 5A. Figure 3 and Table 3 show that CO_2 is very strongly adsorbed on 5A with a very large selectivity over H_2. In particular, the capacity of CO_2 on 5A at lower partial pressures can be very high. However, Fig. 4 shows that desorption of CO_2 from 5A is very difficult, requiring an enormous

Table 2. Physical properties of activated carbons

	Manufacturer	Source	BET area (m²/g)	Pore volume (cc/g)	Bulk density (g/cc)	Skeleton density (g/cc)	Ash (%)
BPL	Calgon	Coal	1100	0.70	0.48	2.1	8.0
RB	Calgon	Coal	1250	1.22	0.41	2.35	23.0
Witcarb 965	Witco	Petroleum	1300	0.65	0.47	–	1.0
Amoco PX21	Amoco	Petroleum	3150	1.8	0.30	–	2.0
PCB	Calgon	Vegetable	1200	0.72	0.44	2.2	6.0
Ambersorb XE 340	Rohm Haas	Polymer	400	0.34	0.60	1.34	<0.5
Molecular sieve carbons							
MSCV	Calgon	Coal	–	0.5	0.67	2.1	–
MSC	Takeda	Vegetable	–	0.43	0.67	2.2	–

123

quantity of H_2 purge gas. This relative balance between capacity and selectivity of adsorption and ease of desorption in an adsorptive separation process often dictates the use of activated carbons as the preferred adsorbent over zeolites or other polar adsorbents for many gas separations.

2. EXAMPLES OF GAS SEPARATION PROCESSES

The data in Table 1 cover a very large spectrum of separation and purification applications using activated carbons. This section describes three specific cases that have been commercially very successful.

2.1 Trace impurity removal by TSA

Packed activated carbon columns are frequently used for removing trace or dilute organic impurities, solvent vapors, odor forming compounds from air or other industrial gases by selectively adsorbing the impurity at near ambient temperature. A clean effluent containing less than 10 ppm impurity can be produced easily. The adsorbed impurity is then desorbed by heating the column countercurrently with an inert gas or steam. The desorbed and concentrated impurity is either vented or collected as a water immiscible liquid by condensing the effluent gas. The process is repeated after cooling the adsorbent by flowing an inert gas through the column at ambient temperature.

Figure 5 shows a schematic flow diagram for a three-column conventional TSA process for removal of a trace impurity (A) from an inert gas (B). A portion of the purified inert gas is used to cool and heat two columns in series while the third column adsorbs component A from fresh feed gas [23].

Activated carbons offer a very special advantage for this application because of their relative hydrophobicity. A significant adsorption capacity for the impurity can be achieved even when the feed gas is wet.

For example, many industrial processes use large quantities of organic solvents and a substantial amount of solvent vapor contaminates the vent streams, which are saturated with water vapor. TSA processes using activated carbons are used to recover and reuse the solvent. The adsorption selectivities of organic compounds over water on activated carbons primarily depend on their polarities and polarizabilities and those of the carbon surface. Table 4 shows several examples of dilute hydrocarbon adsorption capacities on BPL carbon at 298 K from dry and wet gases [2]. The carbon retains a substantial fraction of its capacity for hydrocarbons even in the presence of moisture at 70–80% relative humidity. Most polar adsorbents like zeolites, silica and alumina gels will adsorb very little hydrocarbon under such moist conditions.

2.2 PSA process for production of hydrogen from reformer off gas

The conventional method of producing H_2 is to catalytically reform natural gas or naphtha by

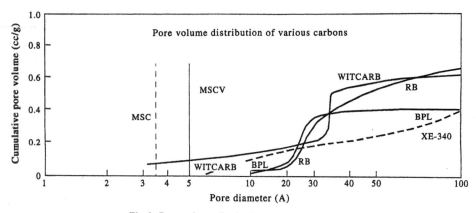

Fig. 2. Pore volume distribution of various carbons.

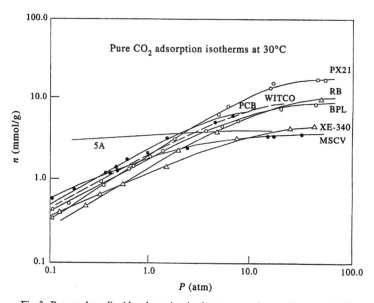

Fig. 3. Pure carbon dioxide adsorption isotherms on various carbons at 303 K.

Table 3. Adsorptive properties of carbon dioxide and hydrogen at 303 K

Carbon	Henry's Law constant (K) mmol/g/atm		Limiting isosteric heat of adsorption of CO_2 (kcal/mol)	Limiting CO_2–H_2 selectivity
	CO_2	H_2		
BPL	3.45	0.038	5.80	91
RB	2.38	0.027	5.40	90
MSC V	7.14	0.026	9.30	275
PCB	4.54	0.046	4.80	98
PX 21	2.94	0.060	5.50	49
Witco 965	3.92	–	6.00	–
XE 340	3.10	–	–	–
5A Zeolite	138.90	0.188	9.40	7400

reacting with steam. The reformed gas is then subjected to a water–gas shift reaction to produce a crude H_2-rich steam containing $\sim 75\%$ H_2, 20% CO_2, 4% CH_4, 1% CO and trace N_2 (dry basis) at a pressure of 150–350 psig. A polybed PSA adsorption process has been developed by Union Carbide

Corporation [3] to produce a $99.999 + \%$ H_2 stream with a H_2 recovery of 75–85% from this crude gas.

Figure 6 shows a schematic flow diagram using nine parallel columns for this purpose [3]. The columns are packed with a layer of activated carbon to selectively adsorb CO_2 from CO, CH_4, N_2 and H_2

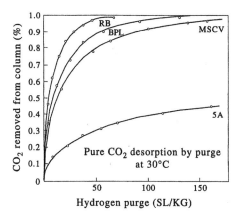

Fig. 4. Isothermal–isobaric desorption of carbon dioxide by purging with pure hydrogen.

measured in our laboratory. Table 5 gives the Henry's Law selectivities between these components at 303 K. It can be seen that the strength of adsorption of these gases on the BPL carbon increases in the order $CO_2 > CH_4 > CO > N_2 > H_2$ while that for the 5A zeolite increases in the order $CO_2 > CO > CH_4 > N_2 > H_2$. The higher selectivity of adsorption of CO over CH_4, N_2 and H_2 and the higher selectivity of adsorption of N_2 over H_2 exhibited by 5A zeolite compared to BPL carbon dictates the use of the zeolite as the layer for removing CO, CH_4 and N_2 from H_2 in the polybed process. On the other hand a carbon is used to remove CO_2 from CO, CH_4, N_2 and H_2 because of the ease of desorption of CO_2.

2.3 PSA process for production of nitrogen from air

A new class of activated carbons called carbon molecular sieves (CMS) has been developed for commercial scale air separation [4]. These are microporous carbons in which the size of the pore mouth has been altered to allow faster diffusion of relatively smaller O_2 molecules (~ 3.46 Å diameter) into the internal pore structure rather than larger N_2 molecules (~ 3.64 Å diameter). Consequently, when a CMS is contacted with air, an O_2-enriched adsorbed phase is produced based on the kinetic selectivity of O_2 over N_2, thus forming the basis of air separation

from the feed gas followed by a layer of 5A zeolite for selectively adsorbing CO, CH_4 and N_2 from H_2. The PSA process has eleven sequential steps consisting of (a) adsorption at feed gas pressure, (b) four co-current depressurization steps, (c) counter-current depressurization, (d) counter-current purge with pure H_2, and (e) four counter-current pressurization steps.

Figures 7 and 8 show the pure gas adsorption isotherms for CO_2, CH_4, CO, N_2 and H_2 on the BPL carbon and 5A zeolite, respectively, at 303 K over a pressure range of 0.1–60 atm. These data were also

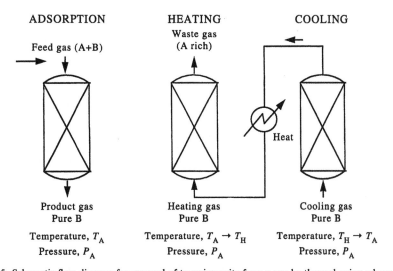

Fig. 5. Schematic flow diagram for removal of trace impurity from a gas by thermal swing adsorption.

Table 4. Adsorption of dilute hydrocarbons on BPL carbon at 298 K

	Partial pressure of hydrocarbon (kPa)	Adsorption capacities (mmol/g)		
		Dry gas	Wet gas	Water RH (%)
n-Hexane	1.29×10^{-4}	0.82	0.50	74.9
	1.58×10^{-3}	1.47	1.03	79.3
	2.00×10^{-2}	2.12	1.99	78.0
Acetone	0.154	2.65	1.92	70.8
	1.275	4.20	3.63	76.0

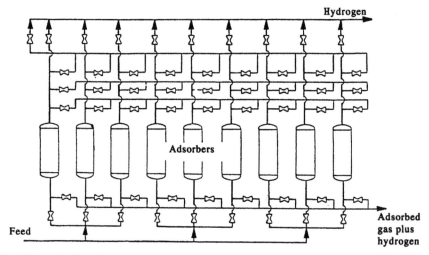

Fig. 6. Schematic flow diagram for production of high purity hydrogen from a steam–methane reformer off gas by pressure swing adsorption.

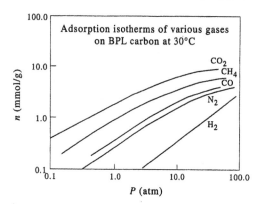

Fig. 7. Adsorption isotherms of various pure gases on BPL activated carbon at 303 K.

Table 5. Henry's Law selectivity on BPL carbon and 5A zeolite at 303 K

Gas mixture	Selectivity	
	BPL	5A
CO_2–CH_4	2.5	195.6
CO_2–CO	7.5	59.1
CO_2–N_2	11.1	330.7
CO_2–H_2	90.8	7400.0
CO–CH_4	0.33	3.3
CO–N_2	1.48	5.6
CO–H_2	12.11	125.0
CH_4–N_2	4.5	1.7
CH_4–H_2	36.6	37.8
N_2–H_2	8.2	22.3

Figure 9 shows an example of the uptake rates of pure O_2 and N_2 by a CMS sample, which were measured in our laboratory. The kinetic adsorption capacity of O_2 increases with increasing contact time with air, but its kinetic selectivity over N_2 decreases. Thus, air contact time during the adsorption step of the PSA cycle is a critical variable.

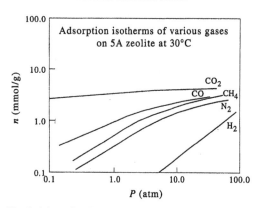

Fig. 8. Adsorption isotherms of various pure gases on 5A zeolite at 303 K.

using a CMS. The kinetic selectivity and capacity of adsorption for O_2 is time dependent. The CMS does not have any appreciable thermodynamic selectivity of O_2 over N_2, and a prolonged contact with air allows the N_2 to eventually diffuse into the CMS pore structure, creating an adsorbed phase of the same composition as air.

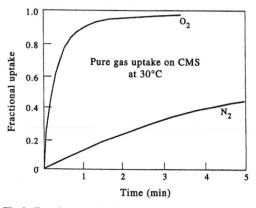

Fig. 9. Transient uptakes of pure oxygen and nitrogen by a carbon molecular sieve.

Numerous PSA processes have been developed for production of N_2 from air by using a CMS [5–9]. A relatively simple version developed by Bergbau Forschung, Germany, uses two parallel adsorbent columns packed with CMS as shown in Fig. 10 [5,6]. The process has four sequential steps consisting of (a) adsorption of air at 80–110 psig, (b) pressure equalization between the columns, (c) countercurrent depressurization and (d) pressure equalization between the columns. A layer of a desiccant is usually placed before the CMS layer in each column in order to dry the feed air.

The performance of the PSA air separation processes using CMS depends on the process cycle design and the adsorptive properties of the CMS. They can be used to produce a N_2-enriched gas containing 98–99.9% N_2. The typical composition of the by-product O_2-enriched gas from these processes is 30–45% O_2. Figure 11 shows an example of the performance of such a process (N_2 productivity against percent O_2 in product N_2) evaluated in a process development unit by Air Products and Chemicals [10].

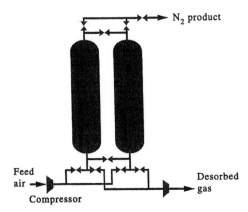

Fig. 10. Schematic flow diagram for production of nitrogen from air by pressure swing adsorption using kinetic selectivity.

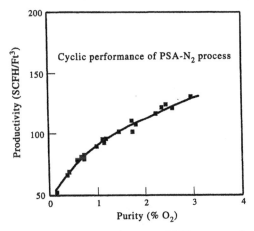

Fig. 11. Air separation performance of PSA process using kinetic selectivity.

3. NOVEL APPLICATIONS OF ACTIVATED CARBONS FOR GAS SEPARATION

The research and development on the use of activated carbons for gas separation is an ongoing effort. This section describes four new directions in that area, which are at different stages of development.

4.1 Nanoporous carbon selective surface flow (SSF®) membrane

A new class of nanoporous activated carbon membrane for gas separation has been developed by Air Products and Chemicals [11,12]. It consists of a thin layer (2–3 μm) of nanoporous (5–7 Å pores) carbon supported on a mesoporous substrate. The carbon membrane is produced by controlled carbonization of a polyvinylidene chloride (PVDC) film at a temperature of 600–1000°C under N_2.

Figure 12 depicts the mechanism of gas separation through the carbon membrane [11]. When a gas mixture is passed over one side of the membrane at an elevated pressure, while maintaining a low pressure on the other side, certain components of the mixture are selectively adsorbed on the pore walls. The adsorbed molecules move towards the low pressure side by surface diffusion and then desorb into the gas phase. The adsorbed molecules also hinder the flow of less selectively adsorbed components through the free space within the pores. As a consequence of selective adsorption and surface diffusion, the SSF membrane offers very high selectivity and permeability for the diffusing species. The less strongly adsorbed components (often the desired product) remain on the high pressure side of the membrane, which is the unique advantage of this membrane. Another advantage is that the SSF membranes can be operated using a relatively low absolute pressure on the high pressure side.

Table 6 shows an example of the pure gas permeabilities of H_2 and C_1–C_4 hydrocarbons through the SSF membrane at 295 K measured using pressures of, respectively, 1.8 and 0.2 atm on the high and low pressure sides [11]. It also shows the permeabilities of these components from a gas mixture at 4.4 atm using ambient pressure helium sweep in the low pressure side. It may be seen that the flow of non-adsorbing small H_2 molecules through the carbon membrane is practically blocked in the presence of more strongly adsorbing, larger hydrocarbons, like C_4H_{10}. The selective permeation of the gases through the membrane (from the mixture) increase in the order $C_4 > C_3 > C_2 > C_1 > H_2$, which is also the order of relative strengths of adsorption of these gases in the carbon.

A unique application of the SSF membrane is recovery of H_2 from a low pressure petroleum refinery waste gas containing a low concentration of H_2. Figure 13 shows a schematic diagram for such an application [11]. The waste gas (41% H_2, 20% CH_4, 20% C_2H_6, 9.5% C_3H_8 and 9.5% C_4H_{10}) at 4.4 atm can be passed over an SSF membrane to produce a

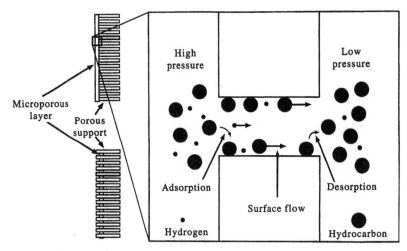

Fig. 12. Mechanism of gas separation by nanoporous carbon membrane.

Table 6. Pure and mixed gas permeabilities (P) through carbon membrane

Gas	P^o, pure gas (Barrer)	$P^o_{HC}/P^o_{H_2}$	P, mixed gas[a] (Barrer)	P_{HC}/P_{H_2}
H_2	129	1.0	1.2	1.0
CH_4	663	5.1	1.3	1.1
C_2H_6	851	6.6	7.2	6.0
C_3H_8	291	2.2	24.1	20.1
C_4H_{10}	156	1.2	120.0	100.0

[a]20.2% CH_4, 9.5% C_2H_6, 9.4% C_3H_8, 19.9% C_4H_{10} and 41.0% H_2.

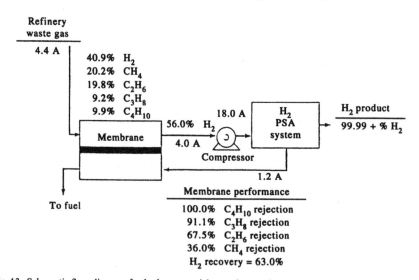

Fig. 13. Schematic flow diagram for hydrogen enrichment from refinery waste gas by carbon membrane.

H_2 enriched stream (56% H_2) at feed pressure with a H_2 recovery of 63%. The membrane can reject 100% C_4H_{10}, 91% C_3H_8, 67% C_2H_6 and 38% CH_4 by selective permeation during the process. The enriched H_2 stream can be compressed to 18.0 atm and purified by a PSA process in order to produce 99.99 + % H_2 with a H_2 recovery of ~75%. The waste gas from the PSA process can be used to purge the low pressure side of the SSF membrane. The SSF membrane allows economic recovery of the valuable H_2

from a waste gas which may not otherwise be practical.

The pore structure of these membranes has been characterized using permeation techniques [11,12]. Table 7 shows the pure H_2 and He permeabilities and the corresponding diffusivities at three temperatures through a carbon membrane [11]. For Knudsen diffusion, the hydrogen diffusivity should be 1.41 times higher than helium at a given temperature. It can be seen from Table 7 that hydrogen permeates 4

Table 7. Permeabilities and diffusivities of pure hydrogen and helium through carbon membrane

T (K)	Permeabilities (Barrer)		Diffusivity $(cm^2/s \times 10^5)$		$P^{\delta}_{H_2}/P^{\delta}_{He}$
	He	H_2	He	H_2	
256.1	280	1520	0.55	3.00	5.4
273.1	340	1700	0.72	3.59	5.0
298.1	440	1850	1.01	4.26	4.2

to 5.5 times faster than helium, indicating that Knudsen diffusion is not occurring through these membranes. Thus, the pore diameter is less than ~ 20 Å. Further evidence of the very small pore size in these membranes is illustrated in Fig. 14, which shows the diffusivity (D) of methane as a function of pore diameter (d). Bulk diffusion of methane occurs through very large pore ($d > \sim 1000$ Å) and Knudsen diffusion occurs at pore diameters between ~ 20 and 1000 Å. The diffusivity for methane becomes a very strong function of pore diameter when $d < 20$ Å (activated diffusion regime). The methane diffusivity data in this pore size range are literature values for diffusion through zeolites of known pore size. Using the curve of Fig. 14 and the measured methane diffusivity through the carbon membrane, we estimated the membrane pore diameters to be in the range of 5–6 Å [12].

3.2 Activated carbon adsorbents for PSA drier

Air or other gas drying by using a PSA or a TSA process is a common practice. Typically NaX zeolite (TSA, PSA) or activated alumina (PSA) are used as adsorbents. Figure 15 shows water adsorption isotherms (amount adsorbed vs relative humidity) at 297 K on (a) NaX zeolite and (b) alumina. The zeolite adsorbs water (Type I isotherm) very strongly, with a high Henry's Law constant (~ 140 g/g) and isosteric heat of adsorption (18.0 kcal/mol). The alumina adsorbs water moderately strongly (Type IV isotherm), with a Henry's Law constant of 2.1 g/g and

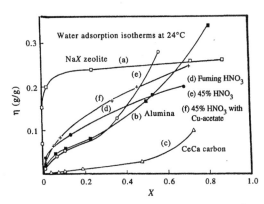

Fig. 15. Pure water vapor adsorption isotherms on various adsorbents at 297 K.

an isosteric heat of 14.0 kcal/mol. Consequently, water is difficult to desorb from zeolite and the process is very energy intensive. The alumina, on the other hand, has low water adsorption capacity at low partial pressures and the adsorption mass transfer zone is very long. The optimum water adsorbent for a PSA drier should be Type I in shape with an adsorption affinity in between NaX and alumina.

Modified activated carbon can fulfill that role. The hydrophobic nature of common activated carbons (CeCa) exhibited by curve (c) in Fig. 15 can be changed by introducing polar oxygen groups on the surface [13]. Curve (f) in Fig. 15 shows a Type I water adsorption isotherm (Henry's Law constant of 5.3 g/g) produced by oxidation of CeCa carbon by heating the carbon in 45% HNO_3 solution at 353 K in the presence of copper acetate catalyst. The isotherm is well suited for gas drying application by PSA.

3.3 Sorption–reaction process for removal of trace VOC from air

Stricter environmental regulations have renewed interest in the removal of trace hydrocarbons from contaminated air. Commonly, these trace hydrocarbons are oxidized to CO_2 and H_2O by thermal or catalytic incineration techniques [14], which consume a significant quantity of fuel (energy) to heat the large volume of contaminated air to the combustion temperature (600–1600 K).

The energy required for air purification can be significantly reduced by using a cyclic sorption–reaction (SR) process patented by Air Products and Chemicals [15]. A schematic diagram of the process is given in Fig. 16. The system consists of two parallel adsorbers containing a physical mixture of an activated carbon and an oxidation catalyst. The adsorbers may consist of shell and tube heat exchangers, so that the adsorbent–catalyst mixture can be heated indirectly. A typical SR cycle would include (a) adsorption of the trace hydrocarbon on the carbon at ambient temperature until impurity breakthrough, (b) in-situ oxidation of the hydrocarbons by directly or indirectly heating the adsorbent-catalyst mixture to about 423 K, and (c) cooling the adsorbent–catalyst

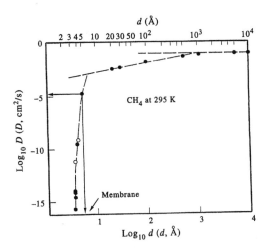

Fig. 14. Diffusivity (D) of pure methane through porous substrates of different pore diameters (d).

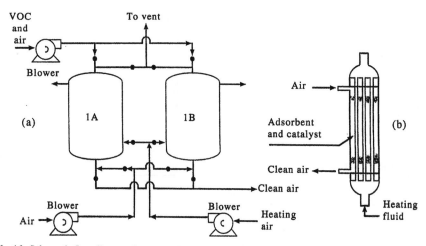

Fig. 16. Schematic flow diagram for removal of trace impurity from air by the sorption–reaction process.

Table 8. Energy saving for sorption–reaction process

	SR process	Catalytic combustion
Energy (MM BTU/h)	0.012	0.41
Adsorbent–catalyst (lbs)	5700	800

mixture to ambient temperature either directly or indirectly and venting the combustion products. Only the adsorber vessel and its contents are heated to the reaction temperature, which drastically reduces the energy necessary for removal and destruction of hydrocarbons.

Table 8 compares the performance of the SR process for cleaning a 1 MMSCFD air stream containing 260 ppm vinyl chloride monomer to a level of 1 ppm with that of a plug flow catalytic reactor using a standard oxidation catalyst at 600 K [1,14]. The adsorbent–catalyst in this case was RB carbon impregnated with 1.5 wt% palladium chloride as the oxidation catalyst. The table shows that savings of an order of magnitude in energy requirement can be achieved using the SR process.

3.4 Storage of natural gas by adsorption on activated carbon

The abundance of natural gas, its relatively lower price, and its potential to be a cleaner fuel has prompted much interest in its use as motor fuel. Numerous vehicles have been adapted to use compressed natural gas (CNG) as fuel. CNG is typically stored at a pressure of 200 atm (2925 psig) in heavy steel cylinders. The net deliverable capacity for CNG at 1.35 atm (5 psig) is 215 standard liter per liter (sl/l) of storage volume. The energy density of CNG, however, is only 29% of that of gasoline.

A large effort has been made to replace CNG by storing methane in a vessel packed with activated carbon. This concept of adsorbed natural gas (ANG) can potentially reduce the highest storage pressure (typical target of 35 atm, 500 psig, which can be obtained by a single stage compressor) so that lightweight cylinders can be used. The key question is whether the net deliverable capacity of such a device can match or exceed that of CNG.

Obviously, one needs a high pore volume, high surface area activated carbon for this purpose. Figure 17 shows the CH_4 adsorption isotherms on PX21 carbon (highest surface area carbon of Table 2) at several temperatures, which were measured in our laboratory. The isosteric heat of adsorption (q) of CH_4 on the carbon is only 4.0 kcal/mol. Table 9 shows the net isothermal deliverable capacity of PX21 (granular form) carbon at 303 K between pressures of 35.0 and 1.35 atm. The net capacity includes adsorbed CH_4 and that in the macropore and external (interparticle) voids (37%). The net capacity is much

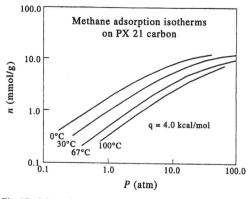

Fig. 17. Adsorption isotherms of pure methane on PX21 carbon at different temperatures.

Table 9. Isothermal deliverable methane capacity of PX21 carbon at 303 K (CNG = 215 Sl/l)

	Bulk density (g/cc)	Capacity (sl/l)
Granular	0.30	82.0
Monolith (no external void)	1.08	210.0

Storage pressure = 35.0 bar; exhaust pressure = 1.35 bar.

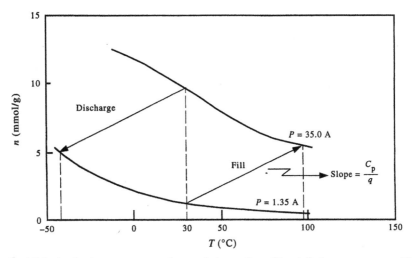

Fig. 18. Adiabatic adsorbent temperature changes during methane fill and discharge processes on PX 21 carbon.

less than that for CNG. On the other hand, if the carbon can be produced in a monolithic form so that the external void in the cylinder is negligible, the net deliverable capacity is significantly increased and it matches the CNG capacity. More recently, activated carbons with BET surface areas of $4000 \ m^2/g$ have been produced by Osaka Gas [16], which moves the goal closer. Reduction of external void, however, is still necessary. The practical feasibility of such carbon packing has not been demonstrated.

The numbers reported in Table 9 are isothermal deliverable capacities. The carbon temperature will increase when filling the cylinder with methane and will decrease when the methane is released as a result of heats of adsorption or desorption. Instantaneous heat removal (addition) to maintain isothermality may not be practically possible. Consequently, the real deliverable capacity of ANG may be much less than those given in Table 9 when quick supply of gas is demanded.

The methane adsorption isobars used to calculate the temperature rise and drop during filling and discharge with AX21 carbon, respectively, are given in Fig. 18. It shows that a temperature rise of 66 K and a drop of 77 K from a base temperature of 303 K will occur during adiabatic filling and discharge of CH_4 from PX21 carbon between pressure levels of 35.0 and 1.35 atm. The adiabatic fill–discharge operating lines in Fig. 18 have slopes equal to the ratio of the heat capacity of the adsorbent (C_p) to the isosteric heat of adsorption of methane (q) on the carbon. The calculated temperature rise is in good agreement with experimental measurements [17]. The adiabatic deliverable CH_4 capacity for granular carbon determined from these isobars is 36.5 sl/l only.

Attempts have been made to incorporate phase change materials with the activated carbon in order to remove (or supply) heat from (or to) the carbon so that the carbon column remains essentially isothermal during the fill–discharge process [17]. This,

however, reduces net carbon inventory in a given volume and increases the cost of the system. Instantaneous thermal exchange between the carbon and the phase change material may also not be possible.

4. CONCLUSION

Activated carbons produced from different precursors provide a large spectrum of pore structures and surface chemistry. They have been successfully employed as adsorbents in many different gas separation and purification applications of practical importance using pressure swing and thermal swing adsorption processes. Molecular engineering of the pore structure and surface chemistry of these carbons have opened up new potential applications, such as gas separation using a nanoporous carbon membrane, gas drying, pollution abatement and natural gas storage.

REFERENCES

1. S. Sircar, Proc. 4th Int. Conf. Fundamentals of Adsorption, Kyoto, Japan, p. 3 (1993).
2. E. N. Rudisill, J. J. Hacskaylo and M. D. Levan, *I&EC Res.* **31**, 1122 (1992).
3. A. Fuderer and E. Rudelstorfer, U.S. Patent 3,896,849 (1976).
4. H. Juntgen, K. Knoblauch and J. Reichenberger, *Monatsschr. Brau.* **30**, 27 (1977).
5. H. Juntgen, K. Knoblauch, J. Reichenberger, H. Heimbach and F. Tarnow, U.S. Patent 4,263,339 (1981).
6. K. Knoblauch, H. Heimbach and B. Harder, U.S. Patent 4,548,799 (1985).
7. R. Jain, U.S. Patent 5,090,973 (1992).
8. A. I. Shirley and A. I. Lacava, U.S. Patent 5,082,474 (1992).
9. N. C. Lemcoff and R. C. Gmelin, U.S. Patent 5,176,722 (1993).
10. T. C. Golden, P. J. Battavio, Y. C. Chen, T. S. Farris and J. N. Armor, *Gas Sep. Purif.* **7**, 274 (1993).
11. M. B. Rao and S. Sircar, *J. Membrane Sci.* **85**, 253 (1993).
12. M. B. Rao and S. Sircar, *Gas Sep. Purif.* **7**, 279 (1993).

13. T. C. Golden and S. Sircar, *Carbon* **28**, 683 (1990).
14. J. J. Spivey, *Ind. Engng Chem. Res.* **26**, 2165 (1987).
15. A. I. Dalton and S. Sircar, U.S. Patent 4,025,605 (1977).
16. Osaka Gas, *Technical Bulletin*, Renoves M Series.
17. C. F. Blazek, W. J. Jasionowski, A. J. Tiller and S. W. Gauthier, Paper presented at 25th Intersociety Energy Conversion Engng Conf., Reno, Nevada (1990).
18. S. Sircar and W. C. Kratz, *Sep. Sci. Technol.* **23**, 2397 (1988).
19. A. Kapoor and R. T. Yang, *Chem. Engng Sci.* **44**, 1723 (1989).
20. P. C. Wankat, *Large Scale Adsorption and Chromatography*, pp. 69–77. CRC Press, Boca Raton, FL (1986).
21. J. W. Leatherdale, In *Carbon Adsorption Handbook* (Edited by P. N. Cheremisinoff and F. Ellerbush), pp. 371–389. Ann Arbor Science, Ann Arbor, MI (1978).
22. J. Izumi, Proc. Symp. Adsorption Proc., Chung-Li, Taiwan, 25 May, pp. 71–84 (1992).
23. G. Keller, R. A. Anderson and C. M. Yon, In *Handbook of Separation Process Technology* (Edited by R. W. Rousseau), pp. 644–696. John Wiley, New York (1987).

133

RESPONDING TO CHANGING CONDITIONS: HOW POWDERED ACTIVATED CARBON SYSTEMS CAN PROVIDE THE OPERATIONAL FLEXIBILITY NECESSARY TO TREAT CONTAMINATED GROUNDWATER AND INDUSTRIAL WASTES*

J. A. MEIDL
U.S. Filter/Zimpro, Rothschild, WI 54474, U.S.A.

(*Received* 30 *June* 1996; *accepted in revised form* 14 *May* 1997)

Abstract—Treatment of contaminated water and wastewater with air stripping, biological systems and/or granular activated carbon are well-documented, conventional approaches. However, the use of powdered activated carbon in these treatment applications is increasing because of the superior performance and operational flexibility that powdered activated carbon can provide.

The growing data base documents the reliability and flexibility of the technology. By merging biological and physical treatment into a single process step, the system is able to buffer toxic loads which might otherwise impair a straight biological system and reduce the amount of carbon otherwise needed by a straight adsorption treatment system.

The PACT system is discussed in detail, with emphasis on groundwater and industrial wastewater treatment. Case studies include treatment of BETX contaminated groundwater and industrial applications for a refinery and two different petrochemical applications. © 1997 Elsevier Science Ltd

Key Words—A. Activated carbon, C. adsorption.

1. INTRODUCTION

The use of activated carbon for wastewater and contaminated groundwater treatment is increasing throughout the world as a greater awareness becomes evident of this planet's limited supply of water. In many areas this one time untainted "universal solvent" was fast becoming an organic solvent in its own right – compliments of the handiwork of man. Cleaning up this mess has fallen on the shoulders of conventional systems such as air stripping, biological systems, granular carbon or a combination of these. However, since its discovery in 1970, the PACT® system has been shown to have both significant performance and cost advantages over conventional treatment combinations.

Applications of PACT technology are wide ranging: organic chemicals, petrochemicals, refineries, textile/dyes, industrial and municipal water re-use, leachates, contaminated groundwater, among others. Major PACT system advantages for industrial wastewater and contaminated groundwater will be covered, including a few case studies.

2. PACT® SYSTEM

2.1 *Process flow diagram*

Typical PACT® system flow designs for single stage continuous and batch systems are shown in

Figs 1 and 2, respectively. Multi-stage systems with combinations of anaerobic, anoxic and aerobic stages exist as well.

2.2 *Advantages*

Major advantages of the system are shown in Table 1.

The importance of some of these advantages are addressed below:

2.2.1 *Organics control by adjusting carbon dose, carbon mixed liquor.* It is well known that increasing carbon dose or using a more "active" carbon will improve organics removal from wastewater. However, it is not usually recognized that by changing solids residence time (SRT) in the system will also impact performance. Since mixed liquor carbon (MLC) and SRT are directly proportional per the equation $MLC = \dfrac{Cdose \times SRT}{HDT}$, a doubling of the SRT will double the surface area in the treatment zone which is provided by the carbon and will also increase the amount of bacteria therein. Figure 3 shows a comparison for TOC performance between the activated sludge and PACT systems. Note that no matter how long the treatment time for the ASU, the effluent TOC never got below 90 ppm. PACT, however, at a constant carbon dose, was able to significantly improve upon its performance solely by increasing SRT.

2.2.2 *Metals control without a separate precipitation step.* Carbon's affinity for metals can, in

---

* American Carbon Society Workshop, Charleston, SC, June 9–12, 1996.

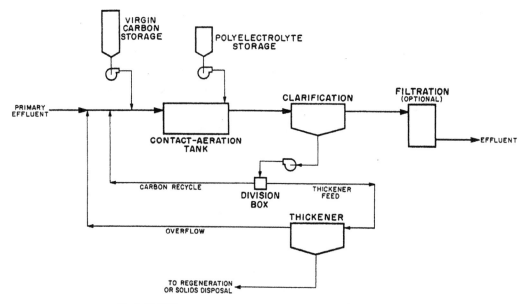

Fig. 1. PACT® wastewater treatment system general process diagram.

many cases, eliminate the need for up-front metals precipitation. This is equally important when toxic metals exist that would otherwise "poison" a conventional biological system or a GAC bed. Some data from treatment of a leachate shows, in Fig. 4, how PACT fares in treating metals with and without iron present. The case history on Schofield, Wisconsin, contained herein is a case in point.

2.2.3 *VOC/odor removal ability may forego adding a separate emissions control system.* Because of the powdered activated carbon (PAC) in the aeration tank, VOC emissions are less of a problem than in conventional bio-systems. Some PACT systems like those at the 54 MGD Kalamazoo, MI, WWTP or the Ciba–Geigy site in Toms River, control odors/VOCs by routing foul gases to the PACT system. A benzene mass balance for a refinery (Fig. 5) shows the fate of that compound in PACT. This type of performance is not possible with an activated sludge (AS) system since there is little adsorptive power in its mixed liquor to hold the benzene in the system long enough for the bacteria to degrade it.

2.2.4 *PAC use controls toxicity within the process and the effluent it discharges.* Because of PAC, biological toxicity is quite rare in the PACT system. More impressive is the system's ability to meet stringent effluent bioassays and tailor effluent quality to actual needs. A comparison against AS is shown in Table 2. Adjusting carbon dose "adjusted" effluent quality. It is also interesting to note that the PACT system gives better bioassay performance than AS + GAC because of GAC's chromatographic behavior and, where ammonia exists, better ammonia removal.

2.2.5 *Simplified residuals management.* For industrial wastewater applications, where sludge needs to be handled, dewatering of PACT system sludge will result in less sludge volume than an AS system due to a drier cake (50% solids vs < 20% solids). Alternatively, the spent carbon can be effectively recovered and regenerated, thus eliminating a major headache. Where regeneration makes sense, wet air oxidation (Fig. 6) is the preferred method due to its ability to regenerate a slurry (no need to dewater) and be easily permitted, even in non-attainment areas of the U.S.

For contaminated groundwater applications where a sanitary sewer is nearby, waste sludge (slurry) may be able to be discharged directly to the sewer, thus totally eliminating waste solids handling altogether. In other cases it may be possible to operate PACT at a very high *SRT* and a low PAC dose where "wasting" of solids occurs via the treated effluent.

2.3 *PACT system case histories*

Currently, the largest applications of the PACT system are found in the refinery, petrochemical, leachate and highly contaminated groundwater markets. An abbreviated list of users is shown in Table 3. A few case histories are included below to help amplify points previously discussed.

2.3.1 *Contaminated water: city garage site.* A small city in Wisconsin had, for years, maintained its own equipment in its garage which was located adjacent to a small river. Years of use led to soil and groundwater contamination that required remediation. The investigation identified 10–12 ppm BTEX groundwater contamination, and a recommendation that air stripping and GAC

BATCH PACT® SYSTEM DIAGRAM

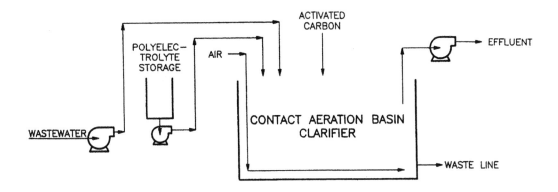

OPERATING SEQUENCE

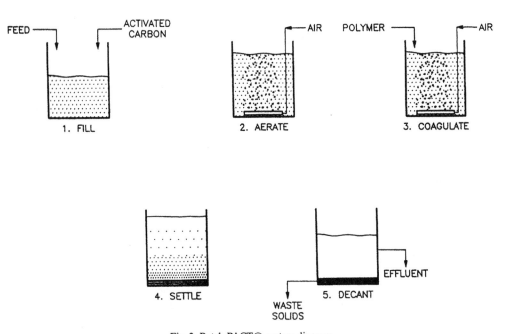

Fig. 2. Batch PACT® system diagram.

polishing be used to treat the groundwater. However, further analytical investigation of the groundwater (Table 4) indicated the contamination to be more than GC–MS identified compounds, but also significant organics (*BOD*, *COD*) and metals present that would pose sliming and scale problems in the air stripper, and blinding off and adsorption problems in the GAC. These were not problems for the PACT system.

A reevaluation of the proposed air stripper + GAC system vs a batch PACT system indicated the GAC's capital cost was, in fact, about 10–15% higher than the batch PACT system. And there was neither a guarantee on performance nor GAC use for the GAC alternative, whereas U.S. Filter/Zimpro was willing to guarantee both for the PACT system without doing testing. And the biggest drawbacks of the GAC alternate were a requirement to regularly clean the air stripper of scale, with all residuals from that cleaning and all spent GAC classified as hazardous waste, requiring special handling and disposal methods.

Table 1. PACT® system advantages

Increased or improved
Organics removal
Color removal
VOC/odor removal
Process stability
Solids residence time
Resistance to shock load
Metals removal
Low temperature performance
Sludge settleability/thickening/dewatering
Anaerobic digestion
Reduced or eliminated
Effluent toxicity
Aerator/effluent foam
Waste sludge
Other advantages
Protection from toxic upsets
Nitrification
Size

Effect of Sludge Age on Effluent Soluble TOC

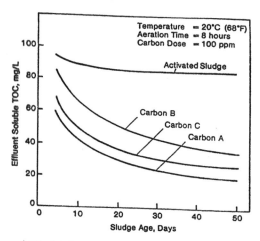

Fig. 3. Effect of sludge on effluent soluble *TOC*.

Thus, the PACT system was found to be not only less capital intensive, but also less operating intensive as well. Since the startup of the PACT system in 1993, it has always met discharge goals. And because the PACT system treats organics (*BOD*, BTEX, naphthalene), its solids are not considered hazardous, easily passing TCLP tests, and are disposed of in the nearby sanitary sewer.

Predicted and actual operating costs for the PACT system are noted in Table 5, with current performance shown in Table 6.

It is interesting to note that PAC was used upon startup in 1993 and for about 1 to 1½ years thereafter. U.S. Filter/Zimpro then began weaning the system off PAC and has noticed only marginal change in treatment performance since then.

2.3.2 Contaminated water: manufacturing site (superfund). A former organic chemicals manufacturing facility in Michigan had contaminated the groundwater below its site with halogenated and aromatic organics. The groundwater, containing about 80 ppm NH$_3$–N and 1500 ppm *COD*, is contaminated with a complex mixture of 50 Appendix IX compounds as noted in Table 7. A number of treatment schemes were analyzed, including a GAC/FB (granular activated carbon/biological fluid bed) system and a two stage PACT system.

Testing of the GAC/FB and PACT systems indicated that although the GAC/FB could achieve high removal of organic compounds, it could not nitrify, whereas the PACT system achieved nearly complete removal of all organic compounds and ammonia. A comparison of performance is shown in Table 8.

PACT system treatment efficiencies were consistent throughout the study. Concentrations of alkalinity and sulfate increased significantly across the first stage of the PACT system as a result of the oxidation of carbon and sulfur containing compounds. Nitrification was stable with decreases in ammonia matching increases in nitrate. Also, a decrease in alkalinity matched well with the theoretical alkalinity required for ammonia nitrogen removed.

Based on performance and overall costs, the PACT system was chosen to treat the site's 800 gpm of contaminated groundwater.

2.3.3 Industrial: refinery. This 20 000 TPD refinery processes "high" and "middle" crude, with two main wastewater streams requiring treatment.

The concentrated wastewater stream consists of sour condensates from the catalytic cracking unit, fluidized catalytic cracking unit and spent caustic from the Kero–Merox unit. The other stream consists of process wastewaters from the crude distillation unit, bituminous blowing unit, and aromatics plant.

The design characteristics of the combined wastewater stream were predicted to be as shown in Table 9. The discharge standards the refinery had to meet are also shown on that table.

The 1.5 MGD (240 m^3 h^{-1}) treatment process is as shown in Fig. 7 and consists of equalization and treatment of the combined equalized waste stream, using a free oil and emulsified oil separation system, followed by the PACT system. The PACT system has an aeration basin with a diffused aeration system and a clarifier. A guard pond stores the treated effluent prior to stream discharge.

Wasting of spent PACT system solids is to a gravity thickener, then to a slurry storage tank. This material is then sent to the 25 gpm (5.7 m^3 h^{-1}) WAR unit. After regeneration, the regenerated carbon is returned to the PACT system.

A small amount of ash material is periodically removed and is dewatered in a centrifuge and trucked off site to a landfill.

The performance of the operating plant is shown in Table 10. From the table it can be seen that all discharge standards are met. Also, an effluent *COD* of 78 mg L^{-1} is achieved even though the influent *COD* is 1494 mg L^{-1}, thus demonstrating the effec-

METALS REMOVAL ENHANCEMENT USING IRON IN THE PACT SYSTEM

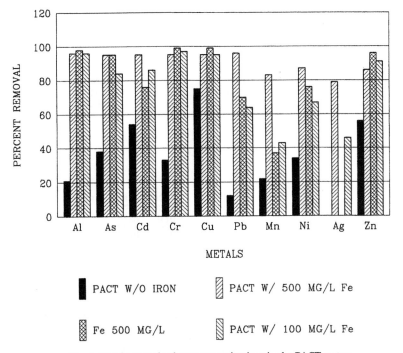

Fig. 4. Metal removal enhancement using iron in the PACT system.

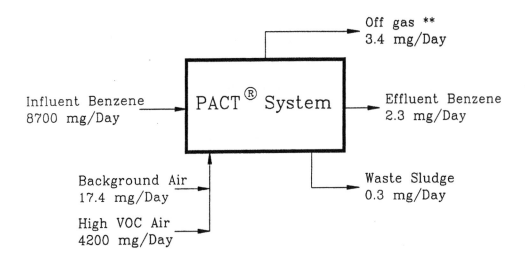

** 0.08% of the Influent benzene was measured in the off gas.

Fig. 5. Benzene material balance.

tiveness of the PACT/WAR system in removing COD, phenol and sulfides.

From the opening history of the plant come the following observations:

(1) The sludge generated for disposal from the PACT/WAR system is almost negligible, being only a sterile ash from the reactor blowdown.

(2) WAR unit operation is thermally self sustaining at a solids concentration of 7%, easily achieved by the gravity thickener.

Table 2. Treatment of a chemical manufacturing wastewater

	BOD (mg L^{-1})	TOC (mg L^{-1})	Color APHA	Cu (mg L^{-1})	Cr (mg L^{-1})	Ni (mg L^{-1})	LC$_{50}^a$ Vol. (%)
Influent	320	245	5365	0.41	0.09	0.52	–
Extended air activated sludge effluent	3	81	3830	0.36	0.06	0.35	11
PACT effluent @ carbon dose							
@ 100 mg L^{-1}	3	53	1650	0.18	0.04	0.27	33
@ 250 mg L^{-1}	2	29	323	0.07	0.02	0.24	>75
@ 500 mg L^{-1}	2	17	125	0.04	<0.02	0.23	>87

[a] Based on Mysid shrimp.

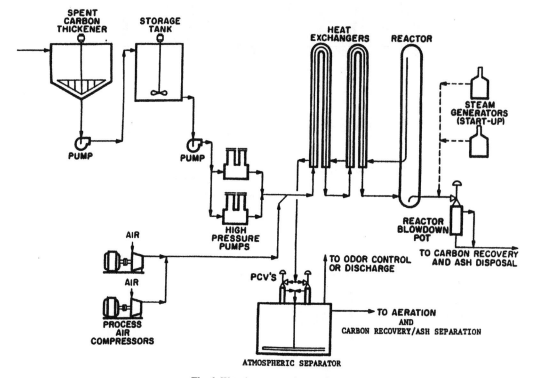

Fig. 6. Wet air regeneration system.

(3) When mixed liquor carbon levels in the aeration basins are allowed to fall too low, the biomass becomes inefficient, thus affecting treatment performance. However, when the carbon levels are returned to normal operating levels, rapid return of performance is seen, thus demonstrating the importance of maintaining sufficient carbon in the system to eliminate bio-toxins.

(4) The plant was able to withstand shock loads as well as spikes very effectively. For example, COD up to 4000 mg L^{-1} (longer duration), sulfide up to 300–500 mg L^{-1} (short duration) and phenol up to 900 mg L^{-1} (short duration) posed no problem in achieving the required treated effluent quality.

(5) Very high influent sulfide levels resulted in sulfur reducing bacteria converting the sulfide to sulfur and storing it in the cells. Subsequent reduction in influent levels then encourages the biomass to oxidize the stored sulfur to sulfate, thus causing a reduction in pH. Thus, it has been found beneficial to maintain sulfides within specified limits at all times.

The treatment of the refinery's wastewater by the PACT/WAR system has shown that a high quality effluent can be produced with minimal residual solids to landfill without a concern for air pollution or the escape of toxics, odors and color to the environment.

2.3.4 *Industrial: petrochemical.* A PACT® system is on-line treating 75 000 to 100 000 gallons a

Table 3. Zimpro environmental Inc. PACT® system users' list

Municipal	Industrial	
Vernon, Connecticut[a]	DuPont (2)	Koch Refinery
Medina, Ohio[a]	General Electric	Elixir Industries
Mt. Holly, New Jersey[a]	Bofors–Nobel[a]	BFI Landfill
Burlington, North Carolina[a]	Exxon (2)	RCC Landfill
Kalamazoo, Michigan[a]	Tenneco	Phillips Petroleum
Sauget, Illinois	Alcoa	Shell Oil
Bedford Heights, Ohio	Crompton–Knowles	Lone Pine Landfill
North Olmsted, Ohio	Moore Business Forms (3)	Thai Ambica Chemicals
El Paso, Texas[a]	Huron Valley Hospital	Hoechst–Celanese
Kimitsu, Japan[a]	Ciba–Geigy	Teknor Apex Inc.
Oga, Japan[a]	Powell–Duffryn	Schofield, WI
Senroku, Japan[a]	Koppers	Puebla Hospital
Oizumi, Japan[a]	Rollins Environmental (2)	Club Toluca
Ibargi, Japan	BKK Landfill	Bristol–Myers Squibb
Burlington County Landfill	Southern Yeast	Cuernavaca Hospital
Citrus County Landfill	Tosco	Kin-Buc Landfill
Greater Lebanon Landfill	Bethlehem Steel	Gadot Petrochemical
Charlotte County Landfill	Unocal[a]	Chambers Development Co.
Western Berks Refuse Authority	BPCL Refinery[a]	Ott/Story/Cordova Site
Hillsborough County Landfill	Yukon Ltd.	Polifin, Ltd.
Frederick County Landfill	Domtar	Al Jubail Fertilizer
Martin County Landfill	Central Services, Inc.	Hagerstown Fibers Ltd.
	Nalco (3)	W.N. Stevenson
	Aldrich Chemical	Sandoz Chemicals
	Safety Kleen	Ta Sheh Industrial Park[a]
	Reilly Industries	Chevron
	BP Oil	
	Waste Management of North America, Inc. Landfill	
	Bostik	

[a] Wet air regeneration system.

Table 4. City garage contaminated groundwater: daily design flow[a] = 17 500 gpd

Parameter	Influent (mg L^{-1})[b]	Required effluent (mg L^{-1}))
COD	50–200	NA
BOD	5–20	NA
TSS	≤10	<40
pH	6.5–7.5	6.5–8.5
Ammonia	1–1.5	NA
Phosphorus	0.8–2	NA
Iron	10–30	NA
Lead	0.01–0.02	<0.162
Manganese	2–3.5	NA
Oil and grease	1–3	<10
Benzene	0.8–1	0.05
BTEX	10–12	0.75
PAH	<0.001	<0.0001
Naphthalene	0.16	NA

[a] Flow consists of 12 gpm of pumped contaminated groundwater.
[b] Influent concentrations are maximum concentrations found in samples.

Table 5. PACT system operating requirements for garage contaminated groundwater

	Predicted	1995/1996 Actual
Power (kW h per day)	60–70	<50
Powdered carbon (lb per day)	50	0
Solids to disposal (lb per day)	7–10[a]	0[b]

[a] Sewered.
[b] Discharge in effluent.

Table 6. PACT system performance results for garage contaminated groundwater (1995)

	Influent	Effluent
SS (ppm)		<5
COD (ppm)	45	22
Benzene (ppb)	717	<0.5
BTEX (ppb)	4434	<5

day of complex, high-strength organic wastewater from the production of organic acids, solvents and aromatics.

The system reduces COD and BOD by more than 95%, meeting strict discharge requirements. Selection of the PACT system followed a six-month pilot test during which PACT outshone biological treatment both in performance and stability, and proved more cost-effective.

The industry sought an effective treatment technology which would be able to handle its wastewater which consists mostly of organic acids (fumaric, phthalic and benzoic) and hydrocarbons (mainly benzene and toluene). The waste varies considerably in composition and volume. Average COD is about 2000 to 3000 mg L^{-1}; BOD averages more than

Table 7. Contaminated groundwater Michigan superfund site

Target parameters	Average concentration	Target parameters	Average concentration
Conventional parameters (mg L^{-1})		Semivolatile organic compounds (SVOCs) (μg L^{-1})	
COD	1795	Anilene	3770
BOD, 5-day	692	Benzoic acid	4639
Nitrogen, ammonia	81	Benzyl alcohol	4174
Volatile organic compounds (VOCs) (μg L^{-1})		Bis(2-ethylhexyl)phthalate	417
Acetone	34259	Butyl benzyl phthalate	417
Benzene	774	Di-N-butylphthalate	417
Carbon tetrachloride	685	Camphor	7619
Chlorobenzene	685	4-Chloroaniline	1670
Chloroethane	685	2-Chlorophenol	436
Chloroform	685	1,2-Dichlorobenzene	879
1,1-Dichloroethane	2000	1,4-Dichlorobenzene	424
1,1-Dichloroethene	922	N,N-Dimethylbenzenamine	3570
1,2-Dichloroethane	83481	N-Ethylbenzenamine	21630
cis-1,2-Dichloroethane	689	2-Methylnaphthalene	418
trans-1,1-Dichloroethene	685	2-Methylphenol	417
1,2-Dichloroethene (total)	1374	3,4-Methylphenol	420
Ethylbenzene	704	N-Nitroso-di-phenylamine	417
Isopropanol	34259	Nitrobenzene	417
4-Methyl-2-pentanone	417	Di-N-Octylphthalate	417
Methylene Chloride	915	1,1,3,3-Tetramethylurea	835
Tetrachloroethene	5163	1,2,4-Trichlorobenzene	417
Tetrahydrofuran	6852	2,4-Dimethylphenol	417
Toluene	5156	2-Ethyl benzenamine	835
1,1,1-Trichloroethane	5678	5-Methyl-s-hexanone	1252
1,1,2-Trichloroethane	685	1,1-Dichloro-2,2-diethoxy-ethane	835
Trichloroethylene	770		
Vinyl chloride	7574		
Xylenes	2337		

Table 8. Performance comparison Michigan superfund site (% removal)

	GAC/FB system	PACT system
BOD	>95	>99
COD	80–90	>90
NH$_3$–N	Unstable	>99
Individual organic compounds[a]	≥95	100[b]

[a] Detection limits for most compounds were 5–10 ppb.
[b] 1,2-DCA was detectable in effluent, removal was >99%.

Table 9. Refinery wastewater characteristics

	Predicted	Discharge standard max. conc. (mg L^{-1})
BOD (ppm)	808	15
Phenolics (ppm)	257	1
Sulfide (ppm)	120	0.5
Oil and grease	–	10
Suspended solids (ppm)	57	20
pH	60	6-8.5
COD (ppm)	1172	250
BOD/COD	0.7	–

1000 mg L^{-1}. In addition, the waste stream contains oil and grease.

During 1990, a six-month pilot test of the PACT system and AS was conducted at the site. Working with U.S. Filter/Zimpro, the petrochemical plant arranged to pilot the PACT system alongside the biological process.

In addition to running at anticipated normal conditions, all pilot systems were "stressed" to simulate operation under shock loading situations. The pilot systems were evaluated for both their ecological and economic implications. Pilot test results are shown in Table 11.

The PACT systems also showed consistently better values throughout the test, and exhibited better ability to cope with the frequent fluctuations in waste composition.

The full-scale system started up in 1993. As Table 12 shows, effluent COD and BOD reflect reductions of more than 95%, and meet new discharge requirements.

Solids wasted from the treatment system are gravity thickened. A filter press is used to dewater the solids to a cake of 40–50%.

Not only was the PACT system selected over conventional biological treatment for environmental reasons, it was also the choice based on cost comparisons. According to a detailed cost analysis, the PACT system was about 10% less in capital costs than biological treatment, primarily because PACT packs more treatment into reactors that are two-thirds the size of those required by an activated sludge system.

On the operating side, despite additional costs for powdered carbon, PACT came in at 10% less expensive. Anticipated cost per cubic meter treated is $0.98

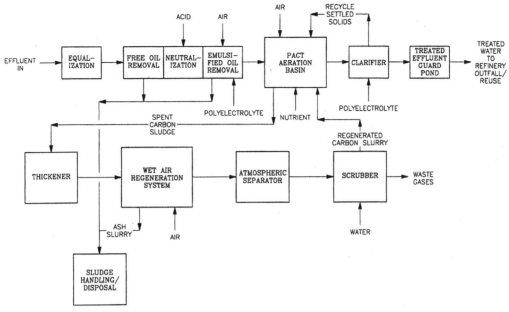

Fig. 7. Refinery waste water treatment plant.

Table 10. PACT/WAR system performance refinery wastewater

	Actual influent	PACT effluent	Discharge Standard
BOD (ppm)	718	7	15
COD (ppm)	1494	78	250
BOD/COD	0.5	–	–
Phenolics (ppm)	70	NIL	1
Sulfide (ppm)	142	NIL	0.5
Oil and grease (ppm)	–	2	10
Suspended solids (ppm)	75	6	20
pH	8	6.6	6–8.5

Table 11. Pilot results petrochemical waste

	Influent	Effluent	
		PACT	AS
COD (mg L^{-1})	3684	100	285
BOD (mg L^{-1})	1517	14	32

Table 12. PACT system performance petrochemical waste

Parameter	Units	Influent	Effluent
Flow	gallon/day	90000	–
SCOD	mg L^{-1}	3600	135
SBOD	mg L^{-1}	1600	<30
Oil and grease	mg L^{-1}	150	5
TSS	mg L^{-1}	650	60

for the PACT system, versus $1.05 for the activated sludge. Significant savings are in man hours to run the system, and in the amount of dewatered sludge produced (0.5 tons per day with PACT; 1.4 tons per day with activated sludge).

2.3.5 Industrial: petrochemical. U.S. Filter/Zimpro conducted an extensive testing programme to develop and optimize an effective approach to the treatment of an industrial vinyl chloride monomer (VCM) wastewater for a South African company to allow complete re-use of the treated water. The PACT® system, along with chemical oxidation processes were evaluated to produce an exceptionally high quality effluent that would allow re-use back in the production facility.

The objective of the testing was to produce a treated effluent with a *TOC* of less than 5 mg L^{-1}. Based on the results, a treated effluent with a *TOC* of less than 5 mg L^{-1} can be produced by a combination of a two-stage PACT system followed by a chemical oxidation process that utilizes UV/hydrogen peroxide.

Pilot testing results when treating the VCM wastewater are shown in Table 13. Further work with UV/oxidation (not reported here) indicated that the

Table 13. PACT system pilot VCM petrochemical waste

Characteristic	Influent	1st Stage effluent	2nd Stage effluent
BOD (mg L^{-1})	1160	205	90
Reduction (%)		82	92
COD (mg L^{-1})	1910	545	345
Reduction (%)		71	82
TOC (mg L^{-1})	910	160	60
Reduction (%)		82	93

PACT system effluent could be reduced to a $TOC < 5$ ppm. The industry had attempted to treat this waste previously with conventional AS, but the AS system was not a reliable alternative, especially in light of the re-use objectives and the amount of ethylene dichloride (EDC) present.

Because of the performance results obtained at U.S. Filter/Zimpro laboratories and the company's willingness to guarantee the PACT system effluent quality, a full scale PACT system with UV/oxidation was furnished. The PACT system is currently treating the VCM wastewater and is achieving < 20 ppm TOC. UV/oxidation is not yet being used by the industry.

Vapor adsorption on coal- and wood-based chemically activated carbons (I) Surface oxidation states and adsorption of H_2O

W.H. Lee, P.J. Reucroft *

Department of Chemical and Materials Engineering, University of Kentucky, Lexington, KY 40506, USA

Received 7 November 1997; accepted 27 May 1998

Abstract

X-ray photoelectron spectroscopy (XPS) was employed to evaluate the surface element distribution/concentration and surface chemical structure of coal- and wood-based chemically activated carbons and to investigate the effect of these features on the adsorption of water vapor at low relative pressure and room temperature. It was found that high surface area carbon samples generally show low concentrations of surface oxygen and low surface area carbon samples generally show high concentrations of surface oxygen. The concentrations of carbon surface oxygen groups such as C–O and C=O generally decrease with increasing heat treatment temperature. Increased water vapor adsorption on the lower surface area activated carbons can be correlated with higher concentrations of surface oxygen functional groups. © 1998 Elsevier Science Ltd. All rights reserved.

Keywords: A. Activated carbon; B. Activation; C. Adsorption; C. Photoelectron spectroscopy; D. Adsorption properties

1. Introduction

Coal- and wood-based activated carbons were prepared using a chemical activation route with varying heat treatment temperature employing KOH and H_3PO_4 as chemical activants [1–3]. To elucidate the effects of microporous structure and surface oxidation states on adsorption capacities in the low and high relative pressure ranges, water vapor adsorption studies were carried out. In current research, the Dubinin–Serpinsky isotherm equation [4] and X-ray photoelectron spectroscopy (XPS) were employed. The XPS technique can provide information on the surface element distribution/concentration and chemical structure in the outmost 3–4 nm of the exposed solid surface. Peak deconvolution of the carbon spectra can also be used to identify/quantify the carbon surface functional components of the activated carbon surface investigated.

* Corresponding author. Tel: +1 606 257 8723;
Fax: +1 606 323 1929; e-mail: reuc@engr.uky.edu

The XPS results have been correlated with the adsorption isotherm data and adsorption capacities, and will be discussed in terms of surface element concentrations and functional components.

2. Experimental

2.1. Materials

Eight different types of activated carbons were prepared at the Center for Applied Energy Research, University of Kentucky. The carbon samples were synthesized from Illinois bituminous coals and white oaks using a chemical activation process that employed potassium hydroxide (KOH) and phosphoric acid (H_3PO_4), respectively, as chemical activation agents, and low and high temperature heat treatments. Details of the synthesis process have been described previously [1–3].

Reprinted from *Carbon* **37** (1), 7-14 (1999)

2.1.1. Activated carbon samples 1–5 (coal-based, activant – KOH)

Coal-based chemically activated carbons using KOH activant at high heat treatment temperatures, 500, 600, 700, 800 and 900°C, respectively, were designated as activated carbon samples 1, 2, 3, 4 and 5.

2.1.2. Activated carbon samples 6–8 (wood-based, activant – H_3PO_4)

Wood-based activated carbons using H_3PO_4 activant at high heat treatment temperatures, 300, 350 and 450°C, respectively, were designated as activated carbon samples 6, 7 and 8.

BET and mesopore surface areas of the carbons are summarized in Table 1.

2.2. Water vapor adsorption

Adsorption isotherms at room temperature (24°C) of all the coal- and wood-based chemically activated carbon samples for water vapor were obtained using a gravimetric adsorption apparatus equipped with a Cahn 2000 electrobalance. About 50 mg of each activated carbon sample was evacuated at 523 K to eliminate volatile vapors absorbed in air and outgassed to a residual pressure of 10^{-3} Torr in the system. The adsorption system was finally evacuated down to a residual pressure of 10^{-5} to 10^{-6} Torr using an ion pump. Samples were exposed to equilibrium relative pressures of water vapor in the relative pressure range of 10^{-3} to 0.9 to obtain isotherms.

Each adsorption isotherm was confirmed by repeating the measurements up to three times, resulting in a deviation range of less than 3% in terms of adsorbed amount at constant relative pressure in $g\,g^{-1}$. Adsorption capacities in terms of limiting micropore volume were evaluated by applying the DR equation [5].

Table 1
BET and mesopore surface areas

Sample no.	S_{BET} ($m^2\,g^{-1}$)	$S_{mesopore}$ ($m^2\,g^{-1}$)
1[a]	580	54
2[a]	835	37
3[a]	1081	42
4[a]	1583	46
5[a]	1605	55
6[b]	783	38
7[b]	1075	38
8[b]	1807	238

[a] Coal-based, KOH activated.
[b] Wood-based, H_3PO_4 activated.

2.3. Surface analysis by X-ray photoelectron spectroscopy

Surface element analysis was conducted by XPS in order to evaluate the distribution of surface elements on the chemically activated carbon samples. A Kratos XSAM 800 combined scanning Auger microprobe (SAM) and X-ray photoelectron spectrometer was used. Samples were carefully mounted on a spectrometer probe tip by means of double-sided adhesive tape. The operating pressure in the XPS was maintained below 1×10^{-8} Torr using an ion pump and a Ti sublimation pump to minimize contributions from contaminants. The spectrometer was operated in fixed retarding ratio (FRR) mode at a pass energy of 11 kV and 13 mA. Under these conditions, full width at half maximum (FWHM) of the $Ag(3d_{5/2})$ peak was about 1.1 eV. A Mg Kα X-ray source was used. All binding energies were referred to the carbon 1s peak. Atomic concentrations were estimated from the XPS element peak area after applying an atomic sensitivity factor [6].

3. Results and Discussion

3.1. Water vapor adsorption

Plots of water vapor adsorption in gram per gram versus relative pressure (P/P_0) for each coal-based KOH chemically activated carbon sample are shown in Fig. 1. In general, typical Type V adsorption isotherms were obtained for most of the coal-based KOH activated carbon samples. It was generally found that increases in the adsorption amount in the relative pressure range

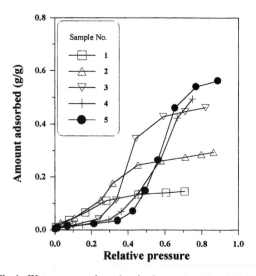

Fig. 1. Water vapor adsorption isotherms (coal-based KOH chemically activated carbons).

above 0.4 reflected surface area/microporosity development in the carbon sample. The water vapor adsorption amount (W) of samples with lower surface areas was somewhat higher than that of those samples with high surface areas at the low relative pressure ranges in most of the KOH activated carbon samples.

Typical plots of water vapor adsorption in gram per gram versus relative pressure (P/P_0) for each wood-based H_3PO_4 chemically activated carbon sample at room temperature are shown in Fig. 2. In general, most of the carbon samples again showed Type V adsorption isotherms with the exception of sample 8. Sample 8 showed a very high water vapor adsorption amount, especially in the low relative pressure range. Sample 8 has a large specific surface area, but also has a high amount of mesoporosity. Mesopores in the adsorbent provide more room to allow multilayer adsorption of water vapor molecules. At low relative pressures in the case of sample 8, the water vapor adsorption mechanism can thus be considered to be due to capillary condensation rather than micropore filling.

The limiting micropore volumes (W_0) obtained from the water vapor adsorption results were deduced from the intercepts on the $\ln(W)$ axis at $A = 0$ in the high relative pressure range ($P/P_0 > 0.4$) because the $\ln(W)$ versus A^2 plots were non-linear over the full relative pressure ($0 < P/P_0 < 0.9$). The non-linear DR plots can be attributed to two different adsorption regions. The resulting W_0 values obtained from water vapor adsorption on the coal-based KOH activated carbon samples are listed in Table 2. The resulting W_0 data were found to be generally in good agreement with the degree of surface area development. This result shows that the micropore structure of the carbon sample is the most

Table 2
Micropore volumes (W_0) and maximum amount of water vapor adsorbed on the specific adsorption sites (a_0) of selected activated carbon samples

Sample no.	W_0 (cm^3 g^{-1})	a_0 (g g^{-1})
2	0.30	0.61
4	0.62	0.11
5	0.68	0.04
6	0.25	0.41
7	0.34	0.10
8	0.54	0.03

important factor which determines the limiting micropore volume. In the case of carbon sample 8, the water vapor adsorption in the low relative pressure range was high compared to that of the other wood-based carbon samples. This adsorption amount represents a large fraction of the entire water vapor adsorption. The relatively low W_0 value of carbon sample 8 can be attributed to multi adsorption in the low water relative pressure range being disregarded in evaluating W_0 values from the DR plots.

Water vapor adsorption phenomena in two significant regions of relative pressures have previously been considered as arising from hydrogen bonding of water molecules onto specific adsorption sites of the adsorbent surface in the low relative pressure range and from a bulk micropore filling process in the high relative pressure range [7–9]. From water vapor adsorption, it was found that the adsorption amount of samples with lower surface areas was somewhat higher than that of samples with higher surface areas in the low relative pressure ranges. This can be due to carbon samples with lower surface areas having more surface oxygen species which can act as dominant active adsorption sites. To examine the effect of surface oxidation on the water adsorption capacity in the case of the chemically activated carbons investigated, the Dubinin–Serpinsky (DS) equation was employed [4].

$$a = a_0 ch/(1 - ch), \qquad (1)$$

where a = gravimetric water adsorption capacity (g g^{-1}), a_0 = maximum amount of water vapor adsorbed on the specific adsorption sites, h = relative pressure (P/P_0) and c = ratio of overall rate of adsorption and desorption.

Typical DS plots of h/a versus h in the case of carbon samples 2 and 5 are shown in Figs. 3 and 4. The best relative pressure regions to fit Eq. (1) were 0.1–0.5 and 0.15–0.55, respectively. Using a linear regression method, the relative values of a_0 for the two carbon samples were estimated and compared. The resulting a_0 values for carbon samples 2 and 5 were 0.61 and 0.04 g g^{-1}, respectively. a_0 of sample 2 is thus much

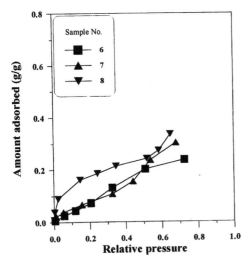

Fig. 2. Water vapor adsorption isotherms (wood-based H_3PO_4 chemically activated carbons).

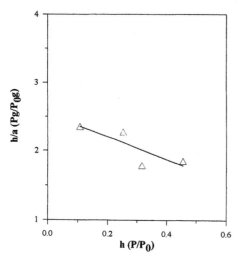

Fig. 3. Dubinin–Serpinsky plot of the water vapor adsorption isotherm (sample 2).

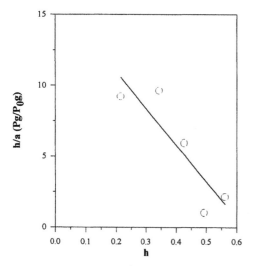

Fig. 4. Dubinin–Serpinsky plot of the water vapor adsorption isotherm (sample 5).

Table 3
XPS surface elemental analysis data expressed in terms of atomic concentration

Sample no.	Atomic concentration (%)		
	C 1s	O 1s	P 2p (K 2p)
2	73.8	26.2	0 (0)
4	82.2	17.8	0 (0)
5	85.4	14.6	0 (0)
6	77.4	22.5	0.2
7	83.7	15.5	0.8
8	83.7	15.5	0.8

3.2. Chemical states on the surface of activated carbons

Wide scan spectra in the binding energy range of 0–800 eV were used to identify the surface elements present and obtain a qualitative analysis. Representative XPS wide scan spectra for two coal-based KOH activated carbons (samples 2 and 5) are shown in Fig. 5. The scale intensity factor for sample 5 is 0.7 times that of sample 2. The spectra show two distinct peaks representing the major constituents, carbon and oxygen. No other elements were detected on the surface of the coal-based KOH activated carbon samples. However, it was found that typical wood-based H_3PO_4 activated carbon samples displayed relatively weak P 2p peaks at a binding energy of 136 eV, which is close to the binding energy of phosphorus in the form of P_4O_{10} [6], in addition to the strong C 1s and O 1s peaks. Elemental phosphorus was oxidized into an insoluble form such as ash and remained in the form of P_4O_{10} when the heat treatment temperature was increased. Fig. 6 shows XPS narrow scan spectra of the C 1s region at 285 eV for two coal-based KOH activated carbons (samples 2 and 5). Fig. 7 shows XPS narrow scan spectra of the O 1s region at 533.5 eV for samples 2 and 5. The spectra show distinct differences in surface oxygen content for these carbon samples.

To determine the surface concentrations (in at%) of the elements, quantitative peak analysis was carried out. The resulting data is listed in Table 3 for coal-based KOH and wood-based H_3PO_4 activated carbon samples. It was found that the atomic concentrations of carbon, the major surface element, were in the 74–85 at% range. The atomic concentrations of oxygen were in the 15 to 26% range. The atomic concentrations of phosphorus on the wood-based H_3PO_4 activated carbon samples were relatively quite small (0.2–0.8 at%). The atomic concentration of potassium could not be determined due to the absence of a K 2p peak on the surface of the coal-based KOH activated carbons samples. This result indicates that the KOH activating agent was completely removed by washing after the activation process.

higher than that of carbon sample 5 and it can be concluded that sample 2 contains more oxygen atoms on the surface.

From water vapor adsorption studies, the microporosity/surface area development was an important factor in determining the sigmoid-shaped water vapor adsorption isotherm at $P/P_0 > 0.4$. The DS analysis indicates that surface chemistry determines the water vapor adsorption amount (W) at low relative pressures ($P/P_0 < 0.4$) rather than the microporosity/surface area development.

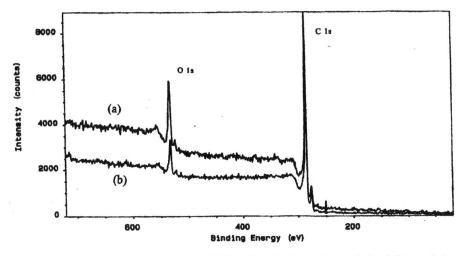

Fig. 5. Wide scan XPS spectra of coal-based KOH activated carbons. (a) Sample 2 and (b) sample 5.

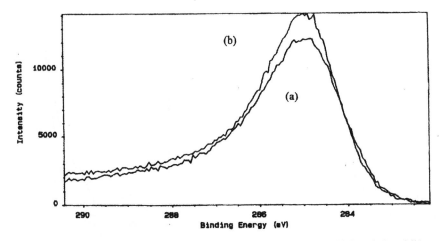

Fig. 6. Carbon 1s narrow scan XPS spectra of coal-based KOH activated carbons. (a) Sample 2 and (b) sample 5.

Generally, the surface concentration of K can be estimated as <0.1 at%, the detectability limit.

To confirm the relation between a_0 and total surface oxygen concentration, the DS equation was correlated with the XPS results, and is shown in Fig. 8. It can be seen that the a_0 values depend on the oxygen concentration. A similar result was found by Barton and Evans [10], who estimated the oxygen concentration by taking the sum of desorbed oxygen species such as CO_2, CO and H_2O. It can be concluded that the surface oxygen concentrations obtained from XPS can also be usefully employed in investigating the relation between a_0 and surface oxygen concentration. However, the physical mean of a_0 is unclear and it can only be used comparatively from carbon to carbon.

The broad carbon peak in the binding energy range from 282 to 292 eV, can be attributed to several carbon-based functional groups which have different binding energies. The C 1s peak of each carbon sample was thus deconvoluted using a peak synthesis procedure which fits the measured peak to several Gaussian peaks, each with a FWHM value of 2 eV and a fixed binding energy [11–13]. To obtain the best fit between experimental and synthesized spectra, the intensity contribution of each functional component peak was estimated by a computer simulation. A representative XPS spectrum of the C 1s peak region at 285 eV deconvoluted into functional group contributions is shown in Fig. 9 in the case of sample 8. Each C 1s peak area was deconvoluted into its functional groups allowing an evaluation of the functional component concentrations. It was found that deconvolution of the carbon 1s peak could be fitted to four line shapes with binding energies at 284, 286.7, 288.4 and 289.7 eV. These binding energy peaks were

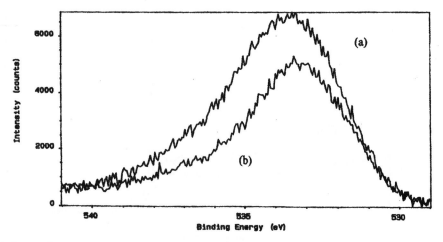

Fig. 7. Oxygens 1s narrow scan XPS spectra of coal-based KOH activated carbons. (a) Sample 2 and (b) sample 5.

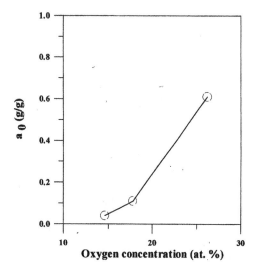

Fig. 8. a_0 versus oxygen concentration (coal-based KOH activated carbons).

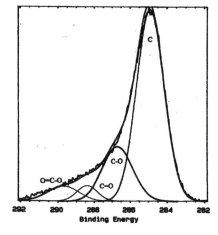

Fig. 9. Deconvolution of C 1s peak area into its functional groups (sample 8).

identified as C or C–H at 285 eV, C–O at 286.7 eV, C=O at 288.4 eV and O–C=O at 289.7 eV following previous results [12,13]. In the case of carbon sample 8, the total area of the C 1s peak region consists of 68% C or C–H, 22% C–O, 4% C=O and 6% O–C=O. Similar results were obtained from deconvolution of the C 1s peaks of the other chemically activated carbon samples and are shown in Table 4.

3.3. Effects of surface functional components on the adsorption capacity

From Table 3, it was found that the atomic concentration of oxygen on the surface decreases with increas-

Table 4
Functional components obtained from the deconvolution of the C 1s peak

Sample no.	Functional components			
	C or C–H	C–O	C=O	O=C–O
2	62.3	25	5.7	4.0
4	67.3	21.8	5.2	5.1
5	68.8	21.4	4.1	4.2
6	59.6	25.2	6.6	6.3
7	64	23.8	5.7	5.2
8	67.5	21.8	4.2	6.0

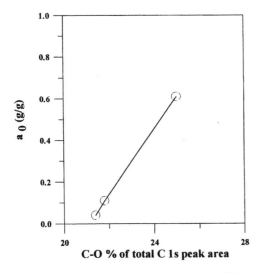

Fig. 10. a_0 versus C–O concentration (coal-based KOH activated carbons).

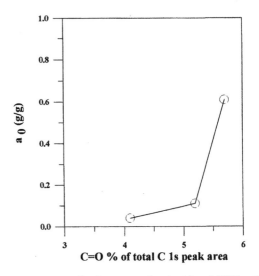

Fig. 11. a_0 versus C=O concentration (coal-based KOH activated carbons).

ing development of surface area. In the process of synthesizing activated carbons samples, the development of surface area was mostly affected by high heat treatment temperature at a fixed concentration of chemical activants [2,14]. In general, the degree of heat treatment temperature enhanced the development of surface area and average micropore size of the activated carbon samples investigated. A high heat treatment temperature during the activation process produced activated carbons with higher carbon contents and relatively lower contents of surface impurities such as chemisorbed oxygen. This produces an inverse relationship between the surface oxygen content and the degree of surface area development. The phenomenon can be attributed to an enhanced desorption rate of surface oxide by the increasing heat treatment temperature.

From the data on surface functional components listed in Table 4, it can be seen that the functional component of highest concentration is C or C–H followed by hydroxyl C–O. The concentration of C or C–H increases with increasing heat treatment temperature. On the other hand, the concentrations of C–O and C=O generally decrease with increasing heat treatment temperature. However, the concentration of O=C–O seems to be independent of the heat treatment temperature.

To investigate the relation between a_0 and surface carbon functional groups, the DS equation was correlated with surface functional group concentrations obtained from the XPS peak deconvolution results of the carbon spectra. Results are shown in Figs. 10 and 11. It can be seen that a_0 values depend on the concentration of C–O and C=O. However, there was not a good correlation with the concentration of O=C–O

because the concentration of O=C–O did not depend on the heat treatment temperature.

In water vapor adsorption isotherms of highly microporous activated carbon samples, a dependence of adsorption capacity on the degree of surface area development was found in the very high relative pressure range ($P/P_0 > 0.4$). High surface area activated carbon samples showed relatively low water vapor adsorption amounts compared to the low surface area samples until the relative pressure range of water vapor reached 0.4. These phenomena lead to the conclusion that differences in the average micropore sizes or degree of surface area development between highly microporous carbon samples are not expected to be a controlling factor in affecting the low pressure water vapor adsorption capacity. In the case of low pressure water vapor adsorption, large specific adsorption sites such as oxygen functional groups on the adsorbent surface are preferable for the adsorption of water molecules. The differences in surface oxygen content between highly microporous activated carbons can affect the actual water vapor adsorption amount in the low relative pressure range ($P/P_0 < 0.4$). It can be concluded that the high adsorption amount of water vapor on the lower surface area carbon samples can be explained by the higher concentration of surface oxygen groups on the surface of these carbon samples compared to the higher surface area activated carbon samples investigated.

From these results, it can be concluded that the influence of the degree of surface area development on the water vapor adsorption amounts was found to be relatively weak in the low relative pressure range. The density of carbon surface oxygen groups is the most important controlling factor which affects the water vapors adsorption capacity in the low relative pressure range.

4. Summary and conclusions

XPS results showed that a high heat temperature during the activation process produces activated carbons with relatively lower contents of surface impurities such as chemisorbed oxygen. This produces an inverse relationship between surface oxygen content and the degree of surface area development. The high adsorption amount of water vapor on the lower surface area carbon samples can be explained by the higher concentration of surface oxygen groups on the surface of these carbon samples compared to the higher surface area activated carbon samples investigated. The a_0 values in the DS equation are correlated with the total oxygen concentration and the percentage of carbon surface oxygen groups such as C–O and C=O obtained from XPS.

Acknowledgements

Activated carbon samples and BET surface area data were provided by M. Jagtoyen and F. Derbyshire, Center for Applied Energy Research, University of Kentucky.

References

[1] Jagtoyen M, Thwaites M, Stencel J, McEnaney B, Derbyshire F. Carbon 1992;30:1089.
[2] Jagtoyen M, Derbyshire F. Carbon 1993;31:1185.
[3] Derbyshire F, Stencel J. In: Technical report on conversion of Illinois coals to activated carbons. Lexington: Center for Applied Energy Research, 1991.
[4] Dubinin M, Zaverna E, Serpinsky V. J. Chem. Soc. 1955;2:1760.
[5] Dubinin M, Radushkevich L. Proc. Acad. Sci. USSR 1947;55:331.
[6] Wagner C, Riggs W, Davis L, Moulder J, Muilenburg G. In: Handbook of X-ray photoelectron spectroscopy. Norwalk: Perkin Elmer Corp., 1979.
[7] Dubinin M, Zaverna E, Serpinsky V. Chem. Soc. 1955;2:1760.
[8] Dubinin M. Carbon 1980;18:355.
[9] Dubinin M, Serpinsky V. Carbon 1981;9:402.
[10] Barton S, Evans M. Carbon 1991;29:1099.
[11] Larsson K, Nording C, Sieghahn K, Stenhagen E. Acta Chem 1966:2880.
[12] Wenming Q, Zhu Q, Ling L. In: Extended abstracts 22th biennial conference on carbon. San Diego, CA, 1995:400.
[13] Silva I, Radovic L. In: Extended abstracts 22th biennial conference on carbon. San Diego, CA, 1955:444.
[14] Laine J, Calafut A, Labady M. Carbon 1989;27:191.

Activated Carbon Compendium
H. Marsh (Editor)

Vapor adsorption on coal- and wood-based chemically activated carbons (II) adsorption of organic vapors

W.H. Lee, P.J. Reucroft *

Department of Chemical and Materials Engineering, University of Kentucky, Lexington, KY 40506, USA

Received 7 November 1997; accepted 27 May 1998

Abstract

The present investigation was undertaken to determine the adsorption properties and evaluate the adsorption capacities of several coal- and wood-based chemically activated carbons using the Dubinin–Radushkevich (DR) characteristic adsorption analysis. Limiting micropore volumes (W_0), determined from CCl_4 and acetone adsorption isotherms at room temperature, were generally in good agreement with the development of surface area. An effect of adsorbate polarity on the adsorption capacity was found in the case of lower surface area activated carbons rather than higher surface area carbons. This can be attributed to higher density of surface oxygen and carbon surface oxygen functional groups on the lower surface area carbon samples. Characteristic adsorption energy (E_0) values obtained from coal-based KOH activated carbons were generally higher than those obtained from wood-based H_3PO_4 activated carbons. These results indicate that the coal-based KOH activated carbons have narrower micropores and uniform micropore size distributions. The average micropore sizes were mostly affected by the degree of surface area development which depended on the heat treatment temperature during the synthesis process. © 1998 Elsevier Science Ltd. All rights reserved.

Keywords: A. Activated carbon; B. Activation; C. Adsorption; D. Adsorption properties

1. Introduction

Microporous adsorbents are widely used to remove chemical species from the gas or liquid phase through their excellent adsorption capacities, which are closely related the large internal surface areas and micropore volumes that are generally associated with microporosity. For practical industrial use, adsorptivity is one of the most important properties of activated carbons with similar surface chemistry.

In recent research, coal- and wood-based activated carbons were prepared using a chemical activation route with varying heat treatment temperatures employing KOH and H_3PO_4 as chemical activants [1–3]. The physical adsorption characteristics of these carbon

* Corresponding author. Tel.: 606-257-8723;
Fax: 606-323-1929; e-mail: reuc@engr.uky.edu

samples with CCl_4 and acetone vapors as adsorbates have been investigated. Limiting micropore volumes and characteristic adsorption energies were obtained by applying the Dubinin–Radushkevich (DR) equation [4].

The resulting adsorption characteristics of the activated carbons investigated will be discussed and correlated with the heat treatment temperature, emphasizing organic vapor adsorption.

2. Experimental

2.1. Materials

Eight different types of activated carbons were prepared at the Center for Applied Energy Research at the University of Kentucky [1–3]. The carbon samples were synthesized from Illinois bituminous coals and white oaks using a chemical activation process with potassium

152

hydroxide (KOH) and phosphoric acid (H_3PO_4), respectively, as chemical activation agents, and low and high temperature heat treatments. The preparative procedure and carbon numbering procedure has been described previously [5]. Their Brunauer, Emmett, Teller (BET) and mesopore surface areas, and average micropore sizes are listed in Table 1.

2.2. Adsorption measurements

2.2.1. Gravimetric adsorption experiments

The gravimetric adsorption procedure has been described previously [5]. Samples were exposed to equilibrium relative pressures of CCl_4 and acetone vapor in the relative pressure range of 10^{-4} to 0.8 to obtain isotherms. The experimental isotherm data were analyzed using the DR equation to obtain adsorption parameters for each carbon sample.

2.2.2. Application of the DR equation to the adsorption isotherms

The DR equation was used to estimate the maximum limiting micropore volume per gram of adsorbent (W_0):

$$\ln(W) = \ln(W_0) - \left(\frac{A}{\beta E_0}\right)^2 \qquad (1)$$

Affinity coefficients for each adsorbate were estimated theoretically from the molar volume of each vapor taking benzene as a reference vapor [6,7]. Micropore volumes (W_0) of the carbon samples for each adsorbate were evaluated from the intercept at $A=0$ from $\ln(W)$ versus $(A/\beta)^2$ plots. Characteristic adsorption energies (E_0) were evaluated from the slope of the $\ln(W)$ versus $(A/\beta)^2$ plots. Average micropore size (L) was estimated from the value $L = k/E_0$ where k was taken to be 17.25 kJ nm mol^{-1} [8–11]. The estimated micropore volumes and the characteristic adsorption energies were then correlated with the BET surface areas.

Table 1
BET and mesopore surface areas, and percentage of the total surface area associated with mesopores

Sample No.	S_{BET} (m^2 g^{-1})	$S_{mesopore}$ (m^2 g^{-1})	Percentage of area due to mesopores
1[a]	580	54	9
2[a]	835	37	4
3[a]	1081	42	4
4[a]	1583	46	3
5[a]	1605	55	3
6[b]	783	38	5
7[b]	1075	38	4
8[b]	1807	238	13

[a]Coal-based, KOH activated.
[b]Wood-based, H_3PO_4 activated.

3. Results and Discussion

3.1. Adsorption isotherms

3.1.1. CCl_4 adsorption isotherms

Weight increases in gram per gram of the five coal-based KOH chemically activated carbon samples on exposure to CCl_4 vapor are plotted versus relative vapor pressure (P/P_0) in Fig. 1. Similar CCl_4 adsorption isotherms for the three wood-based H_3PO_4 chemically activated carbon samples are shown in Fig. 2. The maximum relative pressures investigated in the case of CCl_4 adsorption were in the range of 0.7–0.8.

Typical Type I adsorption isotherms were obtained for the coal- and wood-based chemically activated carbon samples. In the case of sample 8, the saturation plateau is not parallel to the relative pressure axis compared to the other carbon samples. Most of the micropores in sample 8 are filled up below $P/P_0 = 0.2$ and larger mesopores influence the adsorption at $P/P_0 > 0.2$, resulting in increased adsorption as P/P_0 increases. This result implies that sample 8 is a microporous adsorbent which contains a large portion of mesopores. Samples 6 and 7, and the coal-based KOH activated carbons show mostly microporous structures with very little mesoporosity.

3.1.2. Acetone adsorption isotherms

The acetone adsorption results were compared with the CCl_4 adsorption isotherm data in order to ascertain similar adsorption trends even though acetone has polar characteristics while CCl_4 is non-polar. The adsorption properties of each carbon sample were again charac-

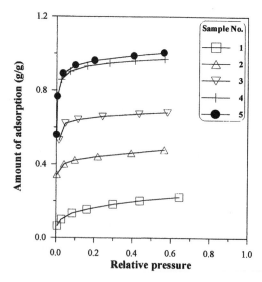

Fig. 1. CCl_4 adsorption isotherms (coal-based KOH chemically activated carbons).

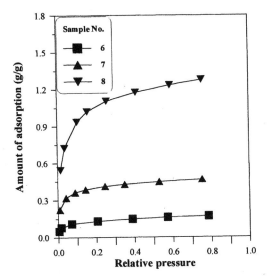

Fig. 2. CCl$_4$ adsorption isotherms (wood-based H$_3$PO$_4$ chemically activated carbons).

terized in terms of microporosity/mesoporosity. Acetone adsorption isotherms on the coal-based KOH activated carbon samples are shown in Fig. 3. Typical Type I adsorption isotherms were again obtained. The acetone adsorption isotherms show similar trends as the CCl$_4$ adsorption isotherms for most of the coal-based activated carbon samples. The adsorption amounts of most of the KOH carbon samples correlate well with the surface area, implying consistent microporous structure for these carbon samples.

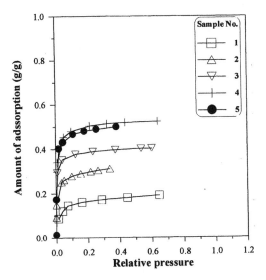

Fig. 3. Acetone adsorption isotherms (coal-based KOH chemically activated carbons).

Acetone adsorption isotherms on the three wood-based H$_3$PO$_4$ chemically activated carbon samples also showed similar trends as the CCl$_4$ adsorption isotherms. As in the case of the CCl$_4$ adsorption isotherms, the saturation capacities increased as P/P_0 increased in the case of Sample 8. Samples 6 and 7 were quite similar to those of the coal-based KOH carbon samples, implying consistent microporous structures.

To further evaluate the effects of porosity on the shape of the adsorption isotherm, the mesoporosity of each activated carbon was obtained [1–3]. The mesopore surface areas were determined by mercury porosimetry. The obtained values of the mesopore surface area ($S_{mesopore}$) are listed in Table 1. The percentage of total surface area on each carbon sample associated with mesopores was determined from ($S_{mesopore}/S_{BET}$) × 100 and is also listed in Table 1. Most of the coal-based KOH activated carbon samples show highly microporous structures with very low mesoporosity (% of $S_{mesopore}$ = 3–9). Most of the wood-based activated carbon samples showed also highly microporous structures with the percentage of surface area associated with mesopores ranging from 2 to 5. On the other hand, sample 8 showed a microporous structure with an increased amount of mesoporosity (13%). The different position of the flat plateau region and increasing saturation capacity with P/P_0 of sample 8 can be therefore be attributed to the larger amount of mesopore surface area. The inflection point which signifies completion of the micropore filling process can be delayed by the effect of mesoporosity.

3.2. Limiting micropore volumes, characteristic adsorption energy and average micropore size from the DR characteristic adsorption isotherms

3.2.1. Limiting micropore volume

The resulting W_0 values for all the coal- and wood-based chemically activated carbon samples are listed in Table 2 for each adsorbate. The standard deviation of

Table 2
Micropore volumes of coal- and wood-based chemically activated carbons (CCl$_4$ and acetone adsorption)

Sample No.	W_0 (cm^3 g^{-1})	
	CCl$_4$	Acetone
1	0.14	0.24
2	0.30	0.39
3	0.43	0.51
4	0.61	0.61
5	0.63	0.64
6	0.11	0.25
7	0.28	0.33
8	0.76	0.78

the W_0 values was in the range of 1–4%. The resulting data show that the W_0 values obtained from CCl_4 and acetone adsorption on each activated carbon sample are generally in good agreement with the development of surface area. As expected, the microporous structure of the activated carbon sample is the most important factor that affects the adsorption properties.

The W_0 values obtained from CCl_4 adsorption were somewhat lower than those obtained from the acetone adsorption in the case of carbon samples 1, 2, 3 and 6. In general, the amount of acetone adsorption (W) is less than that of CCl_4 adsorption for most of the coal-based carbon samples. This difference can be attributed to their different liquid densities (ρ for $CCl_4 > \rho$ for acetone). The effect of adsorbate polarity on the adsorption amount was found to be more significant in the case of adsorbents with lower surface areas. This in turn affects the limiting micropore volume.

To investigate the polarity effect on the adsorption capacity, surface oxygen concentrations of each carbon sample were obtained using the X-ray photon spectroscopy (XPS) technique [5]. The resulting data showed that higher oxygen concentrations were found on the lower surface area carbon samples. It was also found that the lower surface area carbon samples contain a higher density of carbon surface oxygen functional groups such as C–O and C=O, resulting in more active polar adsorption sites. In Fig. 4, it can be seen that the difference in W_0 values for acetone and CCl_4 vapors becomes more significant in the case of lower surface area carbon samples. The effect of polarity on the adsorption capacity can thus be significantly higher in

the lower surface area carbon samples. This can be attributed to the higher density of surface oxygen species and carbon surface oxygen functional groups such as C–O and C=O in the lower surface area carbon samples. However, polarity did not significantly influence the W_0 values in the case of the high surface area samples.

3.2.2. Characteristic adsorption energy and average micropore sizes

Affinity coefficient (β) values for CCl_4 and acetone vapors were estimated to be 1.08 and 0.82 by taking benzene as a reference vapor, respectively. The resulting characteristic adsorption energies for all the coal- and wood-based chemically activated carbon samples are listed in Table 3. The standard deviation of the E_0 values was again in the range of 1–4%.

It was found that the E_0 values obtained from CCl_4 and acetone adsorption generally correlated inversely with the development of surface area in the case of both coal-based KOH and wood-based H_3PO_4 activated carbon samples. Generally, the E_0 values obtained from organic vapor adsorption on the coal-based KOH activated carbon samples were somewhat higher than those obtained on the wood-based H_3PO_4 activated carbon samples. This result implies that the coal-based KOH activated carbons have more highly developed microporous structures, producing higher characteristic adsorption energies.

The average micropore width (L) in nanometers being filled with each adsorbate in the activated carbon samples was evaluated and listed in Table 3. The characteristic adsorption energy data yielded average micropore sizes in the range of 0.72–1.88 nm, when the average value of k was taken as 17.25 kJ nm mol^{-1}. These values are in good agreement with the size of typical micropores (<2 nm). The average micropore sizes of the coal-based KOH activated carbon samples were found to be lower (0.72–1.02 nm) than those of the wood-based H_3PO_4 activated carbon samples

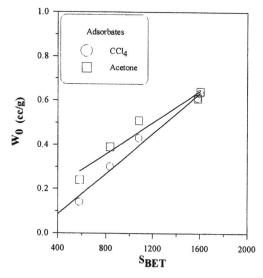

Fig. 4. Plots of W_0 versus S_{BET} obtained from the CCl_4 and acetone adsorption isotherms (coal-based KOH activated carbons).

Table 3
Characteristic adsorption energy and average micropore size of coal- and wood-based chemically activated carbons (CCl_4 and acetone adsorption)

Sample No.	E_0 (kJ mol^{-1}) [L (nm)]	
	CCl_4	Acetone
1	23.9 (0.72)	23.6 (0.73)
2	23.9 (0.72)	23.5 (0.73)
3	20.2 (0.85)	22.7 (0.76)
4	19.9 (0.87)	21.8 (0.79)
5	16.9 (1.02)	18.9 (0.91)
6	11.6 (1.49)	12.0 (1.44)
7	10.4 (1.66)	12.7 (1.36)
8	8.2 (2.10)	9.2 (1.88)

(1.36–2.1 nm). It can thus be again concluded that the coal-based KOH activated carbon samples have highly developed microporous structures with smaller average micropore sizes than the wood-based H_3PO_4 activated carbon samples. This also results in the higher characteristic adsorption energies.

3.3. Correlation of the surface area with micropore volume, characteristic adsorption energy and average micropore size

3.3.1. Correlation of the surface area development with limiting micropore volume

To evaluate the relationship between surface area and limiting micropore volume, Fig. 4 plots BET surface area versus W_0 for CCl_4 and acetone adsorption on the coal-based KOH activated carbon samples. The straight lines indicate that there is generally a good correlation between surface area development and micropore volume. It can be concluded that the limiting micropore volume is closely related to the surface area development. This produces more adsorption sites for the micropore filling process. In Fig. 4, adsorbate polarity leads to an increase in the value of the limiting micropore volume, especially in the case of activated carbon samples with low surface area. The wood-based H_3PO_4 activated carbon samples showed similar S_{BET} versus W_0 plots. The different slopes associated with the S_{BET} versus W_0 plots for the carbon samples can be attributed to the effect of polarity on the W_0 values in the case of lower surface area carbon samples which contain higher oxygen concentrations.

3.3.2. Correlation of the surface area development with characteristic adsorption energy and average micropore size

Plots of the BET surface area versus characteristic adsorption energy obtained from CCl_4 and acetone adsorption on the KOH activated carbon samples are shown in Fig. 5. The results indicate that the E_0 values for both CCl_4 and acetone adsorption decrease slightly with increasing BET surface area. Similar results were obtained in the case of the wood-based activated carbon samples. This result implies that the average micropore size of the carbon samples increases with increasing the degree of heat treatment temperature in the chemical activation synthesis process.

It was found that the heat treatment temperature was a critical parameter in changing the properties of the carbon sample in synthesizing the activated carbon samples investigated. A high heat treatment temperature generally increased the development of surface area and also the limiting micropore volume of each activated carbon sample. From current research, it can be seen that the average micropore sizes are also affected by the heat treatment temperature. Higher heat treatment tem-

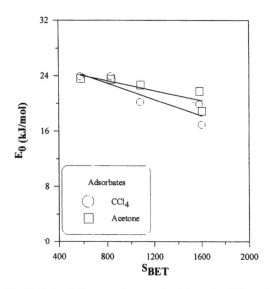

Fig. 5. Plots of E_0 versus S_{BET} obtained from the CCl_4 and acetone adsorption isotherms (coal-based KOH activated carbons).

peratures generally leads to micropore widening phenomenon.

4. Summary and conclusions

W_0 values of coal- and wood-based activated carbons determined from CCl_4 adsorption isotherms were generally found to be lower than those obtained from the acetone adsorption isotherm data in the case of lower surface area activated carbons. Significant influence of adsorbate polarity on the adsorption capacity especially in the lower surface area carbon samples was attributed to higher oxygen concentrations, resulting in more polar adsorption sites. However, adsorbate polarity did not significantly influence the W_0 values in the case of high surface area carbon samples.

E_0 values obtained for acetone adsorption were in good agreement with those obtained from CCl_4 adsorption. E_0 values obtained from KOH activated carbons were generally higher than those obtained from wood-based H_3PO_4 activated carbons. This can be attributed to the more uniform micropore distribution and narrower micropore sizes obtained with coal-based KOH activated carbons. A higher heat treatment temperature generally increases the surface area, micropore volume and average micropore size of carbon samples investigated in the chemical activation synthesis process.

156

Acknowledgements

Activated carbon samples and BET surface area data were provided by M. Jagtoyen and F. Derbyshire, Center for Applied Energy Research, University of Kentucky.

References

[1] Jagtoyen M, Thwaites M, Stencel J, McEnaney B, Derbyshire F. Carbon 1992;30:1089.
[2] Jagtoyen M, Derbyshire F. Carbon 1993;31:1185.
[3] Derbyshire F, Stencel J. In: Technical Report on Conversion of Illinois Coals to Activated Carbons, Center for Applied Energy Research, Lexington, 1991.
[4] Dubinin M, Radushkevich L. Proc. Acad. Sci. USSR 1947;55:331.
[5] Lee WH, Reucroft PJ. Vapor adsorption on coal- and wood-based activated carbons: I surface oxidation states and adsorption of H_2O. Carbon submitted.
[6] Dubinin M, Timofeyev D. Acad. Sci. USSR 1946;54:701; 1947; 55:137.
[7] Noll K, Wang D, Shen T. Carbon 1989;27:239.
[8] Bradley R, Rand D. Carbon 1991;29:1165.
[9] Kaehenbuehl F, Steockli H, Addoun A, Shrburger P, Connet J. Carbon 1986;24:483.
[10] Bansal R, Donnet J, Steockli H. Carbon 1982;20:376.
[11] Dubinin M, Plavnik G, Zaverina D. Carbon 1968;6:183.

Activated Carbon Compendium
H. Marsh (Editor)

Vapor adsorption on coal- and wood-based chemically activated carbons
(III) NH_3 and H_2S adsorption in the low relative pressure range

W.H. Lee, P.J. Reucroft *

Department of Chemical and Materials Engineering, University of Kentucky, Lexington, KY 40506, USA

Received 7 November 1997; accepted 27 May 1998

Abstract

NH_3 and H_2S adsorption studies on coal- and wood-based chemically activated carbons were carried out at room temperature using a gravimetric adsorption technique. The adsorption capacity of NH_3 and H_2S in the very low relative pressure range was independent of the surface area development. Activated carbons with lower surface area generally adsorbed more of these vapors than those with higher surface area. This can be attributed to the presence of smaller average micropore sizes and a greater number of active adsorption sites on the low surface area chemically activated carbons. © 1998 Elsevier Science Ltd. All rights reserved.

Keywords: A. Activated carbon; B. Activation; C. Adsorption; D. Adsorption properties

1. Introduction

Coal- and wood-based activated carbons were prepared using a chemical activation route with varying heat treatment temperature employing KOH and H_3PO_4 as chemical activants [1–3]. The physical adsorption characteristics of these carbon samples with CCl_4, acetone and water vapors as adsorbates up to the relative pressure range $(P/P_0) = 1$ have been investigated in terms of limiting micropore volumes, characteristic adsorption energies, average micropore size and surface oxygen concentrations [4,5].

Ammonia and H_2S adsorption studies were carried out to investigate the adsorption characteristics in the low relative pressure range at room temperature. These adsorptives were selected in order to investigate the effects of micropore structure and surface chemistry on the adsorption capacity in the low relative pressure range. The Dubinin–Radushkevich (DR) isotherm equation was again used to estimate micropore volumes and characteristic adsorption energies. The resulting data were correlated with the average micropore size, surface oxygen functional groups and the heat treatment temperature in the chemical activation synthesis process.

2. Experimental

2.1. Materials

Eight different types of activated carbons were prepared at the Center for Applied Energy Research at the University of Kentucky [1–3]. The carbon samples were synthesized from Illinois bituminous coals and white oaks using a chemical activation process with potassium hydroxide (KOH) and phosphoric acid (H_3PO_4), respectively, as chemical activation agents, and low and high temperature heat treatments. The preparative procedure and carbon numbering procedure has been described

* Corresponding author. Tel: +1 606 257 8723;
Fax: +1 606 323 1929; e-mail: reuc@engr.uky.edu

Reprinted from *Carbon* **37** (1), 21-26 (1999)

previously [5]. Their BET and mesopore surface areas, and average micropore sizes are listed in Table 1.

2.2. Adsorption measurements

2.2.1. Gravimetric adsorption experiments

The gravimetric adsorption procedure has been described previously [5]. Since the NH_3 and the H_2S adsorption was at room temperature, the maximum allowed pressure of these vapors in the current gravimetric adsorption system was 735 Torr. The maximum relative pressures for NH_3 and H_2S were thus limited to 0.1 and 0.05, respectively. Each adsorption isotherm was confirmed by repeating the measurements up to three times, resulting in a deviation range of less than 3% in terms of adsorbed amount of constant relative pressure in g g^{-1}.

2.2.2. Analysis of the adsorption isotherms using the DR equation

Adsorption capacities in terms of limiting micropore volume were evaluated by applying the DR equation to the measured adsorption isotherms obtained for each carbon sample. The characteristic adsorption energies were obtained by analyzing the DR characteristic curves. Affinity coefficients for each adsorbate were estimated theoretically from the molar volume of each vapor taking benzene as reference vapor [6,7]. The estimated β values for NH_3 and H_2S were 0.315 and 0.484, respectively, taking β for benzene as 1.

Micropore volumes of the carbon samples for each adsorbate were evaluated from the intercept at $A = 0$ of $\ln(W)$ versus $(A/\beta)^2$ plots. The characteristic adsorption energies were evaluated from the slopes of the $\ln(W)$ versus $(A/\beta)^2$ plots.

Table 1
BET and mesopore surface areas, and average micropore sizes

Sample no.	S_{BET} (m^2 g^{-1})	$S_{mesopore}$ (m^2 g^{-1})	Average micropore size (nm)
1[a]	580	54	0.72
2[a]	835	37	0.72
3[a]	1081	42	0.85
4[a]	1583	46	0.87
5[a]	1605	55	1.02
6[b]	783	38	1.49
7[b]	1075	38	1.66
8[b]	1807	238	2.10

[a] Coal-based, KOH activated.
[b] Wood-based, H_3PO_4 activated.

3. Results and Discussion

3.1. Adsorption isotherms

3.1.1. Ammonia adsorption isotherms

The main investigation focused on comparing adsorption amounts for the various activated carbon samples in the low relative pressure range. Adsorption properties of each carbon sample were characterized in terms of microporosity and the results were compared with CCl_4 and acetone adsorption isotherm data.

Typical ammonia adsorption isotherms for each coal-based KOH and wood-based H_3PO_4 activated carbon sample are shown in Figs. 1 and 2, respectively. In general, ammonia adsorption amounts on carbon samples with higher surface areas were somewhat lower than on the lower surface area carbon samples. Adsorbents with higher surface areas generally show more adsorption due to the higher adsorption capacity but the ammonia adsorption results show the reverse effect, especially at low relative pressures.

The main factors affecting ammonia adsorption in the low relative pressure range were subdivided into micropore volume, micropore size and surface chemistry. From the CCl_4 and acetone adsorption results, it has been concluded that the main controlling adsorption mechanism was a micropore filling process, which generally depends on the surface area development over the full relative pressure range. The degree of surface area/microporosity development and the micropore volume do not seem to control adsorption phenomena in the very low relative pressure range. Surface chemistry and the micropore size from carbon sample to carbon

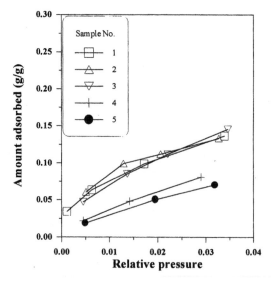

Fig. 1. Ammonia vapor adsorption isotherms (coal-based KOH chemically activated carbons at $P/P_0 < 0.04$).

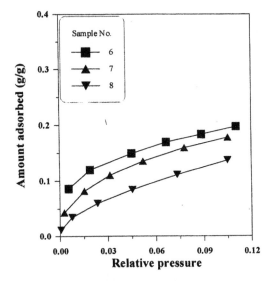

Fig. 2. Ammonia vapor adsorption isotherms (wood-based H_3PO_4 chemically activated carbons at $P/P_0 < 0.12$).

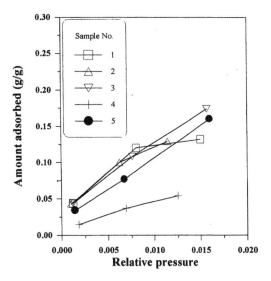

Fig. 3. H_2S adsorption isotherms (coal-based KOH chemically activated carbons at $P/P_0 < 0.02$).

sample can be possible factors affecting the adsorption amount in the very low relative pressure range.

3.1.2. H_2S adsorption isotherms

The main objective in the H_2S adsorption isotherm studies has been to assess differences in the adsorption amount from carbon to carbon sample. In addition, the adsorption properties of each carbon sample obtained from this investigation have been evaluated in terms of micropore structure and the adsorption data have been compared with the ammonia, CCl_4 and acetone adsorption results.

Typical plots of the weight increase due to H_2S adsorption in gram per gram for each coal-based KOH and wood-based H_3PO_4 activated carbon sample versus relative pressure (P/P_0) at room temperature, are shown in Figs. 3 and 4, respectively. The H_2S adsorption, in g g^{-1} of carbon sample in the case of the higher surface area carbon samples was again in general lower than that of the lower surface area carbon samples. The primary controlling factors appear to be surface chemistry and/or micropore size at very low relative pressure range.

3.2. Determination of limiting micropore volume and characteristic adsorption energy from DR characteristic adsorption isotherms

3.2.1. Limiting micropore volume

The resulting W_0 values of the coal-based KOH and wood-based H_3PO_4 activated carbon samples are listed in Table 2. The W_0 value of the KOH activated carbon

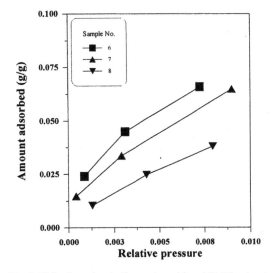

Fig. 4. H_2S adsorption isotherms (wood-based H_3PO_4 chemically activated carbons at $P/P_0 < 0.009$).

samples obtained from ammonia and H_2S adsorption are generally in good agreement with the development of surface area. On the other hand, the W_0 values obtained from ammonia and H_2S adsorption on the wood-based H_3PO_4 activated carbon samples are relatively constant and independent of the development of surface area, in comparison to the organic vapor adsorption. Especially, sample 8 shows significantly low values of W_0 for ammonia and H_2S adsorption. In the case of the coal-based KOH activated carbons, earlier initiation

Table 2

Micropore volumes and characteristic adsorption energies of activated carbon samples for ammonia and H$_2$S adsorption

Sample no.	W_0 (cm^3 g^{-1})		E_0 (kJ mol^{-1})	
	Ammonia	H$_2$S	Ammonia	H$_2$S
1	0.36	0.26	32.3	27.8
2	0.37	0.37	30.1	25.7
3	0.55	0.53	26.6	23.8
4	0.63	0.64	24.7	20.9
5	0.63	0.62	21.5	19.1
6	0.37	0.22	33.2	25.3
7	0.38	0.28	29.1	23.3
8	0.39	0.30	24.1	18.3

of the flat plateau region ($P/P_0 = 0.1$) was found from the full relative pressure of CCl$_4$ and acetone adsorption isotherms. This relative pressure range for initiation of the flat plateau (the end of micropore filling) in the case of CCl$_4$ and acetone adsorption on the KOH activated carbon samples is thus close to the same relative pressure that produced this effect in the ammonia and H$_2$S adsorption results.

In the case of sample 8, initiation of the flat plateau in the organic vapor adsorption was at a relative pressure of about 0.2, which is out of the range of the relative pressure that was investigated in the ammonia and H$_2$S adsorption. The large amount of mesoporosity in sample 8 affected the late initiation of the flat plateau, resulting in incomplete assessment of the limiting micropore volume in the low relative pressure range. It was found that the nature of the micropore distribution rather than the development of surface area also affects the difference in values of the limiting micropore volume obtained from the low relative pressure range adsorption.

3.2.2. Characteristic adsorption energy

The resulting E_0 values obtained from the coal-based KOH activated carbon samples for each adsorbate are listed in Table 2. The E_0 values obtained from ammonia adsorption were higher than those obtained from CCl$_4$, acetone and H$_2$S adsorption. E_0 values obtained from the wood-based H$_3$PO$_4$ activated carbon samples for ammonia adsorption, also shown in Table 2, were again generally higher than those obtained from the CCl$_4$, acetone and H$_2$S adsorption data. The result indicates that there is generally a strong adsorption interaction between the carbon surface and ammonia vapor.

It can be concluded that the structure of the activated carbon samples is probably the most important factor affecting the primary adsorption process but the surface chemistry is another important factor which can affect characteristic adsorption energy differences between organic and inorganic adsorption processes.

3.2.3. Correlation of the adsorption capacity with average micropore size and surface oxygen concentrations

In analyzing ammonia and H$_2$S adsorption isotherms for both coal- and wood-based activated carbons samples, it was observed that the adsorption especially at very low relative pressures, was not always well correlated with BET surface area. In general, adsorption amount was higher for high surface areas when measurements were made over the full relative pressure range. Ammonia and H$_2$S adsorption showed the reverse effect in the very low relative pressure range investigated. This result can be discussed in terms of the average micropore size.

In synthesizing the activated carbon samples, heat treatment temperature was a critical parameter in changing the properties of the carbon sample. A high heat treatment temperature generally increased the development of surface area and also the limiting micropore volume of each activated carbon sample. In the case of ammonia adsorption on the coal-based KOH activated carbon samples, adsorption on carbon samples with high surface area was somewhat lower than on lower surface area carbon samples at very low relative pressures ($P/P_0 < 0.04$). Adsorption at very low relative pressures was not significantly influenced by the development of surface area. However, the average micropore size affects the adsorption amount. The reverse correlation between the development of surface area and the adsorption amount at very low pressures can be attributed to differences in the average micropore size.

Smaller micropore size produced higher adsorption, due to the higher adsorption potential produced by the narrower walls of the micropores. As the micropore filling process proceeded at high relative pressures, the adsorption became increasingly influenced by the development of surface area, which correlates with the limiting micropore volume. From these results, it can be concluded that the adsorption amount is significantly influenced by the average micropore size at low relative pressures due to the narrower walls of the micropores, and by the limiting micropore volume in the high relative pressure range.

In the case of H$_2$S adsorption at very low relative pressure, adsorption is generally affected more by the average micropore size than by the development of surface area and limiting micropore volume. Similar trends were found in the case of wood-based H$_3$PO$_4$ activated carbons. It can thus be concluded that the degree of surface area development and limiting micropore volume control the adsorption process in the high relative pressure range, and the average micropore size affects the adsorption process in the very low relative pressure range through the micropore filling mechanism.

Higher values of E_0 and W_0 for ammonia adsorption compared to other adsorption results in the presence of surface oxygen on the activated carbon samples.

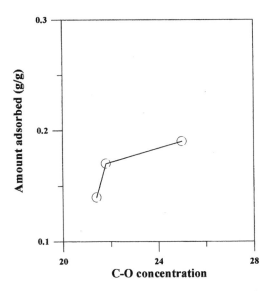

Fig. 5. Plot of amount adsorbed versus C–O concentration (NH₃ on wood-based H₃PO₄ chemically activated carbons at $P/P_0 = 0.1$).

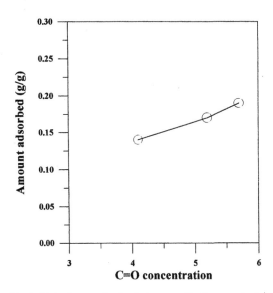

Fig. 6. Plot of amount adsorbed versus C=O concentration (NH₃ on wood-based H₃PO₄ chemically activated carbons at $P/P_0 = 0.1$).

Significant differences in these values were found on the samples 1 and 6 which have lower surface areas. This can possibly be explained in terms of surface chemistry. Nitrogen has a high electronegativity (3.0). The hydrogen atoms in NH_3 can strongly interact with oxygen due to high electrostatic attraction. Their hydrogen bonding (N–H⋯O) energy has been reported to be as much as 3 kcal mol^{-1} [8]. Hydrogen atoms in the form of NH_3 can therefore preferably adsorb on specific adsorption sites such as oxygen functional groups rather than on pure carbon sites.

It was found that NH_3 adsorption in the low relative pressure region increases with increasing oxygen concentration on the carbon surface obtained from XPS [5]. The adsorption amount (W) is also clearly related to the percentage of carbon surface oxygen functional groups such as C–O and C=O. Typical relations between amount adsorbed (W) in g g^{-1} and C–O and C=O concentrations in the case of the wood-based activated carbon samples are shown in Figs. 5 and 6, respectively. Lower surface area carbon samples which contain higher densities of surface oxygen functional groups show higher W values in the low relative pressure range. Oxygen in the form of hydroxyl and carbonyl thus provides specific adsorption sites for ammonia adsorption (N–H⋯O) in the very low relative pressure range, resulting in large adsorption amounts on lower surface area carbon samples. It can be concluded that higher values of E_0 and W_0 on lower surface area carbon samples for ammonia may be attributed to the existence of a high density of surface oxygen.

4. Conclusions

E_0 values obtained from ammonia adsorption were higher than those obtained from H_2S adsorption due to the strong adsorptive attraction between surface oxygen atoms and the hydrogen atoms in the ammonia molecules. The adsorption capacity of ammonia and H_2S in the very low relative pressure range was independent of the degree of surface area development. This can be attributed to the presence of smaller average micropore sizes and the presence of a higher density of surface oxygen on the low surface area activated carbons. The degree of surface area development and the limiting micropore volume generally control the adsorption process in the high relative pressure range through the micropore filling mechanism. The average micropore size and surface chemistry have dominant effects on the adsorption process in very low relative pressure ranges.

Acknowledgements

Activated carbon samples and BET surface area data were provided by M. Jagtoyen and F. Derbyshire, Center for Applied Energy Research, University of Kentucky.

References

[1] Jagtoyen M, Thwaites M, Stencel J, McEnaney B, Derbyshire F. Carbon 1992;30:1089.
[2] Jagtoyen M, Derbyshire F. Carbon 1993;30:1185.

162

[3] Derbyshire F, Stencel J. In: Technical report on conversion of Illinois coals to activated carbons. Lexington: Center for Applied Energy Research, 1991.

[4] Lee WH, Reucroft PJ. Vapor adsorption on coal- and wood-based chemically activated carbons: (II) adsorption of organic vapors. Submitted to Carbon.

[5] Lee WH, Reucroft PJ, Vapor adsorption on coal- and wood-based chemically activated carbons: (I) surface oxidation states and adsorption of H_2O. Submitted to Carbon.

[6] Dubinin M, Radushkevich L. Proc. Acad. Sci. USSR 1947;55:331.

[7] Noll K, Wang D, Shen T. Carbon 1989;27:239.

[8] Mortimer C. In: Chemistry. Seoul: Chung-Moon Kak, 1983:333.

Removal of SO_x and NO_x over activated carbon fibers

Isao Mochida[a], Yozo Korai[a,*], Masuaki Shirahama[a], Shizuo Kawano[a],
Tomohiro Hada[a], Yorimasa Seo[b], Masaaki Yoshikawa[c], Akinori Yasutake[d]

[a]*Institute of Advanced Material Study, Kyushu University, Kasuga, Fukuoka 816-8580, Japan*
[b]*Center for Coal Utilization, 6-2-31, Roppongi, Minato-ku, Tokyo 106-0032, Japan*
[c]*Research and Development Center, Osaka Gas Co., Ltd., 6-19-9, Torishima, Konohana-ku, Osaka 554-0051, Japan*
[d]*Mitsubishi Heavy Industries., Ltd., Technical Headquarters Nagasaki Research and Development Center Chemical Research
Laboratory 5-717-1, Fukahori-machi, Nagasaki 851-0301, Japan*

Received 23 April 1999; accepted 14 June 1999

Abstract

The recent development of de-SO_x and de-NO_x using activated carbon fibers (ACF) in Japan is introduced in this review comparing it with the conventional commercialized system. Pitch-based ACFs showed higher de-SO_x activity than other ACFs from different precursors. De-SO_x activity can be further modified by the heat-treatment, continuous and complete removal of SO_x. ACF is also effective for removing NO_x in the presence of NH_4. The mechanisms of de-SO_x and de-NO_x on the surface of ACFs are proposed and discussed. Based on fundamental research results de-SO_x, de-NO_x and their combined processes are designed and proposed. © 2000 Elsevier Science Ltd. All rights reserved.

Keywords: A. Activated carbon; B. Activation; Heat treatment; C. Adsorption; D. Catalytic properties

1. The necessity of technology development for SO_x and NO_x removal

Energy consumption has increased with an increase in the population and activity of human beings since the industrial revolution in the 18th century, especially after the second world war. The total energy supply in Japan was 4.7×10^{15} kcal in the year 1991, which was equivalent to 530 million kl of petroleum [1]. The shares of coal, petroleum, natural gas, atomic nuclear energy, and others including water were 17.0, 56.1, 10.9, 10.0, and 6.0%, respectively. As shown here, a major part of the energy used in our country is supplied from the fossil fuels, i.e., coal, petroleum, and natural gas.

Recently, new energy, such as solar heat, light, wind, waves of the sea, and geothermal energy, has tried to supplement present energy sources. However, they contribute very little or are difficult to commercialize because of the problems of security, cost, supply stability, technical limitation, low energy density, and seasonal fluctuation. Therefore, we will have to depend on nuclear energy for

another source. However, the safety and waste disposal of fossil fuels will be of major concern the next 30 years or more.

The combustion of such fossil fuels in massive amounts produces air pollutants and solid wastes such as SO_2, NO_x, CO, VOC (volatile organic compounds), carbon particulates, and ashes. Furthermore, CO_2, which produces the green-house effect, is also produced by the combustion of fossil fuels.

Some air pollutant gases such as SO_2 and NO_x are also produced in nature. Production rates of some important pollutants are summarized in Table 1 [1,2]. Sulfur, nitrogen, and some heavy metals are necessary the life of plants and animals. Sulfuric and nitric acids in the rain supply

*Corresponding author. Tel.: +81-92-583-7800; fax: +81-92-583-7797.
E-mail address: korai@cm.kyushu-u.ac.jp (Y. Korai)

Table 1
The amounts of artificially and naturally produced gases that influence the global environment[a]

	Artificial	Natural
Flon	1	0
SO_2	140	100
NO_x	100	100
VOC	60	1000
CO_2	25 000	75 0000

[a] Ten thousand tons/year.

such important elements, especially in areas where the soil lacks sulfur. Such species are, thus, useful or even essential for the nature of the earth. However, they cause pollutants that harm human life when they are produced in excess [2].

The amounts of SO_2 produced by human activity and nature are 1.4 and 1 million tons per year, respectively. In this case of the excess of SO_2 produced by humans beings becomes a self inflicted poison to themselves. The amounts of NO_x produced by both human beings and nature are 1 million tons per year [2]. Recently, the amount of exhausted SO_2 in the USA, Westen Europe, and Japan has decreased, while that from developing nations has increased. The amounts from Europe (including post Soviet Union), North America (including Canada), and Asia are 50, 30, and 30 million tons/year, respectively [2].

A large percentage of VOC is generated from the forest, while human activity mainly generates VOC in urban and industrialized areas [2].

In North Europe and North America, acid rain has acidified thousands of lakes and killed fish in the lakes. Acid rain and fog are believed to have damaged forests for the past ten years. The average pH of rain in North Europe is about 4.1, in Japan it is 4.5–5.0. Although, it is also reported that many trees can grow well even when the average pH of rain in a year is 3.5, the stronger acidity of the rain and fog destroys the leaves. Ozone has been revealed also to cause damage [3–5]. Ozone is produced by the photochemical reaction of VOC and NO_x as the main components of the oxidant. The oxidant accelerates the oxidation of SO_2 and NO_x into toxic sulfuric and nitric acids, respectively. The removal of VOC and NO_x is very important to reduce the concentration of ozone.

Sulfuric and nitric oxides are the major causes of acid rain and fog. Sulfur trioxide is another source. The properties and effects of SO_2 and SO_3 in acid rain or fog are summarized in Table 2. The concentration of SO_3 in the flue gas derived from the combustion of coal or heavy oil is 2–7% of SO_2, while that from the calcination furnace in iron and cement factories is sometimes much higher. The concentration of sulfuric acid mist derived from SO_2

in the flue gas is smaller than 0.05% (of gas pH is higher than 2), because SO_2 is oxidized very slowly to SO_3 which is hydrated into sulfuric acid in the atmosphere. On the other hand, SO_3 forms a mist, carrying sulfuric acid of high concentration (its concentration is often higher than 30%). In particular, its concentration in the exhaust mist from combustion and calcination furnaces is higher than 50% and, therefore, its influence is really serious in the neighborhood of the exhaust site. Since the size of the mist droplet is 0.1–1 nm, it is hardly removed by washing or wet desulfurization. Acid mist is observable even at the road side because SO_2 of low concentrate exhaust from light fuel oil burned in the diesel engines of bus and track can be concentrated in the mist [6].

Sulfur dioxide had been believed to cause Yokkaichi asthma. In truth, a large amount of SO_3 in the exhaust gas from chemical factories caused the problem [4,6]. Emission of SO_2 and NO_x in the flue gas is now regulated at acceptable levels in the developed countries. However, asthma continues to increases, probably because there is no regulation for SO_3.

More than 1800 wet desulfurization units of high efficiency are working in Japan. They remove more than 95% of SO_2 because that is what they are designed for SO_2 removal. However, they can remove only 50% of SO_3. What is worse is that the damage by acid mist is enhanced after passing through a wet desulfurization process because the process lowers the temperature of the flue gas and increases the formation of the acid mist. Therefore, SO_3 removal coupled with SO_2 removal is very important in Japan. Improved combustion, dry desulfurization, and electric dust collectors are believed to be effective for the removal of SO_3 mist.

2. Current processes of SO_x and NO_x removal

2.1. SO_x

Wet desulfurization using limestone has been applied most widely, and it will be employed further in the future because limestone is the cheapest and most resourceful absorbent. Dry and semi-dry desulfurization methods using limestone or lime have been also developed.

The capital and running costs for a current SO_2 removal plant are 150 000~200 000 Japanese yen per 1 kw and 100 000~150 000 Japanese yen/t-SO_2, respectively. Many optional ways were studied in the Clean Coal Technology Projects in USA to decrease the cost of SO_2 removal. Fig. 1 illustrates the schemes of desulfurization processes using limestone or lime [7]. Types A and B are dry methods, C and D are semi-dry methods, while G is the most popular wet limestone–gypsum method. Type A is the furnace limestone injection method (FLI), where limestone is thrown into the furnace at around 1000°C. Type B is the duct lime injection method (DLI), where $Ca(OH)_2$ is put

Table 2
The properties and influences of acid rain and fog derived from SO_2 and SO_3

Source	SO_2	SO_3
Contribution to H_2SO_4 in acid fog	less than 0.05%	more than 30%
pH of acid fog	larger than 2	less than 0
Contribution to H_2SO_4 in acid rain	more than 0.005%	more than 0.1%
pH of acid rain	more than 3	less than ca.2
Spreading area	10–1000 Km	less than 10 Km
Main influence	acidification of lakes	asthma, damage for the plant

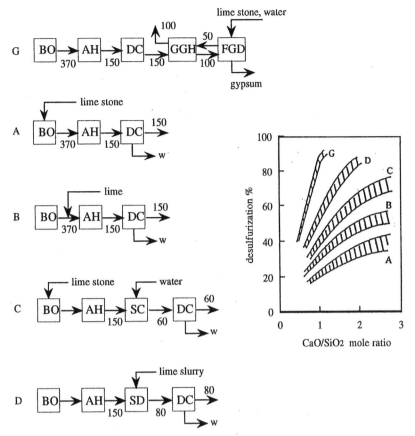

Fig. 1. Construction of desulfurization system using lime or limestone [2]. BO: Boiler, AH: air heater, DC: dust collector, GGH: gas-gas heater: SC: spray cooler, SD: spray dryer, W: waste.

into the duct at around 350°C. The desulfurization efficiency of both types A and B is lower than 60%. Application of these systems has been limited in number in spite of their low capital cost. Type C is a modified scheme of type A by attaching a water spray tower after the air-heater to enhance the desulfurization efficiency. Type D is a lime slurry spray-drying method that has been employed recently in Europe. The capital cost of types C and D is about 6000 and 13 000 yen/kw, respectively.

The comparisons of the running cost and desulfurization efficiency of these methods are summarized in Fig. 2. The cost of G1 is estimated by assuming that the by-product gypsum can be sold as a construction material and that of G2 includes the cost of disposal of the gypsum. When SO_2 concentrations of the inlet are 1000 and 3000 ppm, the cheapest ways are type C and G1, respectively. At higher SO_2 concentrations, wet desulfurization is more favorable because of the high utilization rate of limestone. To enhance the reactivity of lime and decrease the running cost of the spray-dryer system, the addition of fly ash or sodium hydroxide to lime, the conditions for the prepara-

tion of lime slurry, and the effects of the humidity in the duct have been examined by many researchers [8–11].

The most common desulfurization method in Japan is the wet type. However, it has some disadvantages. First of all, this system requires large space and high capital costs, especially for the wastewater purifier. Hence, it will be difficult to introduce this method in developing nations. It can not remove SO_2 completely, with about 50 ppm of SO_2 leaking out. It requires large amount of water and energy. Furthermore, the cost for carrying limestone is not negligible since it is mined far away from the desulfurization plant. The by-product gypsum must be disposed of when there is no demand.

2.2. NO_x

In 1970s, the discharge of NO_x was controlled and regulated since it is recognized that it causes acid rain and photochemical smog. Vanadium pentoxide (V_2O_5) supported on titania was used to reduce NO with NH_3 in the fixed bead reactor. To solve such problems, a number of

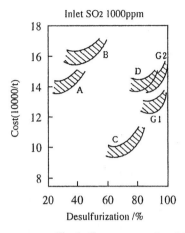 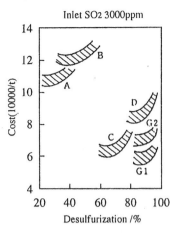

Fig. 2. Cost to remove 1 t of SO$_2$ by various Limestone methods.

other de-NO$_x$ methods have been proposed. Dry NO$_x$ adsorption by a carbon adsorbent is one of the most expected ways. This system has been called SCR (selective catalytic reduction) because NH$_3$ did not produce N$_2$O through the reduction of NO. A three-component catalyst system was installed in a gasoline-powered vehicle to reduce NO$_x$. Thus, the total reduction of NO$_x$ was almost achieved for such combustion systems. However, NO$_x$ in the atmosphere still increases because more NO is exhausted from an increasing number of sources. As a consequence, there are many regions where the NO$_x$ concentration in the atmosphere does not satisfy the environmental standards because combustion equipment such as diesel cars, small-scale combustion equipment, sintering furnaces, and incinerators still exhaust NO$_x$. Recently, gasoline-powered vehicles have adopted a lean burn system for combustion efficiency. In such a system NO$_x$ is not sufficiently reduced because there is not a sufficient amount of reducing agent in the exhaust gas. An adsorbent for the leaked NO$_x$ and/or the pulse addition of the reducing agent are needed.

Some metal oxides have been examined for the desulfurization or simultaneous removal of SO$_2$ and NO$_x$. In this system, SO$_2$ is removed as a metal sulfate or sulfuric acid and NO$_x$ is reduced with NH$_3$ on the sulfate catalyst. The saturated adsorbent is regenerated to the original metal oxide using NH$_3$ or CH$_4$. Alumina supported cupric oxide has been focused as a potential sorbent/catalyst for the removal of SO$_2$ (copper acting as a sorbent by forming CuSO$_4$) and for the selective catalytic reaction of NO$_x$ (CuSO$_4$ acting as the catalyst) [12]. The copper system is attractive because CuO readily reacts with SO$_2$ and O$_2$ to form its sulfate, and its regeneration by reduction is much easier than those of any other metal sulfates [12,13]. Similar processes using CuO–Al$_2$O$_3$–TiO$_2$ [14], CuO–SiO$_2$–TiO$_2$ [15], Cr$_2$O3–TiO$_2$, Pt/ZSM-5 [16], and V$_2$O$_5$–MoO$_3$–TiO$_2$ [17] catalysts have been also pro-

posed. The largest problem of the catalyst for commercial use is the high regeneration cost of sulfates to oxides.

The simultaneous removal of NO$_x$ and SO$_x$ by irradiation with an electron beam has been proposed [18,19]. Flue gas is cleaned by a dust collector and cooled to about 70°C by a water spray and then the gas, mixed with sufficient amount of NH$_3$, is irradiated by an electron beam. SO$_2$ and NO are oxidized immediately by the activated oxygen species through irradiation to react with NH$_3$, producing ammonium salts. This method is attractive because of its high efficiency for the removal of both SO$_2$ and NO$_x$ and there is no need for a catalyst. Its low reliability and high capital cost are the problems.

3. SO$_x$ and NO$_x$ removal over carbon

The Bergbau Forschung Co. in Germany (now DMT) proposed a simultaneous SO$_2$ and NO$_x$ removal process by using a two-stage moving bed using activated coke [20]. NO$_x$ is reduced with NH$_3$ over activated coke at the first bed and SO$_2$ is removed at the second bed. The flue gas is first desulfurized and then denitrified. Sulfur dioxide, adsorbed in H$_2$SO$_4$, on the activated coke is recovered as SO$_2$ at 400–450°C by regenerating the adsorption abilities of the activated coke which is then recycled to the first bed. The process has been developed and commercialized in Germany and Japan (Mitsui Mining and Sumito Heavy Machinary, Japan), respectively [21–23].

It has been revealed that activated carbon is very effective for dry desulfurization of flue gas for long period. Tamura et al. [24] proposed the mechanism of SO$_2$ adsorption onto activated carbon in the presence of oxygen and water vapor at about 100°C as follows:

$$SO_2 \rightarrow SO_{2\,(ad)} \tag{1}$$

$$H_2O \rightarrow H_2O_{(ad0} \tag{2}$$

$$1/2O_2 \rightarrow O_{(ad)} \tag{3}$$

$$SO_{2(ad)} + O_{(ad)} \rightarrow SO_{3(ad)} \tag{4}$$

$$SO_{3(ad)} + H_2O_{(ad)} \rightarrow H_2SO_{4(ad)} \tag{5}$$

$$H_2SO_{4(ad)} + nH_2O_{(ad)} \rightarrow H_2SO_4 \cdot nH_2O_{(ad)} \tag{6}$$

First of all, adsorbed SO_2 as H_2SO_4 has to be removed to regenerate the adsorption ability of activated carbon. Yamamoto et al. [25–28], Gupner [29], and Mauvin et al. [30] tried to regenerate its ability by water scrubbing. However, this required large amounts of water since H_2SO_4, produced in the pores of activated carbon, must be extracted. Furthermore, strongly adsorbed H_2SO_4 tends to remain even after washing and prohibits complete regeneration.

Heating regeneration of carbon adsorbents has been widely examined [20–23,31–35]. Adsorbed SO_2 as H_2SO_4 is recovered as concentrated SO_2 by the following reaction:

$$[C]-H_2SO_4 \cdot nH_2O \rightarrow [C]-SO_3 + (n+1)H_2O \tag{7}$$

$$[C]-SO_3 \rightarrow SO_2 + [CO] \tag{8}$$

$$[CO] + SO_3 \rightarrow SO_2 + CO_2 \tag{9}$$

where [] means the surface of the activated carbon.

As shown above, H_2SO_4 is dehydrated and reduced to SO_2 by consuming carbon. Carbon loss is not negligible in the heating regeneration because of mechanical loss due to reduced strength as well as chemical loss. Thus, the cost of the carbon adsorbent is the key issue to commercialize this system.

Gomi et al. [34] has found that co-existence of a small amount of NH_3 in the flue gas is effective to decrease carbon loss. They proposed the regeneration reaction as follows:

$$(NH_4)_2SO_4 \rightarrow NH_4HSO_4 + NH_3 \tag{10}$$

$$NH_4HSO_4 \rightarrow NH_3 + SO_3 + H_2O \tag{11}$$

$$2/3NH_3 + SO_3 \rightarrow SO_2 + H_2O + 1/3N_2 \tag{12}$$

$$2/3NH_3 + [CO] \rightarrow [C] + H_2O + 1/3N_2 \tag{13}$$

They stated that NH_3 in the flue gas that may have reacted with H_2SO_4 to produce ammonium sulfate decomposed to NH_3 and SO_3 and reduced SO_3 and surface oxygen functionalities, which are produced by the reaction [8]. Therefore, the more NH_3 in the flue gas, the smaller the carbon loss. Plugging of the activated coke bed and the high temperature needed for the complete decomposition and recovery are major disadvantages. Plugging is reported to cause fire because of heat accumulation.

A number of mechanisms have been proposed for the catalytic oxidation of SO_2 on activated carbons. Daytyan [36] has indicated that oxygen reversibly adsorbs on activated carbon and oxidizes SO_2 to SO_3. Siedlewski [37] has shown that free radicals on activated carbon act as active centers for the chemisorption of SO_2. Kitagawa et al. [38] have found that the capacity for SO_2 removal does not depend on the surface area of the activated carbon but on the number of oxygen atoms in activated carbon. Yamamoto et al. [25,26] have also shown that the capacity depends on the preparation conditions and raw materials for activated carbon rather than on the surface area of the activated carbon. Kamino et al. [39,40] estimated that basic surface groups were active sites for the oxidative chemisorption of SO_2. Sano and Odawa [41] have shown that certain impurities in the activated carbon, such as surface nitrogen species, seem to have favorable influences on activity for SO_2 removal. Davini [42] has shown that basic oxygen functionalities on activated carbon are the chemisorption sites for SO_2 and that they play an important role to govern the SO_2 adsorption capacity. Mochida et al. [43] have reported that polyacrylonitrile-based activated carbon fibers (PAN–ACF) exhibited very a high SO_2 adsorption rates and large capacities among the carbon adsorbents in the co-existence of oxygen and water vapor at 100–180°C. Surface functional groups on PAN–ACF as well as its unique pore distribution may be responsible for its SO_2 adsorption capacity.

4. Preparation and characteristics of activated carbon fiber

Activated carbon is a superior adsorbent and there is support for its use as a catalyst because of its large number of micropores and high surface area. Therefore, it has been widely applied for gas purification and separation, deodorization, and water purification, although, a powder or granular form of activated carbon has traditionally been applied. Its fiber form, activated carbon fibers, (ACFs) has been produced commercially and applied recently. Some properties of ACFs are summarized in Table 3 [44].

ACFs have higher adsorption and desorption rates and a larger adsorption ability compared with granular activated carbons. In spite of such a high performance, the application of ACFs has been limited because of is high cost for preparation. Preparation schemes of PAN-based and pitch-based ACFs are shown in Fig. 3 [45,46]. Cellulose and phenolic resin based ACFs are also available.

Synthetic organic fiber is expensive compared to granular activated carbon. Furthermore, the slow stabilization step increases the cost of pitch-based ACFs.

The unique property of ACFs may be their pore size distribution. Fig. 4 illustrates pore structures of ACFs and granular activated carbon. The granular activated carbon has macropores (pore diameter is $10^2 - 10^5$ nm), transi-

168

Table 3
Some properties of ACF

Precursor	PAN	Cellulose	Phenol resin
Diameter of fiber (μ)	6–11	15–18	9–11
Total surface area (m^2/g)	700–1200	1000–1500	1500–2000
Outer surface area (m^2/g)	1.5–2.0	0.2–0.7	–
Pore diameter (Å)	20–30	10–16	15–30
Elemental analysis (wt%)			
C	88–91	92.0–94.5	–
H	0.7–0.9	0.6–0.8	–
N	2.5–5.5	–	–
O	2.5–8.8	2.9–3.5	–

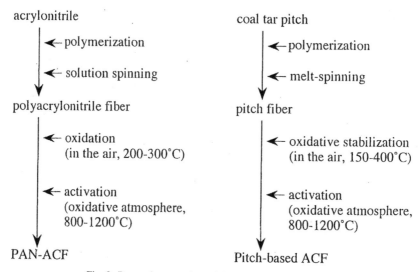

Fig. 3. Preparation procedure of PAN-and Pitch-based ACF.

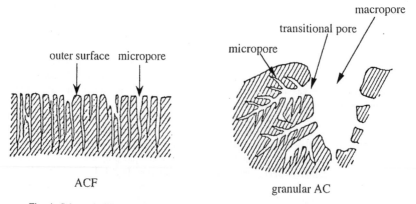

Fig. 4. Schematic illustration of the pore structure of ACF and granular active carbon.

tional pores ($2-10^2$ nm), and micropores (<2 nm). As shown in Fig. 4, an adsorbate molecule must pass through the macro and transitional pores to the adsorption site in the micropores. In contrast, micropores in ACFs are believed to open directly to the outer surface. Therefore, adsorption rates over granular activated carbon and ACFs strongly depend on the diffusion rate of adsorbing gas molecules in the pores [47].

Mochida et al. [43] reported that PAN–ACF showed a very high SO_2 adsorption rate and large capacity. However, the reasons for its enhanced adsorption capacity and durability in the repeated runs were not yet clarified. Basic research to clarify the reasons are important not only for the commercialization of dry SO_2 removal but also for the elucidation of the roles of the surface functionalities of the carbon in the SO_2 adsorption on the activated carbons.

5. Activity of ACFs for SO$_x$ removal at room temperature

Fig. 5 shows de-SO$_x$ characteristics of as-received ACFs prepared from different precursors. All ACFs used had almost same specific surface area of about 1000 m^2/g. The reactions were performed under the same conditions, inlet SO_2 concentration of 1000 ppm, water content of 10% (over-saturated), and reaction temperature of 25°C. The contact time (min) in weight of ACF (g) per gas flow rate (ml/min, W/F) was set to 0.005 g min/ml. C/C_0 in the figure indicates the de-SO_2 ratio (outlet SO_2 concentration/ inlet SO_2 concentration). The figure shows the de-SO$_x$ activities of these ACFs in the following decreasing order: pitch-, PAN-, cellulose-, and phenol-based ACFs. The experiments also revealed that the de-SO$_x$ ratio was kept stable at least for 15 h.

As compared with ACFs, a granular activated carbon made from coconut husk (900 m^2/g) had a relatively lower activity and was proved to be unsuitable for this method. From the results of ion chromatography for the products in the trap at the outlet (condensed water), only sulfate ions (SO_4^{2-}) were detected, indicating that no sulfurous acid was released.

Fig. 6 shows de-SO$_x$ activity of ACFs heat-treated in the temperature range from 600 to 900°C in nitrogen having a specific surface area of 1060 m^2/g. The reaction conditions were nearly the same as for Fig. 5, but the contact time was reduced by half to 0.0025 g min/ml. The figure definitely indicates that heat-treatment improved the efficiency of ACF for the de-SO$_x$ reaction.

Fig. 7 shows activities of six ACFs of different specific surface areas (480~2150 m^2/g) as a function of heat-treatment temperature in an atmosphere of nitrogen. In this figure the de-SO$_x$ ratio after reaching the steady state is shown. The figure reveals that all de-SO$_x$ conversions increase proportionally to the heat-treatment temperature and that the de-SO$_x$ ratio becomes higher with increasing specific surface area. The ACFs with the maximum

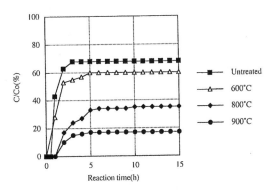

Fig. 6. Behavior of the de-SO$_x$ reaction using heat treated ACF with specific surface area of 1060 m^2/g, SO_2: 1000 ppm, O_2: 5 vol%, H_2O: 10 vol%, W/F: 0.0025 g min/ml, T: 30°C.

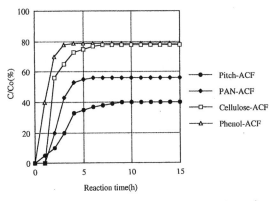

Fig. 5. Breakthrough profiles of de-SO$_x$ reactions using various kinds of ACFs. SO_2: 1000 ppm, O_2: 5 vol%, H_2O: 10 vol%, W/F: 0.005 g min/ml, T: 25°C.

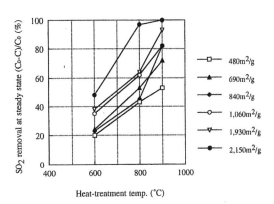

Fig. 7. Correlation between the heat-treatment temperature of ACF and the de-SO$_x$ ratio.

specific surface area of 2150 m^2/g showed 100% de-SO$_x$ at this contact time after the heat-treatment at 900°C.

6. Activity of ACFs for NO$_x$ removal, reduction and oxidation

The authors found that the ammonia reduction of NO$_x$ at 100~150°C successfully proceeded over activated coke [48]. De-NO$_x$ activity of the coke remarkably increased by the repeated use of the activated coke for de-NO$_x$ and de-SO$_x$ in the Bergbau Forschung process. The activity increase appeared to reflect the increase in the oxygen and nitrogen functional groups on the surface. The authors also reported activity of PAN-based ACF (PAN–ACF); activation of PAN–ACF using sulfuric acid provided much higher de-NO$_x$ activity by increasing oxygen functional groups on the surface. The authors work showed de-NO$_x$ activity with NH$_3$ at room temperature over the PAN–ACF since adsorbed NH$_3$ on the ACF was found to reduce NO [49]. However, inhibition by humidity becomes a problem especially when the humidity exceeds 60% [49].

De-NO$_x$ activity at room temperature was remarkably improved by the decomposition of surface oxygen functional groups as observed in SO$_2$ removal [50,51]. Increasing oxygen functional groups on the PAN–ACF surface was effective in increasing the de-NO$_x$ activity at 100~150°C. In a marked contrast the heat-treatment of the pitch-based ACF of very large surface area provided much higher de-NO$_x$ activity at room temperature.

The amount of NO$_2$ evolved by the oxidation of NO on the surface of the ACFs is illustrated in Fig. 8. The concentration of NO influences the oxidation rate because the oxidation of NO in the vapor phase is second order in NO. Therefore, when the NO concentration is lower, the oxidation rate of NO decreases sharply. A major part of low concentration NO is liberated to the atmosphere without oxidation and causes photochemical smog and acid rain through slow oxidation in the atmosphere.

Breakthrough of NO is not observed at the initial stage

of reaction since NO absorbs on the ACF, neither CO nor CO$_2$ being observable at the outlet. After some period, breakthrough of NO and elution of NO$_2$ start. The NO$_2$ concentration at the outlet gradually increases and reaches a steady concentration, while the NO concentration increases initially to the maximum, and then decreases significantly to obtain the steady state conversion. Although, this tendency is recognized over any carbon surface, the maximum decreases with a decrease in the oxygen concentration, suggesting that the balance of adsorption and oxidation with different reaction orders in NO concentration provides such a unique maximum.

7. Mechanism of SO$_x$ removal over carbon

It is well known that the activated surface of carbon adsorbs SO$_2$ as described above. The adsorbed SO$_2$ is oxidized to SO$_3$, which becomes sulfuric acid by hydration. Therefore, the procedures for the continuous removal of SO$_2$ through its oxidative adsorption depend on how to recover sulfuric acid and how to regenerate the adsorption ability of the carbon surface. Recovery of H$_2$SO$_4$ and regeneration of ACF are performed by two ways as illustrated in Fig. 9: (1) reduction of SO$_3$ to SO$_2$ over the carbon at 400°C, (2) hydration and elution as sulfuric acid.

Reduction of SO$_3$ leaves one oxygen over the carbon surface that produces the oxygen functional group on the surface. Therefore, a half or one carbon atom is lost by the recovery of every SO$_2$ molecule. Generally, such oxygen functional groups decompose into CO$_2$ or CO by the heat-treatment. Carbon dioxide is generated at around 400°C, while CO at around 800°C. The adsorption and oxidizing activities of the carbon surface were found to decrease with the amount of oxygen functionality produced on the surface. Hence, decomposition of the oxygen functionality is required for the regeneration of the ad-

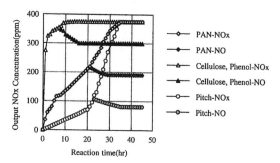

Fig. 8. NO and NO$_x$ concentrations after NO oxidation reaction over various kinds of ACFs. NO: 380 ppm, O$_2$: 4 vol%, H$_2$O: 3 vol%, W/F: 0.03 g min/ml, T: 25°C.

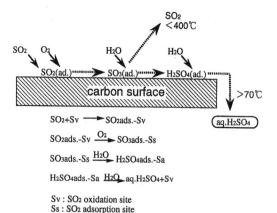

Fig. 9. Two de-SO$_x$ mechanisms.

sorption ability. The carboxyl and lactone groups are believed to produce CO_2, while ketone, aldehyde, and phenols produce CO. Which carbon on the surface is responsible for such functional groups when heated is yet known. Activated carbons of higher surface area tend to generate larger amount of CO. Therefore, the temperature to regenerate adsorption activity depends on the nature of the carbon.

Sulfuric acid produced on the surface must be sufficiently hydrated for its recovery. In this recovery method, there is basically no loss of carbon. In such a system, extraction efficiency by liquid water is low for the hydration of H_2SO_4 in the pores, while the condensation of the steam in the pores is effective for hydration, depending strongly on the relative humidity. Hence, a lower temperature is favorable. ACF is more suitable for the recovery of sulfuric acid compared to granular active carbon because the diffusion of sulfuric acid is much easier in the pores of the fiber.

The active site for oxidizing SO_2 on the carbon surface was believed one of the oxygen functional groups. Among the oxygen functional groups, a basic and/or oxidizing one was considered to be active for the oxidative adsorption of SO_2. However, desulfurization reactivity was remarkably enhanced by heat-treatment to remove almost all surface oxygen functional groups from the carbon surface. Fig. 10 illustrates the schematic structure of the carbon surface. The surface defects induced by the decomposition of the oxygen functional groups appear to be responsible for SO_2 adsorption and oxygen adsorption and activation. The surface dangling bond may be formed through the decomposition of an oxygen functional group. However, there is no evidence that the free radical detected by ESR is an active site. The hexagonal carbon planes carry two prismatic edges of zig-zag and arm-chair types. So far, there is no evidence to identify which edge is responsible for the activity. The benzene-type bond formed by the elimination of the oxygen functional groups may be preferably located on the zig-zag edge. A larger hexagonal plane may stabilize such sites, suggesting higher activity of pitch-based ACF than PAN-based ACF. PAN–ACF can contain nitrogen atoms in the carbon skeleton. Such sites are believed to form adsorption sites for acidic SO_2 and electronically influence the oxidative active site. In addition to the active site, hydrophobicity of the surface may be another important factor for desulfurization activity

because hydration and elution of the sulfuric acid from the active site determines the reaction rate. Both high-temperature treatment of the carbon and a higher degree of graphitization are thus favorable for continuous desulfurization. Such a set of results may provide us a guideline for preparing ACF surfaces of higher activity for SO_2 removal.

8. Mechanism in NO_x removal over ACF

In the past ten years, de-NO_x is reported to proceed on a carbon surface which does not contain any metals or metal oxides. The de-NO_x process can be divided into the following two categories [33]: (a) selective reduction by NH_3 (SRC) and (b) oxidation of NO_x to NO_2. The fomer is very influenced by the reaction temperature.

8.1. Ammonia reduction at room temperature

At room temperature, a reaction proceeds between adsorbed NH_3 and adsorbed NO_2. The latter species is formed on the surface through the oxidation of the vapor phase NO after adsorption. The contribution of gas phase NH_3 through the Rideal mechanism appears small. Nitric oxide is assumed to adsorb on the defect site that is formed through the decomposition of oxygen functional groups on the carbon surface because the adsorption of NO followed by oxidation to NO_2 drastically increases over the ACF with the higher heat-treatment temperature as observed in the oxidative adsorption of SO_2.

The adsorption ability of ACF against ammonia at room temperature does not have a large effect on the activity because NO_3 may be adsorbed on a polar site on the surface and even dissolved in the adsorbed water to participate in the reaction.

Strong inhibition of humidity and promotion of oxygen were recognized. Oxidative adsorption of NO explains the promotion by oxygen in the reaction. The heat-treatment removes the oxygen functional group, reducing the inhibition by humidity. High hydrophobicity of pitch-based ACF due to large carbon layers may explain its higher resistance against humidity especially after the heat-treatment.

8.2. The high temperature de-NO_x

De-NO_x at 100~150°C appears to be governed by the adsorbed ammonia. Hence, the amount of adsorbed ammonia is proportional to the activity in this temperature range. The introduction of acidic oxygen functional groups is very effective in improving the activity of ACF in this range.

The oxidative site of NO may also influence de-NO_x since NO is certainly oxidized to NO_2 before the reduction even in this temperature range. However, the adsorption of NO sharply decreases at higher temperatures. Complexes

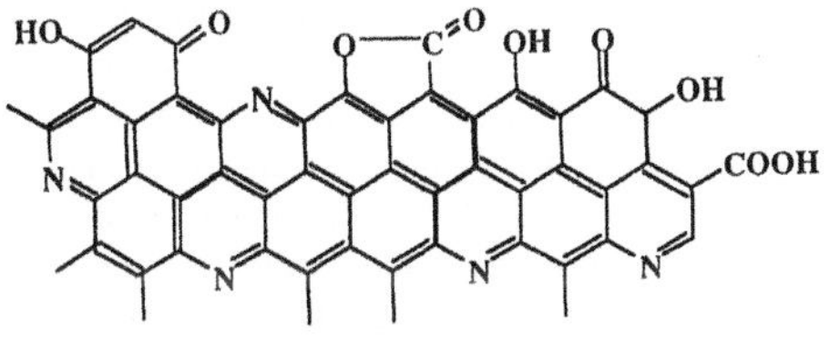

Fig. 10. Surface functional groups of PAN-based ACF.

of the type NO–O_2 on the surface may react with adsorbed ammonia. Hence, the reaction order in NO is first order. Inhibition by H_2O is not marked since its adsorption on the carbon surface is much more restricted in this temperature range.

The reaction below 100°C may follow the transient form of the above two mechanisms. Under these conditions, high temperature tends to reduce the reaction rate simply because of less adsorption in dry air. Nitric oxide adsorption and inhibition of H_2O may compete to give the highest activity at around 50~60°C in humide air.

8.3. Oxidation of NO

The reaction of NO over ACF in the presence of O_2 shown in Figs. 8 and 11 is characterized in the initial period as follows. Nitric oxide is adsorbed until its saturation over the surface and breakthrough to increase its outlet concentration. It must be noted that the breakthrough takes place before the saturation of adsorption, indicating a smaller adsorption rate than the gas flow rate. Nitrogen dioxide is found at the outlet after the breakthrough of NO and increases its concentration to a constant value. The concentration of NO shows a maximum and then decreases to a constant value where NO disappearance and NO_2 production are balanced. The adsorbed NO_x species is

Oxidation of NO into NO_2

$$NO + 1/2 O_2 \longrightarrow NO_2$$

Adsorption of NO

$$NO \longrightarrow NOad \longrightarrow NO_2ad$$
$$O_2 \rightleftharpoons 2O$$

Oxidation

NO   NO

$$NOad \longrightarrow NO_2ad.neigh \longrightarrow NO_2 + NOad$$
$$1/2\ O_2$$

$$NOad + 1/2 O_2 \longrightarrow NO_2ad$$

Reaction Oeders:
1st in O_2
< 1st in NO

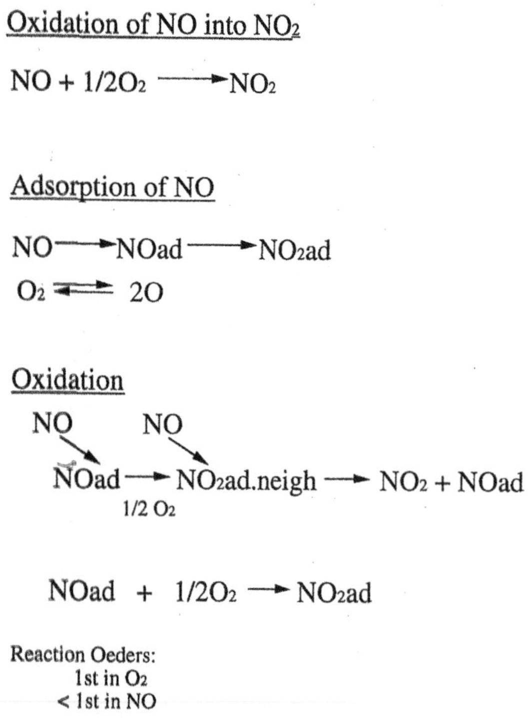

Fig. 11. Reaction mechanism of NO over ACF in the presence of O_2.

initially NO as detected by TPD and then NO_2 becomes dominant. Adsorbed NO is oxidized to NO_2 in the course of time.

Such a unique profile of the NO reaction over the ACF surface suggests some insights into the mechanism for NO oxidation. Nitric oxide in the gas phase adsorbs onto the ACF surface to be oxidized into NO_2, which stays on the surface. Adsorbed NO_2 does not appear to desorb until its coverage reaches a certain level, for example, 60% of the saturated adsorption. The decrease of NO adsorption appears more rapid than NO_2 formation in the initial stage, increasing the outlet concentration of NO. However, NO_2 production becomes rapid along with an increase of adsorbed NO_2. Thus, the outlet NO concentration becomes a maximum just before the rapid increase of NO_2 production, and then starts to decrease rapidly to reach the stationary state. A reaction order larger than unity is suggested for NO_2 formation over the ACF surface. It has been reported that NO oxidation in the gas phase is second order in NO, with N_2O_2 being proposed as the intermediate. An adsorbed NO_2 complex may be an intermediate with NO for the NO oxidation over the ACF surface to produce gaseous NO_2. The stationary conversion of NO into NO_2 appears to be achieved when the coverage of NO_2 adsorption reaches 80%. Adsorption of one NO molecule may liberate one NO_2 by producing another intermediate molecule.

9. SO_x and NO_x removal processes using ACF

9.1. Recently developed process

The removal processes for SO_x and NO_x can be designed using ACF for the environment of busy traffic crossings, parking spaces, and large halls as well as for exhaust gases such as the flue gas from a power plant, catalyst regeneration for fluidized catalytic cracking process (FCC), and ventilated gas from motorway tunnels. Activation of NO is strongly inhibited by SO_2 and humidity, the reactant such as NH_3 can react with SO_2 to produce the sulfate, which may remain on the ACF, covering and plugging its surfaces and pores. In addition, pure SO_2 and H_2SO_4 are desirable resources for recovery. Because of these reasons, separate removal of SO_x and NO_x is designed for the practical application, although, simultaneous removal for the simple process may appear very attractive. Fig. 12 illustrates a basic process flow diagram where NO_x is reduced by a conventional SRC process at 300–400°C, and then SO_2 is removed after the gas is cooled to the designed temperature. The higher temperature up to 100°C is desirable for de-SO_x because the treated gas must be liberated from the chimney.

Another scheme consists of de-SO_x followed by de-NO_x. In this case, de-SO_x is operated below 100°C. Hence,

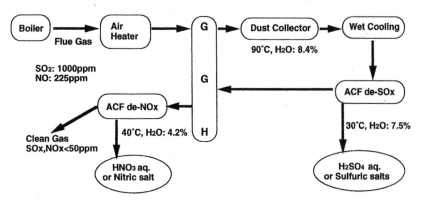

[Basic test conditions of de-SOx]

Design temperature: 30˚C

Inlet SO2 concentration: 1,000ppm

Water content required: 7.5vol%

[Basic test conditions of de-NOx]

Design temperature: 40˚C

Inlet NO concentration: 225ppm

Inlet water content: Saturated at 30˚C (40˚C R.H. 60%)

Fig. 12. De-SO$_x$ and de-NO$_x$ process.

low temperature de-NO$_x$ is required for this process. Optimization of this process is now in progress.

The ACF is moulded with binder into paper which is fabricated into a honey-comb or laminate as shown in Fig. 13. A higher concentration of ACF in the paper form is essential to achieve higher conversion with the lower pressure drop. At present, the gas flows down along the paper surface, H$_2$SO$_4$ eluting concurrently with the gas. Eluted H$_2$SO$_4$ is trapped in the vessel and de-SO$_x$ed gas moves to the chimney or the second reactor. Table 4 summarizes the specifications and performance of de-SO$_x$ tower.

A counter-current reactor may also be designed, where aqueous H$_2$SO$_4$ and gas flow down and up respectively.

9.2. Additional facts for design of de-SO$_x$ and de-NO$_x$ by ACF

9.2.1. Coexistence of NO and SO$_2$

When NO and SO$_2$ coexist, NO oxidation is inhibited by the adsorption of SO$_2$. The de-SO$_x$ reaction is slightly inhibited by coexisting NO, although, NO is scarcely adsorbed. Hence, pure sulfuric acid free from nitric acid is obtained even when both NO and SO$_2$ are contained in the flue gas.

9.2.2. Influence of reaction temperature

The de-SO$_x$ activity decreases very sharply above 60°C at a fixed humidity of flue gas. De-SO$_x$ at 70~100°C is expected to be developed.

Impregnation of ACF by silica, ferric sulfate, and magnesium chloride to around 10% hardly reduces the de-SO$_x$ activity. A cycle of removal of SO$_x$ at 70°C and regeneration of adsorption ability at 30°C was reported to work.

Dust in flue gas is captured by the ACF bed and proved not to inhibit the De-SO$_x$.

10. Forming of ACF

ACF of either toe or mat form can be used for the de-SO$_x$ or de-NO$_x$ processes. However, the density of ACF tends to be low in the packed bed. To increase the efficiency of ACF per volume, high density forming of ACF is very critical. There are two approaches; (a) moulding of ACF with the smallest amount of binder, (b) binderless moulding of stabilized CF followed by carbonization and activation.

The first approach must select the best binder according to the conditions of application. Some plastics can be used for low temperature applications. Very facile preparation, flexible shaping and high density are advantages, while the

Fig. 13. The honeycomb filter.

low softening point limits the temperature of application. The C–C composite can be used at higher temperatures while tedious preparation and brittleness are disadvantages. The density of ACF is also limited.

The second procedure allows a high density of ACF. However, activation after shaping limits the mass production of an uniform shape.

11. Future prospects of ACF

De-SO$_x$ and de-NO$_x$ using ACF have been developed for commercialization in the near future. Nevertheless, there is still room for further improvement of performance. The activity of ACF and the forming and process designs are of value for continued research. As for ACF preparation for high activity, the selection of precursors in terms of better yield, controlled surface elements, graphitizability, spinning into thinner fibers, efficient activation for the higher surface area, introduction of more surface oxygen functional groups at a higher yield, and post-treatment to control the porosity, surface functional groups and graphitization extent are all keys for further research. A catalyst can be supported on ACF to introduce other catalytic functions.

As for forming, a higher density of ACF packing is the important target. The shape and size of the fiber are also of value for research to achieve more compact packing. Thinner paper can moderate the pressure drop in the packed bed.

Process design is concerned with how to combine the de-SO$_x$ and de-NO$_x$ processes and how to enhance the

Table 4
Specifications of de-SO$_x$ tower

Actual scale standard condition	
Amount of gas to be treated	1×10^6 Nm^3/h
Composition of flue gas	
SO$_2$ concentration	1000 ppm
H$_2$O concentration	8.4 vol%
Specifications	
Amount of ACF to be filled	2.5×10^{-3} g min/cc $\rightarrow$ 41.75t
	Bulk density = 0.1 t/m^3 $\rightarrow$ 417.5 m^3
Shape of ACF	Straight plate or corrugated plate
Gas flow rate	1 m/s
Dimension of reactor	$16.7 \times 16.7 \times 1.5$
Design temperature	30°C (saturated water content)
Amount of water to be supplied	None (or little)
Necessary attached equipment	Storage tank for sulfuric acid which is attached to the reactor, various kinds of pumps, water pumps
Achievable de-SO$_x$ ratio	95%
Amount of sulfuric acid produced[a]	4.17 t/h as 100% H$_2$SO$_4$ (concentration of diluted sulfuric acid produced 11.0 wt%)

[a] Calculated from the reaction temperature and amount of water to be supplied. Further concentration by re-circulation is under consideration.

contact of SO_2, NO, O_2 and H_2O with the ACF surface. A thinner bounding layer is reported favorable under rapid gas flow. Interdisciplinary co-operation and progressive ideas are still wanted.

Acknowledgements

This research was partly supported by the New Energy and Industrial Technology Development Organization (NEDO) through the Center Coal Utilization, Japan (CCUJ).

References

[1] Kudo H. Petrotech 1993;16:19–21.
[2] Ando J. Sangyou-kougai 1992;28:230–3.
[3] Hendricks CN. In: NAPAP interim assessment, Washington DC: NAPAP, 1987, pp. 150–5, Executive Summary.
[4] Ando J. Sangyou-kougai 1987;23:7–9.
[5] Ando J. Shokubai 1989;31:548–50.
[6] Kitagawa T. Taikiosen-gakkaishi 1978;13:2–5.
[7] Ando J. In: Esett 91, UN and ENEA, Milano, Italy, October 1991, Abstracts.
[8] Moller MJ. Environ Prog 1986;5:171–5.
[9] Jozewicz W, Rochelle GT. Environ Prog 1986;5:219–24.
[10] Martinez JC, Izquierdo JF, Cunill F, Tejero J, Querol J. Ind Eng Chem Res 1991;30:2143–50.
[11] Ho CS, Shin SM. Ind Eng Chem Res 1992;31:1130–3.
[12] Lowell PS, Schwitzgebel K, Parsons TB, Sladek KJ. Ind Eng Chem Process Des Dev 1971;10:384–8.
[13] Uysal BZ, Aksakin J, Yucel H. Ind Eng Chem Res 1988;24:434–6.
[14] Kasaoka S, Sasaoka E, Funahara M, Nakashima S. J Fuel Soc Jpn 1982;61:126–30.
[15] Kasaoka S, Sasaoka E, Nakashima S. J Fuel Soc Jpn 1984;63:54–60.
[16] Kasaoka S, Sasaoka E, Iwasaki H. J Fuel Soc Jpn 1989;68:45–7.
[17] Teraoka Y, Shimane K, Yamazoe N. Chem Lett 1987;1987:2047–51.
[18] Matsing H. In: Prigogine I, Rice SA, editors, Advanced in chemical physics, Vol. LXXX, New York: John Wiley and Sons, Inc, 1991, pp. 315–20.
[19] Kawamura K. Kagaku-kougaku 1989;53:820–3.
[20] Knoblauch K, Richiter E, Juntgen H. Fuel 1981;60:832–6.
[21] Komatsubara Y, Shiraishi I, Yano M, Ida S. J Fuel Soc Jpn 1985;64:225–31.
[22] Komatsubara Y, Tsuji K, Ida S. J Fuel Soc Jpn 1985;64:339–41.
[23] Komatsubara Y, Tsuji K, Shiraishi I, Ida S, Mochida I. J Fuel Soc Jpn 1985;64:840–5.
[24] Tamura Y, Hishinuma Y, Hisamura T, Mori A. Hitachi Hyouron 1968;1968:78–82.
[25] Yamamoto K, Seki M. Kogyo Kagaku Zasshi 1971;74:78–83.
[26] Yamamoto K, Kaneko K, Seki M. Kogyo kagaku zasshi 1971;74:84–8.
[27] Yamamoto K, Seki M, Kawazoe K. Nippon Kagaku Kaishi 1972;1972:1046–50.
[28] Sugiyama I, Kawazoe K, Yamamoto K, Seki M. Nippon Kagaku Kaishi 1972;1972:1052–8.
[29] Gupner O. VDI-Berichte Nr 1970;149:127–31.
[30] Mauvin PG, Jonakin J. Chem Eng 1970;77:173–7.
[31] Dratwa H, Juntgen H, Peters W. Chem Ing Techn 1967;39:949–55.
[32] Gomi K, Komuro T, Arashi N, Hishinuma Y, Nozawa S, Obata T. J Fuel Soc Jpn 1984;63:275–9.
[33] Komuro T, Gomi K, Arashi N, Hishinuma Y, Kanda O, Kuroda H. J Fuel Soc Jpn 1985;64:264–70.
[34] Gomi K, Komuro T, Arashi N, Hishinuma Y, Kanda O, Kuroda H. J Fuel Soc Jpn 1985;64:760–3.
[35] Komuro T, Gomi K, Arashi N, Hisanuma Y, Kanda O, Nishimura T, Kuroda H. J Fuel Soc Jpn 1986;65:204–10.
[36] Daytyan OK. Zh Fiz Khim 1961;35:992–8.
[37] Siedlewski J. Int Chem Eng 1965;5:608–13.
[38] Kitagawa H, Yuki N, Sanada Y, Watari S, Honda H. Kogyo Kagaku Zasshi 1969;72:2260–5.
[39] Kamino Y, Onitsuka S, Yasuda K. Bull Jpn Petrol Inst 1972;14:141–6.
[40] Kamino Y, Onitsuka S, Yasuda K. Bull Jpn Petrol Inst 1972;14:147–51.
[41] Sano H, Ogawa H. Sangyo Kogai 1974;10:2245–50.
[42] Davini P. Carbon 1990;28:565–9.
[43] Mochida I, Masumura Y, Hirayama T, Fujitsu H, Kawano S, Goto K. Nippon Kagaku Kaishi 1991;1991:269–73.
[44] Ishizaki N. Kagaku to Kogyo 1985;59:171–5.
[45] Okuda K, Fujimaki Y. In: Inagaki M, editor, Kaitei tanso zairyo nyumon, Tokyo: Tanso Zairyo Gakkai, 1984, pp. 199–205.
[46] Shindo N, Tai K, Matsumura Y. Chem Eng 1987;10:28–33.
[47] Ishizaki N. Chem Eng 1984;7:24–30.
[48] Mochida I, Ogaki M, Fujitsu H, Komatsubara Y, Ida S. Nippon Kagaku Kaishi 1985;4:680–4.
[49] Mochida I, Kawano S, Kisamori I, Hironaka H, Matsumura Y, Yoshikawa M. Energy and Fuels 1994;8:1341–4.
[50] Mochida I, Kawano S, Kisamori S, Fujitsu H, Maeda T. Carbon 1994;32:175–89.
[51] Mochida I, Kawano S, Yatsunami S, Hironaka M, Matsumura Y, Yoshikawa M. Energy and Fuels 1995;9:659–64.

(E) Theoretical

CHANGES IN PSD OF PROGRESSIVELY ACTIVATED CARBONS OBTAINED FROM THEIR SUPERCRITICAL METHANE ISOTHERMS

K. A. SOSIN,* D. F. QUINN and J. A. F. MACDONALD
Department of Chemistry and Chemical Engineering, Royal Military College of Canada, Kingston, Ontario K7K 5L0, Canada

(*Received 15 August 1995; accepted in revised form 12 April 1996*)

Abstract—When relatively high pressure methane isotherms at room temperature are measured on carbons, differences in the overall uptake are very obvious. They are believed to be due to differences in the total pore volume and, also, the pore size distributions (PSDs) of these carbons. A recently developed [1] method for determining a PSD) of a carbon adsorbent based on the high pressure methane isotherm was used to analyze two series of PVDC carbons, progressively activated, one with steam, and the other with CO_2.

Notable similarities in the properties of both series were observed. Methane uptakes (at 600 psi) per unit mass rose continuously from ~90 mg CH_4/g in unactivated samples to ~190 mg CH_4/g at 65% weight loss. Considered on the "per unit mass of the original sample" basis, the uptakes exhibited a maximum of ~98 mg CH_4/1 g original sample, at about 15% weight loss, after which they decreased steadily down to ~63 mg CH_4/1 g original sample, at 65% weight loss. This would suggest that, in samples activated beyond 15% weight loss, the extent of changes to the porous structure is outweighed by the loss of adsorbent material during activation.

The size of the average micropore also increased during activation, from an initial value of ~6.5 to ~14 Å at 65% weight loss. Interestingly, at 15% weight loss it was found to be ~8 Å, which is close to the optimum determined by Matranga *et al.* [2]. The analysis also suggests that all the changes were due to alterations to the existing pores and that no new pores were created during the activation. Copyright © 1996 Elsevier Science Ltd

Key Words—Steam activation, CO_2 activation, pore size distribution.

1. INTRODUCTION

It is widely believed that physical adsorption in porous solids could markedly increase the capabilities of natural gas storage systems. This belief has been the driving force behind research efforts in the adsorption of methane (serving as a standard substitute for natural gas) and in the development of porous materials best suited for this task. One class of materials considered for this purpose are active carbons. A study by Matranga *et al.* [2] led to the realization that the optimal carbon pore with respect to methane storage should have a width of 11.2 Å if measured between the planes determined by the centers of carbon atoms constituting two adjacent parallel pore walls of ~7.8 Å according to the convention introduced by Everett and Powl [3].

In general, the purpose of the activation process is to increase the pore volume and to enlarge the pores created during the carbonization process by the use of various chemical and/or physical means [4]. It may involve a number of process parameters (temperature, time, gas flow rates, etc.) which should be optimized to produce a desired material. This, in turn, requires effective methods of characterization of carbons. Among those methods, the determination of the pore size distribution (PSD) offers an intuitive way of visualization and understanding these materials.

2. METHOD

When relatively high pressure methane isotherms at room temperature are measured on carbons, although all of them can classify as type 1 according to the Brunauer classification [5], there are very noticeable differences in the overall shape of isotherms from different carbons (Fig. 1). These are caused partially by differences in the overall micropore volume, but additionally and importantly are due to different PSDs of the carbons. Recently, a procedure was developed to analyze the high pressure, high temperature methane isotherm and provide a PSD for the carbon adsorbent [1]. In this work this procedure was used as a means to monitor changes in PSDs of progressively activated carbons. Since the method has been already published, we give here merely an outline.

It is assumed that pores in an active carbon are infinite parallel-wall slits so that the pore wall separation is the only inhomogeneity parameter of the structure. Under these conditions, the density v of the adsorbed gas is exclusively a function of the external (bulk) pressure p and the pore wall separa-

---

*Corresponding author.

Reprinted from *Carbon* **34** (**11**), 1335-1341 (1996)

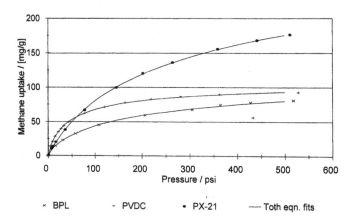

Fig. 1. High pressure methane isotherms @ 298 K of three active carbons having very different pore structures, and their Toth equation fits. (x) BPL; (+) PVDC; (solid circles) PX-21; (line) Toth equation fits.

tion w: $\upsilon(w,p)$. The general adsorption isotherm is given by [6,7]:

$$\theta(p) = \int_0^\infty \vartheta(w,p)f(w)dw \qquad (1)$$

where $\theta(p)$ can be determined experimentally, the function $\upsilon(w,p)$ (the local isotherm) is a characteristic of the carbon–methane interaction in general, and $f(w)$ is an unknown PSD having the following meaning: the differential $f(w)dw$ is the volume of pores contained in a unit mass of the adsorbent and having walls separated by no less than w and no more than $w+dw$.

Formally, eqn (1) provides the means for determining the PSD. The problem, however, is ill posed, i.e. the solution function $f(w)$ is unstable with respect to even small changes in $Q(p)$ [5,8]. We overcome this difficulty by:

1) searching for PSD in the form of a histogram over 11 pore size ranges instead of trying to determine it as a function $f(w)$. The selected ranges are (in Å): 3.8–6, 6–10, 10–15, 15–20, 20–40, 40–60, 60–100, 100–150, 150–200, 200–500 and 500–1000. This is equivalent to reducing the number of unknowns from infinity to 11.

2) taking advantage of the experimentally known fact that high pressure methane isotherms measured at 25°C can be very effectively curve-fit using the Toth equation [9]. Fig. 1 provides examples of such fits. We use the equation in the form adapted by Valanzuela and Myers [10]:

$$\theta(p) = \frac{mp}{(b+p^t)^{1/t}} \qquad (2)$$

where m, b and t are fitting parameters. "Optimizer", a Quattro Pro® and 5.0 for Windows® spreadsheet tool and an IBM compatible PC are used to determine the parameters. After they have been determined based on a limited number of experimental data points, the value $\theta(p)$ can be calculated at any

pressure. For each adsorbent we calculate θ at 100 pressure points: 5, 10,..., 495, 500 psi. It is assumed that there is no adsorption at zero pressure; $\theta(0)=0$.

Using the results of Tan and Gubbins [11] and Cracknell et al. [12], average densities of methane adsorbed in carbon pores of various wall separations were found for the selected gas pressures [1]. Those were later averaged over each of the 11 pore size ranges listed above, using the formula

$$d_{p_k}^{(i)} = \frac{\int_{w_i}^{w^i} J(w,p_k)dw}{w^i - w_i} \qquad (3)$$

where $d_{p_k}^{(i)}$ ($i=1,...,$ 11, $p_k=0, 5,...,$ 500 psi) are average methane densities in pores of i-th pore size range at the pressure p_k, w_i and w^i are, respectively, the lower and the upper limit of the i-th pore size range. The densities are presented in Figs 2 and 3.

It is possible now to transform eqn (1) into a set of 101 linear equations of the form

$$\theta(p_k) = \sum_{i=1}^{11} v^{(i)} d_{p_k}^{(i)} \qquad (4)$$

where $v^{(i)}$ ($i=1,...,$ 11) is the (unknown) volume of i-th pore size range per 1 g of carbon and $Q(p_k)$ ($p_k=0, 5,...,$ 500 psi) is the mass of methane adsorbed per 1 g of carbon at pressure p_k, calculated from the Toth eqn (2). Again, the "Optimizer" is used to solve these equations for volumes $v^{(i)}$.

Also, a knowledge of $v^{(i)}$ allowed us to calculate total internal surface areas, TISA, by extending the concept of the internal micropore surface area [13] over all pore sizes. We used the following formula, which accounted for the fact that each pore size range contained a variety of pore widths [1]:

$$TISA = 2 \sum_{i=1}^{11} \frac{v^{(i)}}{w^i - w_i} \ln \frac{w^i}{w_i} \qquad (5)$$

In the above method, the definition of pore width used by Everett and Powl [3] has been applied. Thus

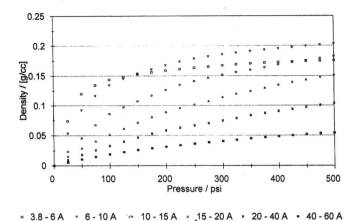

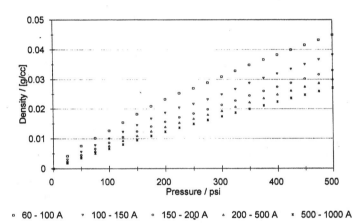

Fig. 2. Average densities of methane adsorbed in parallel-wall carbon pores at 298 K (I). Squares 3.8–6 Å; inv. triangles 6–10 Å; circles 10–15 Å; triangles 15–20 hourglasses 20–40 Å; crossed squares 40–60 Å.

Fig. 3. Average densities of methane adsorbed in parallel-wall carbon pores at 298 K (II). Squares 60–100 Å; inv. triangles 100–150 Å; circles 150–200 Å; triangles 200–500 Å; hourglasses 500–1000 Å.

the optimal pore width as determined by Matranga et al. [2], under this definition should be reduced by 3.4 Å from 11.2 to 7.8 Å.

All calculations have been implemented as a macro in a generic spreadsheet software, Quattro Pro® 5.0 for Windows®.

The method was tested for three carbons having very different PSDs: Calgon BPL, Amoco PX-21 and a PVDC carbon [1]. The results were found to be slightly different than those derived from subcritical argon isotherms using the density functional theory and the Horvath–Kawazoe method, but still comparable. In this study the method was applied to results from two series of Saran carbon samples activated using steam and CO_2.

3. EXPERIMENTAL

Samples for this study were made out of 100% Saran A, Dow Chemical Company homopolymer polyvinylidene chloride (PVDC). Cylindrical pellets (3/4 inch in diameter, approximately 1/4 inch thick) were pressed in a hydraulic press under pressures exceeding 15000 psi, and slowly pyrolyzed under nitrogen to 700°C. The process resulted in highly crosslinked char artifacts which were almost completely microporous. The pellets were degassed at 200°C under vacuum and weighed, also under vacuum. Activation was carried out in a quartz tube furnace. For steam activation, a nitrogen–steam mixture was passed at the rate of 50–100 ml per minute over the carbon heated to 800°C. For CO_2 activation, carbon dioxide was passed over the samples at 850°C. For both activation methods weight losses were found to be approximately proportional to the time of activation. Carbon samples with weight losses reaching $\sim 45\%$ for steam, and 65% for CO_2, were obtained.

Methane isotherms at 25°C were measured up to pressures exceeding 600 psi using a high pressure volumetric apparatus described elsewhere [14]. For each isotherm, a pore size distribution (PSD) was determined. Also, the total micropore volume, the total mesopore volume and the total internal surface area (TISA) were calculated.

Nitrogen isotherms at 77 K were carried out to saturation on all the samples, using a Micromeritics ASAP 2000. Micropore volumes were determined using the Dubinin–Radushkevich equation [15].

4. RESULTS AND DISCUSSION

Both activation methods resulted in a significant increase in methane adsorption at elevated pressures, expressed on the "per mass" basis. Methane uptakes were approximately proportional to weight losses, although some deviation from linearity could be noticed in the range from 35–40% weight loss scale. Figure 4 presents methane uptakes at 600 psi plotted against the degree of activation expressed as the relative weight loss. Triangles represent uptakes in milligrams of methane adsorbed per unit mass of an activated sample. It is evident that 1) both activating agents act in a similar manner, and 2) the activation increased significantly the carbons' methane adsorption capabilities on the mass per mass basis.

Although there is no doubt that activation results in the carbon adsorbing more methane per unit mass, it is also true that the process removes some adsorbent carbon. It is interesting to see to what extent, if any, the loss of the original material is being compensated by the improvement in the adsorbing capabilities of the remaining carbon. This can be done by using the product of methane uptakes in mg/g multiplied by (1 − relative weight loss) [16]. Throughout the paper, these values are referred to as "per 1 g of original sample". They are marked with ovals in Fig. 4. They show a slight increase over the initial stage of the activation process, up to about 15% weight loss, and then start decreasing. At ∼30% weight loss, the values again reach the initial uptake of ∼87 mg/g. This suggests that the changes to the Saran carbon structure, most favorable with respect to methane adsorption, take place in those early stages of the activation processes.

Since methane adsorption at 25°C takes place mainly in micropores and small mesopores, these significant changes in methane uptake must have been caused by notable alterations to the structure of the samples. Indeed, the D–R micropore volumes determined by N_2 adsorption show a systematic increase with weight loss, from about 0.45 ml/g in the untreated carbon, to 1.08 ml/g at about 65% weight loss (Fig. 5). Micropore volumes determined from methane adsorption show a similar trend, although the volumes are slightly greater in the low weight loss region and they do not rise with weight loss as quickly as those based on the adsorption of N_2. This may be explained by the fact that the methane isotherms were measured at 25°C, a temperature much higher than methane's critical temperature, −82°C, which facilitated its entry into narrow micropores. Similar behavior, i.e. N_2 micropore volumes values being smaller, has been observed for CO_2 adsorption at 273 K, and restricted diffusion of N_2 at 77 K into very narrow micropores was suggested as the reason for the difference [17]. This tendency of activation to increase micropore volumes in active carbons has been observed by a number of authors [17,18].

Since methane is adsorbed mainly in micropores, it is interesting to see whether the increase in micropore volume (determined from CH_4 isotherms) caused by activation is large enough to compensate for the loss of carbon during the process. Again, we do this by looking at the products of micropore volumes of activated carbons, multiplied by (1 − relative weight loss). The data for both activation processes are shown in Fig. 6. It can be seen that the micropore volume remains almost constant over the first 30% of weight loss, in spite of the fact that carbon is being removed from the samples. This may be interpreted by taking into account that, while small micropores may be enlarged and their size may remain under 2 nm so that their volume still contributes to the

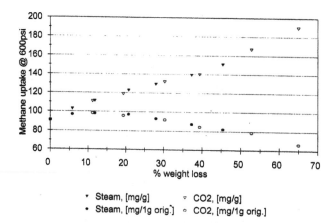

Fig. 4. Methane uptake @ 600 psi and 298 K of Saran carbons activated with steam and carbon dioxide. Inverted triangles – uptakes in [mg/g]; circles – uptakes in [mg/1 g of original sample]; filled symbols – steam activated samples; open symbols – CO_2 activated samples.

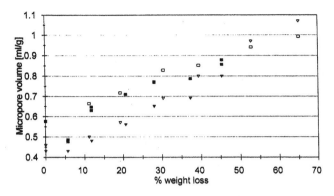

Fig. 5. Micropore volume of Saran carbons activated with steam and carbon dioxide, shown on the "per unit mass" basis. Squares – results from high pressure methane isotherms @ 298 K; inv. triangles – D–R results from N_2 isotherms @ 77 K; filled symbols – steam activated samples; open symbols – CO_2 activated samples.

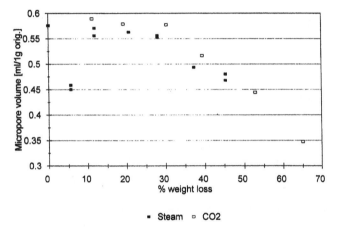

Fig. 6. Micropore volume, calculated from high pressure methane isotherms @ 298 K, of Saran carbons activated with steam (filled squares) and carbon dioxide (open squares), referred to the unit mass of the original sample.

micropore volume, large micropores become mesopores upon enlargement and their volume ceases to be counted as micropore. At the weight loss of ∼ 30% the micropore volume starts decreasing rapidly. This may indicate that, by then, the number of smaller micropores has fallen to the point where their enlargement cannot compensate for the process of larger micropores becoming mesoporous.

Figure 7 illustrates how the average (weighted by volume) micropore size, determined from methane adsorption, increased with the activation. Other authors have made similar observations [19,20]. Again, there is no indication that the average size of micropores depends on the activating agent. It is interesting, however, that the average micropore size at ∼ 15% weight loss (which corresponds with the maximum methane uptake on the "per unit mass of the original sample" basis) is ∼ 8 Å, i.e. very close to the optimal pore size calculated by Matranga et al. [2]. The tendency of enlarging the pores during the activation is also demonstrated in Fig. 8, illustrating

pore size distributions of three Saran carbon samples: one before activation and two others, steam-activated to 28 and 45% weight losses, respectively. From this figure it is evident that activation decreases the volume of very small pores, of the order of 5 Å while the volumes of larger micro- and mesopores increase. This seems consistent with our earlier comments on micropore volume (Fig. 6).

In the model of parallel slit shaped pores, which we assumed for our PSD calculations, the total internal surface area (TISA) considered per unit mass of the original sample, serves as an indicator of whether pores are being created, maintained or destroyed. In this model, no change to the width of a pore would result in a change in its contribution to the TISA, while the creation of a new pore or the destruction of an existing one would have to be accompanied by an increase or, respectively, a decrease in the TISA value. Figure 9 illustrates TISA values plotted against % weight loss. Based on the model, neither activation method produced new

184

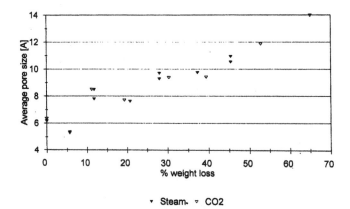

Fig. 7. Width of an average micropore, calculated from high pressure methane isotherms @ 298 K, of Saran carbons activated with steam (filled symbols) and carbon dioxide (open symbols).

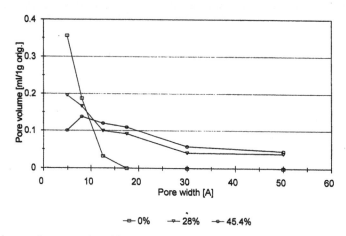

—□— 0%   —▽— 28%   —●— 45.4%

Fig. 8. Pore development in steam activated Saran carbon. Results from high pressure methane isotherms @ 298 K.

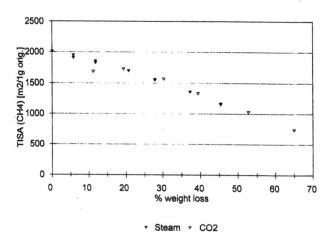

▾ Steam ▿ CO2

Fig. 9. Total internal surface area (*TISA*), from high pressure methane isotherms @ 298 K, of Saran carbons activated with steam (filled symbols) and carbon dioxide (open symbols).

pores, and both resulted in a gradual deterioration of the porous structure of the carbons. As in all the previous cases, the type of activating agent did not have an impact on the process.

The fact that TISA decreases with the progress of activation is not surprising. It can be interpreted by taking into account that, while carbon atoms are being removed from the structure, some inter-pore barriers are being destroyed so that their surface ceases to contribute to TISA. From Fig. 9 it is also evident that the rate at which the TISA decreases is nearly proportional to the weight loss (although the slope changes slightly at about 30% weight loss).

5. CONCLUSIONS

The picture of the activation process emerging from the presented analysis seems to be reasonable and self-consistent. It points to the fact that the effects of steam activation on Saran carbon are very similar to those of carbon dioxide. This, indeed, is to be expected, since the van der Waals diameters of both activating molecules are rather similar (3.2 Å for H_2O and 2.9 Å for CO_2), and both are significantly smaller than the diameter of the methane molecule used as a probe (3.8 Å).

The analysis of TISA indicates that both activating agents act in existing pores only and no new pores are being created in the process. In fact, Fig. 9 suggests that the porous structure of the carbon is being continuously destroyed with the progress of activation. On the other hand, since carbon is being removed from the structure, the size of the pores increases which manifests itself as a steady increase of the micropore volume (Fig. 5) and the average micropore size (Fig. 7) with activation. This leads to a rapid decrease of the volume of very small micropores (~ 5 Å), and a corresponding increase in wider pores, as seen in Fig. 8.

From the point of view of methane storage, where the uptake of methane per unit volume of carbon must be considered and not per unit mass, benefits of activation are rather small, and limited to the early stages of the process (0–15% weight loss). Beyond this point, the gain in the storage capacity of the activated material is too small to compensate for the loss of material itself.

Acknowledgements—The authors wish to thank AGLARG, Atlanta Gas Light Adsorbent Research Group, for their financial assistance for this study. NOTE: Diskettes with the spreadsheet to calculate PSDs of carbon adsorbents based on their high pressure methane isotherms at 298 K are available from K. Sosin upon request.

REFERENCES

1. K. A. Sosin and D. F. Quinn, *Journal of Porous Materials* **1**, 111 (1995).
2. K. R. Matranga, A. L. Myers and E. D. Glandt, *Chemical Engineering Science* **47**, 1569 (1992).
3. D. H Everett and J. C. Powl, *J. Chem. Soc, Faraday Trans.* **72**, 619 (1976).
4. R. C. Bansal, J.-B. Bonnet and F. Stoeckli, *Active Carbon*, Mercel Dekker, New York and Basel (1988) p. 7.
5. S. Brunauer, *The Adsorption of Gases and Liquids.* Princeton University Press, Princeton, NJ (1945) pp. 1–271.
6. B. McEnaney, T. L Mays and P. D. Causton, *Langmuir* **3**, 695 (1987).
7. N. A. Seaton, Walton, J. P. R. B. N. Quirke, *Carbon* **27**, 853 (1989).
8. W. Rudzinski and D. H. Everett, *Adsorption of Gases on Heterogeneous Surfaces.* Academic Press, London (1992) ch. 11.
9. J. Toth. In *Fundamentals of Adsorption I* (Edited by A. L. Myers and G. Belfort), p. 657. Engineering Foundation (1984).
10. D. P. Valanzuela and A. L. Myers, *Adsorption Equilibrium Data Handbook.* Prentice Hall, Englewood Cliffs, NJ (1989) p. 8.
11. Z. Tan and K. E. Gubbins, *Journal of Physical Chemistry* **94**, 6061 (1990).
12. R. F. Cracknell, P. Gordon and K. E. Gubbins, *Journal of Physical Chemistry* **97**, 494 (1993).
13. Ref. [4], p. 121.
14. S. S. Barton, J. R. Dacey and D. F. Quinn. In *Fundamentals of Adsorption I* (Edited by A. L. Myers and G. Belfort), p. 65. Engineering Foundation (1984).
15. M. M. Dubinin and L. V. Radushkevitch, *Proc. Acad. Sci. USSR* **55**, 331 (1947).
16. J. J. Kipling and B. McEnaney. In *Proceedings of the Second Conference on Industrial Carbon and Graphite*, p. 380. London (1966).
17. J. Garrido, A. Linares-Solano, J. M. Martin-Martinez, M. Molina-Sabio, F. Rodriguez-Reinoso and R. Terregrosa, *Langmuir* **3**, 76 (1987).
18. F. Rodriguez-Reinoso. In *Fundamental Issues on Control of Carbon Gasification Reactivity* (Edited by J. Lahaye and P. Ehrenburger), p. 533. Kluwer Academic Publishers, Dordrecht (1991).
19. H. F. Stoeckli, L. Ballerini and S. De Bernardini, *Carbon* **27**, 501 (1989).
20. F. Stoeckli, T. A. Centeno, J.-B. Dennet, N. Pusset and E. Papirer, *Fuel* **74**, 1582 (1995).

Activated Carbon Compendium
H. Marsh (Editor)

THEORETICAL STUDY OF THE PERFORMANCE OF ACTIVATED CARBON IN THE PRESENCE OF BINARY VAPOR MIXTURES

N. Vahdat

Chemical Engineering Department, Tuskegee University, Tuskegee, AL 36088, U.S.A.

(*Received* 22 *October* 1996; *accepted in revised form* 29 *April* 1997)

Abstract—The performance of activated carbons in an environment containing two contaminants was evaluated theoretically. A mathematical model developed earlier was used to determine the effect of different parameters on the breakthrough curves of binary mixtures through activated carbon. Due to the difference between the adsorption capacity of carbon for the two components of a binary mixture, one of the compounds travels faster in the carbon bed and has a shorter breakthrough time. The results of this study show that under certain conditions, changing the concentration of the compounds can cause the more strongly adsorbed component to become less strongly adsorbed and travel faster in the bed.

A database containing Langmuir constants for adsorption isotherms of many compounds for eleven different types of activated carbon was generated. The data base can be used to predict breakthrough times of pure or binary mixtures through activated carbons. © 1997 Elsevier Science Ltd

Key Words—A. Activated carbon, D. adsorption properties.

1. INTRODUCTION

Activated carbon is used extensively to remove contaminants from air in respirator cartridges and in air monitoring devices. Air is passed through a bed of activated carbon where the molecules of the contaminants are transferred to the solid phase. This process (known as adsorption) is a surface phenomenon, that is, the contaminant molecules are accumulated on the surface of solid particles. The number of molecules adsorbed per unit area is usually small; therefore, highly porous solids with very large internal area per unit volume such as activated carbon are used.

There are many parameters which influence an adsorption process, and thus affect the performance of activated carbon. They are: type of carbon, properties and concentration of contaminants in air, properties of the carbon bed (cross sectional area and weight of carbon), atmospheric conditions (temperature and relative humidity), and flow rate of air. The performance of activated carbons has been evaluated by many investigators during the last twenty years. Nelson *et al.* studied the effect of concentration, flow rate, different solvent vapors and water vapor on the performance of carbons [1–6]. Other investigators have evaluated the performance of these activated carbons against many pollutants and in different conditions [7–11]. Many attempts have also been made to predict the adsorption capacity and breakthrough time of activated carbon for different pollutants based on theoretical models. Several equations have been used to describe the effect of vapor concentration on adsorption capacity. Wood and Moyer [12] reviewed some of these correlations (Langmuir,

Dubinin/Radushkevich and Freundlich) and concluded that all of them fit experimental data equally well. Vahdat *et al.* [13] used successfully the Langmuir equation to correlate adsorption capacity of several organic vapor/adsorbent pairs. Wood [14] developed a generalized correlation based on the Dubinin/Radushkevich equation which can be used to predict adsorption capacities of activated carbon.

Of the several models that have been proposed for predicting breakthrough time of compounds, the modified Wheeler equation [15] is probably the most widely used. Wood [16,17] developed a generalized correlation for estimating breakthrough time based on the Wheeler equation.

During the last few years multi-component adsorption has been attracting much attention and several papers were published [18–22]. The atmosphere in many workplaces contains more than one contaminant. As a result, users of respirators and air monitors must have the necessary information to estimate the performance of these devices in the presence of vapor mixtures. Yoon *et al.* [20] developed an empirical method to predict the breakthrough curves for binary mixtures, and applied it to the adsorption of acetone *m*-xylene mixture on activated carbons. The results showed good agreement between the experimental and theoretical breakthrough curves; however, the correlation contains several parameters which should be calculated from binary experimental data. Vahdat *et al.* [22] developed a model for predicting breakthrough curves for the adsorption of binary mixtures on a solid adsorbent. In this model, adsorption capacities of the adsorbent for the two components are first computed using the ideal adsorbed solution theory (IAST) developed by Myers and Prausnitz

[23]. The calculated adsorption capacities are then used in expressions similar to the modified Wheeler equation for pure compounds to determine the breakthrough curves for each component. This model uses only data for pure-component adsorption equilibrium to calculate the breakthrough curves. The model was successfully applied to several chemical mixtures/activated carbon systems [22].

In this paper, Vahdat's correlations are used to study the adsorption behavior of activated carbon for binary mixtures. The effect of different parameters on the adsorption capacity and breakthrough curves are investigated.

2. BACKGROUND

Breakthrough curves for the adsorption of binary mixtures on a solid adsorbent are more complicated than the corresponding curves for pure compounds [22]. The complication is mainly due to the difference in the adsorption capacity of the bed for the two compounds. The weakly adsorbed compound 1 is replaced by the more strongly adsorbed compound 2, and consequently it travels faster in the adsorption bed. A typical set of breakthrough curves for a binary mixture usually consists of four zones (Fig. 1). In zone I, compound 1 breaks through. The concentration of the compound in the gas leaving the bed increases with time, and reaches a maximum concentration, C_1^{max}, which is higher than the concentration of the compound in the gas stream entering the bed. In zone II the concentration of the compound remains constant (C_1^{max}). In the third zone the second compound breaks through. The concentration of compound 2 leaving the bed increases, and that of compound 1 decreases with time in this zone. Finally, in zone IV, the gas leaving the solid bed has the same concentration as the gas entering the bed which indicates that the adsorbent is completely saturated with the compounds.

In the model presented by Vahdat et al. [22], a correlation was developed for the breakthrough curves for each zone in Fig. 1. A brief description of

the theoretical model is given here. For more information, readers should refer to ref. [22].

The adsorption capacity of porous adsorbents for pure compounds may be represented empirically by the Langmuir equation [24]

$$W_e = \frac{mkC}{1+kC} \tag{1}$$

Where m and k are Langmuir constants, and C is concentration. Based on the ideal adsorbed solution theory (IAST) [23], the amounts of the components of a binary mixture adsorbed on a solid bed are given by the following equations:

$$\frac{1}{W_{et}} = \frac{X_1}{W_{e1}^\circ} + \frac{1-X_1}{W_{e2}^\circ} \tag{2}$$

$$W_{e1} = W_{et}X_1 \tag{3}$$

$$W_{e2} = W_{et}(1-X_1) \tag{4}$$

Where: W_{e1}°, W_{e2}° = adsorption capacity of pure component 1 and 2, respectively (g g^{-1}),

W_{e1}, W_{e2} = adsorption capacities of the components in a binary mixture (g g^{-1}),

W_{et} = total adsorption capacity (g g^{-1}),

X_1 = mole fraction of component 1 in adsorbed phase.

The mole fraction X_1 is calculated from

$$X_1 = \frac{Py_1}{P_1^\circ} \tag{5}$$

Where y_1 is the mole fraction of component 1 in the gaseous mixture, P is the pressure and P_1° is a parameter (known as pure component pressure of component 1 at spreading pressure of mixture) which is computed by solving the following simultaneous equations

$$\frac{1+k_1P_1^\circ}{1+k_2P_2^\circ} = \exp\left(\frac{m_2}{m_1}\right) \tag{6}$$

$$X_1 + X_2 = 1 \tag{7}$$

where: m_1, k_1, and m_2, k_2 are the Langmuir constants for components 1 and 2, respectively.

The breakthrough curves for adsorption of pure compounds on solid adsorbents can be represented by the modified Wheeler equation [15]

$$t_b = \frac{W_e}{C^\circ Q}\left[W - \frac{\rho_B Q}{k_v}\ln\left(\frac{C^\circ - C}{C}\right)\right] \tag{8}$$

where:

t_b = breakthrough time (minutes),
C = exit concentration (g cm^{-3}),
C° = inlet concentration (g cm^{-3}),
Q = volumetric flow rate (cm^3 min^{-1}),
W = weight of adsorbent (g),
ρ_B = bulk density of packed bed (g cm^{-3}),
k_v = rate coefficient (min^{-1}),
W_e = adsorption capacity (g g^{-1}).

For a binary system, separate equations were devel-

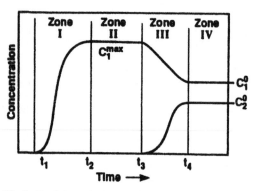

Fig. 1. Breakthrough curves and adsorption zones for a binary mixture.

oped for each zone of Fig. 1. For zone III the breakthrough curve for component 2 (more strongly adsorbed component) is found from eqn (8) by substituting the adsorption capacity of component 2 in the mixture (W_{e2}) for W_e. For component 1, the breakthrough curve is given by the following relation [22]

$$C_1 = C_1^{\max} - (C_1^{\max} - C_1^\circ)\frac{C_2}{C_2^\circ} \qquad (9)$$

where the maximum concentration of component 1, $C_1^{\max}$ is [22]

$$C_1^{\max} = \frac{2[QC_1^\circ t_4 - WW_{e1} - \epsilon VC_1^\circ - QC_1^\circ(t_4 - t_3)/2]}{Q[t_3 + t_4 - t_2 - t_1]} \qquad (10)$$

ϵ is bed porosity which can be measured experimentally or estimated [22], and

t_1 = breakthrough time of component 1 at $C_1/C_1^{\max} = 0.001$,

t_2 = breakthrough time of component 1 at $C_1/C_1^{\max} = 0.999$,

t_3 = breakthrough time of component 2 at $C_1/C_1^\circ = 0.001$,

t_4 = breakthrough time of component 2 at $C_1/C_1^\circ = 0.999$.

In zone II the concentration of component 2 is zero, and component 1 is at a constant concentration of $C_1^{\max}$ (eqn (10)). In zone 1, the breakthrough curve for component 1 is found from the following equation [22]

$$t_{b1} = \frac{W_{e1}^{\max}}{\dfrac{C_1^\circ Q}{W - \dfrac{\rho_B Q}{K_v}\ln\dfrac{C_1^{\max} - C_1}{C_1}} + \dfrac{C_2^\circ Q\rho_B(W_{e1}^{\max} - W_{e1})}{W(W_{e2}\rho_B + \epsilon C_2^\circ)}} \qquad (11)$$

where $W_{e1}^{\max}$ is the adsorption capacity of pure compo-

nent 1 at maximum concentration $C_1^{\max}$. In all the breakthrough calculations (Section 4), a carbon bed with a weight of 50 g and density of 0.45 g cm^{-3} was assumed. The porosity of the bed was estimated from equation 20 in ref. [22].

3. ADSORPTION CAPACITY OF ACTIVATED CARBON FOR VAPORS

Wood [14] collected adsorption capacity data and affinity coefficients for wide varieties of pure vapors and activated carbons, and correlated the data with the Dubinin/Radushkevich equation. Adsorption capacities of carbons for pure vapors can be estimated from this correlation using molar polarizations, liquid densities and vapor pressures. However, Wood's correlation cannot be used for gaseous mixtures, because the Radushkevich equation is not a thermodynamically consistent equation (it is indeterminant at low concentrations). The equation does not reduce to Henry's law as concentration approaches zero, which is a requirement for any thermodynamically consistent isotherm equation. The correlation cannot therefore be used along with the ideal adsorbed solution theory. The simplest adsorption isotherm which is thermodynamically consistent is the Langmuir equation [24]. For this reason, the Langmuir equation was selected to correlate the experimental data for activated carbon/vapor pairs.

Figure 2 shows the Langmuir isotherms of pure compounds for three different sets of constants. For a binary mixture, depending upon the relative values of the Langmuir constants, two cases can be identified. The first case is when $m_i k_i \geq m_j k_j$ and $m_i \geq m_j$, where i and j are the two components. For this case the isotherm for compound i is above the isotherm for compound j (curves A and B or A and C) for the entire range of concentration (if $m_i = m_j$, the curves approach each other at large concentrations). The second case is when $m_i k_i < m_j k_j$ and $m_i > m_j$. For

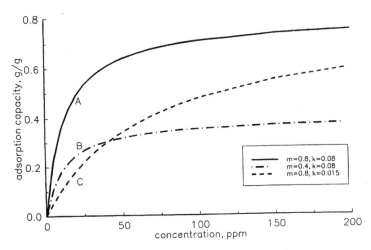

Fig. 2. Adsorption isotherms for pure compounds.

190

Table 1. Langmuir constants $W_e = \dfrac{mkC}{1+kC}$

Compound	Carbon type (refer to Table 2)	Temperature (°C)	m (g g^{-1})	k (ppm^{-1})	R^2	Ref.
Acetone	I	25	0.401	0.000670	0.983	[6]
Acetone	II	25	0.272	0.00295	0.997	[6]
Benzene	I	25	0.389	0.00492	0.995	[6]
Carbon tetrachoride	I	25	0.959	0.00206	1.000	[6]
Dichloromethane	I	25	0.298	0.000716	0.999	[6]
Diethylamine	I	25	0.461	0.00139	0.999	[6]
Hexane	I	25	0.337	0.00924	0.999	[6]
Hexane	II	25	0.357	0.0196	0.999	[6]
Isopropanol	I	25	0.456	0.00185	1.000	[6]
Methylacetate	I	25	0.304	0.00170	0.995	[6]
Methylchoroform	I	25	0.666	0.00443	0.998	[6]
p-xylene	III	25	0.404	0.0419	1.000	[25]
Acetone	III	25	0.309	0.000900	0.991	[25]
Tetrachloroethylene	III	25	0.764	0.0281	1.000	[25]
Butanol	III	25	0.371	0.0232	1.000	[25]
Benzene	III	25	0.372	0.00435	0.998	[25]
Ethylacetate	III	25	0.391	0.00481	0.997	[25]
Carbon tetrachloride	III	25	0.616	0.00670	0.999	[25]
Pyridine	III	25	0.467	0.00847	0.999	[25]
Toluene	V	25	0.447	0.00686	1.000	[26]
o-xylene	V	25.9	0.446	0.00912	1.000	[26]
Nitrobenzene	V	44.4	0.568	0.00154	1.000	[26]
Methanol	V	25	0.394	0.000349	0.998	[26]
Ethylacetate	V	27	0.451	0.00165	1.000	[26]
Acetone	V	25	0.389	0.000570	0.999	[26]
Trichloroethylene	V	25	0.703	0.00265	1.000	[26]
Carbon tetrachloride	V	23	0.805	0.00187	1.000	[26]
Benzene	VIII	26.1	0.438	0.00145	0.999	[26]
Toluene	VIII	25	0.441	0.00629	0.999	[26]
o-xylene	VIII	25.9	0.448	0.00804	1.000	[26]
Methanol	VIII	25	0.396	0.000270	0.997	[26]
Ethylalcetate	VIII	27	0.450	0.00158	1.000	[26]
Acetone	VIII	25	0.391	0.000491	0.999	[26]
Trichloroethylene	VIII	25	0.708	0.00204	1.000	[26]
Carbon tetrachloride	VIII	23	0.805	0.00187	1.000	[26]
Chloroform	XI	25	0.614	0.000802	0.999	[27]
Benzene	XI	25	0.348	0.00182	1.000	[27]
Hexane	XI	25	0.288	0.00211	1.000	[27]
Carbon tetrachloride	XI	25	0.629	0.00203	1.000	[27]
Dioxane	XI	25	0.481	0.00226	0.999	[27]
Tetrachloroethane	XI	25	0.505	0.0285	1.000	[27]
Acetone	XI	25	0.306	0.000398	0.999	[27]
Acetaldehyde	XI	25	0.272	0.000083	0.999	[27]
Methanol	XI	25	0.350	0.000030	1.000	[27]
Acetonitrile	XI	25	0.294	0.000302	0.999	[27]
Ethylacetate	XI	25	0.354	0.00157	0.999	[27]
Dichloromethane	XI	160	0.487	0.000197	0.999	[28]
Dichloromethane	XI	180	0.379	0.000126	0.999	[28]
Dichloromethane	XI	200	0.0828	0.0000273	0.999	[28]
Benzene	XI	160	0.120	0.0652	0.999	[28]
Benzene	XI	180	0.111	0.0165	0.999	[28]
Benzene	XI	200	0.100	0.00933	0.999	[28]
Benzene	XI	220	0.00920	0.00499	0.999	[28]
Toluene	III	25	0.402	0.0170	1.00	[25]
Toluene	III	40	0.378	0.0133	0.998	[25]
Toluene	III	60	0.361	0.00544	0.994	[25]
Trichloroethylene	III	25	0.607	0.857	0.999	[25]
Trichloroethylene	III	40	0.580	0.367	1.00	[25]
Trichloroethylene	III	60	0.568	0.176	0.991	[25]
Benzene	IV	25	0.419	0.00141	0.999	[26]
Toluene	IV	25	0.408	0.00703	1.000	[26]
o-xylene	IV	25	0.416	0.0235	1.00	[26]
Nitrobenzene	IV	25	0.571	0.145	1.00	[26]
Methanol	IV	25	0.364	0.000347	1.00	[26]
Ethanol	IV	25	0.367	0.00136	0.999	[26]
Formic Acid	IV	25	0.383	0.00572	0.999	[26]

Table 1. (*continued*)

Compound	Carbon type (refer to Table 2)	Temperature (°C)	m (g g^{-1})	k (ppm^{-1})	R^2	Ref.
Ethyl Acetate	IV	25	0.417	0.00174	1.000	[26]
Acetone	IV	25	0.054	0.00363	0.999	[26]
MEK	IV	25	0.377	0.00157	1.000	[26]
Chloroform	IV	25	0.468	0.000700	0.999	[26]
Carbon tetrachloride	IV	25	0.746	0.00179	1.00	[26]
Trichloroethylene	IV	25	0.700	0.00197	1.000	[26]
Benzene	VI	25	0.393	0.00157	1.000	[26]
Toluene	VI	25	0.384	0.00873	1.00	[26]
Methanol	VI	25	0.344	0.000446	0.999	[26]
Ethanol	VI	25	0.353	0.00145	1.000	[26]
Trichlorethylene	VI	25	0.650	0.00331	1.000	[26]
Benzene	VII	25	0.124	0.00379	1.000	[26]
Toluene	VII	25	0.681	0.00371	1.000	[26]
o-xylene	VII	25	0.378	0.0238	1.000	[26]
Nitrobenzene	VII	25	0.518	0.140	1.000	[26]
Methanol	VII	25	0.334	0.000317	0.999	[26]
Ethanol	VII	25	0.333	0.00133	0.999	[26]
Acetone	VII	25	0.330	0.000520	0.999	[26]
MEK	VII	25	0.342	0.00149	0.999	[26]
Chloroform	VII	25	0.631	0.000685	0.999	[26]
Carbon tetrachloride	VII	25	0.677	0.00177	1.000	[26]
Trichloroethylene	VII	25	0.625	0.00242	1.000	[26]
Benzene	IX	25	0.318	0.00177	1.000	[26]
Toluene	IX	25	0.316	0.00775	1.000	[26]
Methanol	IX	25	0.285	0.000365	0.999	[26]
Ethylacetate	IX	25	0.284	0.00159	0.999	[26]
Trichlorethylene	IX	25	0.549	0.00145	1.000	[26]
Benzene	IX	25	0.365	0.00176	1.000	[26]
Toluene	X	25	0.316	0.00775	1.000	[26]
Methanol	IX	25	0.285	0.000365	0.999	[26]
Ethylacetate	IX	25	0.284	0.00159	0.999	[26]
Trichloroethylene	IX	25	0.549	0.00145	1.000	[26]
Benzene	X	25	0.365	0.00176	1.000	[26]
Toluene	X	25	0.363	0.00785	1.000	[26]
o-xylene	X	25	0.370	0.0261	1.000	[26]
Nitrobenzene	X	25	0.507	0.160	1.000	[26]
Methanol	X	25	0.325	0.000379	0.999	[26]
Ethanol	X	25	0.326	0.00158	0.999	[26]
Formic acid	X	25	0.510	0.00427	0.999	[26]
Ethylacetate	X	25	0.374	0.00156	1.000	[26]
Acetone	X	25	0.323	0.000599	0.999	[26]
MEK	X	25	0.335	0.00176	1.000	[26]
Chloroform	X	25	0.619	0.000619	1.000	[26]
Carbon tetrachloride	X	25	0.663	0.00202	1.000	[26]
Trichloroethylene	X	25	0.592	0.0131	1.000	[26]

this case the two pure compound isotherms intersect (curves B and C in Fig. 2). The concentration at the point of intersection depends on the relative sizes of the Langmuir constants. This case represents very interesting phenomenon for the breakthrough curves of the binary mixtures, which will be discussed in the next section.

Adsorption capacity data available in the literature were used to calculate Langmuir constants for many compound/activated carbon systems. The results are presented in Table 1. When available, reported values are given in the table. In other cases, the adsorption capacity data were fitted to the Langmuir equation and the optimum values of the constants were computed using the SYSTAT software package. The square of correlation coefficients (R^2) are also given

in Table 1. Eleven types of activated carbon were used in the literature data. Specification of these adsorbents are given in Table 2 at room temperature. The value of the Langmuir constant m varies from about 0.25 to 0.96 g of contaminant/g of carbon. As expected, at higher temperatures, the value of m drops drastically. The variation in parameter k is much higher, it varies by more than four orders of magnitude.

The adsorption isotherms of gaseous mixtures for a solid adsorbent can be estimated from the pure compound isotherms using the ideal adsorbed solution theory (IAST) [23]. This method was successfully applied to the adsorption of several binary mixtures on activated carbon [22]. A computer program was prepared to generate adsorption isotherms

Table 2. Properties of various types of activated carbon

No.	Name	Origin	Density (g cm^{-3})	Surface area (m^2 g^{-1})	Pore volume (cm^3 g^{-1})	Particle mesh size	Manufacturer	Ref.
I	N/A[a]	Coconut	0.38–0.44	1500–1625	0.9–1.0	8–16	North American Carbon Inc.	[6]
II	N/A[a]	Petroleum	0.41–0.45	1400–1500	N/A[a]	10–16	Witco Chemical Co.	[6]
III	BAC	N/A[a]	N/A[a]	1200	N/A[a]	N/A[a]	Union Carbide	[25]
IV	4GS-S	Coal & Coconut	0.43	1170	0.82	N/A[a]	N/A[a]	[26]
V	HC-8	Coconut	0.44	1270	0.70	N/A[a]	N/A[a]	[26]
VI	SX	Coconut	0.41	1090	0.94	N/A[a]	N/A[a]	[26]
VII	Y-20	Coconut	0.45	1098	0.57	N/A[a]	N/A[a]	[26]
VIII	B-CG	Coconut	0.43	1240	0.65	N/A[a]	N/A[a]	[26]
IX	A	Coal	0.42	840	0.97	N/A[a]	N/A[a]	[26]
X	G-BAC	Oil pitch	0.51	1000	0.56	N/A[a]	N/A[a]	[26]
XI	BPL	Coal	N/A[a]	1058	N/A[a]	12–30	N/A[a]	[27]

[a] Not available.

of multicomponent mixtures on carbon. The input data required for this program are the concentrations and the pure compound Langmuir constants which are given in Table 1. A typical set of isotherms for a binary system is shown in Figs 3, 4 and 5. The isotherms for the individual compounds are given in Figs 3 and 4, and the total adsorption capacity as a function of concentration is shown in Fig. 5. Addition of a second component will cause a reduction in the equilibrium adsorption of the first component. The amount of reduction depends on which one of the compounds is more strongly adsorbed by the solid adsorbent. When comparing two compounds, the one which has a higher adsorption capacity (for the same concentration) is more strongly adsorbed by the adsorbent. The important factor, is therefore, W_e/C. The compound with a higher W_e/C will be

adsorbed more strongly. From eqn (1)

$$\frac{W_{ei}}{C} = \frac{m_i k_i}{1 + k_i C} \qquad (12)$$

If compound 2 is more strongly adsorbed, the adsorption capacity of compound 1 will be reduced drastically by adding only a small amount of compound 2 to the system (see Fig. 3), on the other hand, if compound 1 is more strongly adsorbed, addition of compound 2 has little effect on the equilibrium isotherm of the first compound (see Fig. 4). For a binary mixture in which $m_2 k_2 \geq m_1 k_1$ and $m_2 \geq m_1$, the isotherm for compound 2 is above the isotherm for compound 1 (see Fig. 2) for the entire range of concentration. For this case compound 2 is more strongly adsorbed by the solid

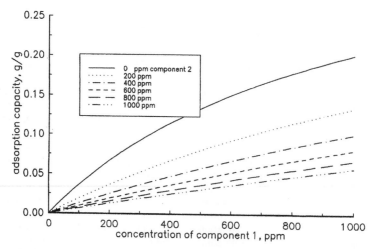

Fig. 3. Adsorption isotherms of component 1 in a binary mixture. Langmuir constants: for component 1, $m_1 = 0.4$ g g^{-1}, $k_1 = 0.001$ ppm^{-1}; for component 2, $m_2 = 0.4$ g g^{-1}, $k_2 = 0.005$ ppm^{-1}.

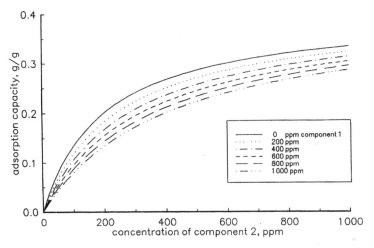

Fig. 4. Adsorption isotherms of component 2 in a binary mixture. Langmuir constants: for component 1, $m_1 = 0.4$ g g^{-1}, $k_1 = 0.001$ ppm^{-1}; for component 2, $m_2 = 0.4$ g g^{-1}, $k_2 = 0.005$ ppm^{-1}.

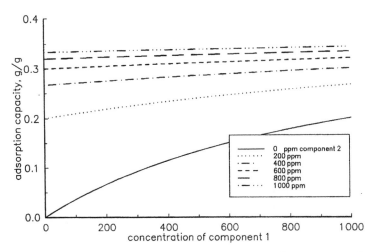

Fig. 5. Total adsorption isotherms of a binary mixture. Langmuir constants: for component 1, $m_1 = 0.4$ g g^{-1}, $k_1 = 0.001$ ppm^{-1}; for component 2, $m_2 = 0.4$ g g^{-1}, $k_2 = 0.005$ ppm^{-1}.

adsorbent at any concentration. This is clearly shown in Fig. 6 where W_e/C is plotted versus concentration for the same three compounds given in Fig. 2. Binary systems A–B and A–C represent mixtures in which component A is more strongly adsorbed. For a binary mixture in which $m_2 k_2 < m_1 k_1$ and $m_2 > m_1$, isotherms will intersect (Fig. 2). For this case compound 2 is more strongly adsorbed for low concentrations, but at higher concentration, component 1 is the more strongly adsorbed compound (Fig. 6).

4. BREAKTHROUGH CURVES OF BINARY VAPOR MIXTURES FOR ACTIVATED CARBON

The breakthrough curves of binary mixtures depend on several parameters. The influence of these parameters on the shape of the curves is discussed in this section.

4.1 Concentration

The effect of concentration of the more strongly adsorbed component (compound 2) on the breakthrough curves is shown in Figs 7 and 8. As the concentration of component 2 increases, breakthrough times of the two components decrease (the curves shift to the left), the size of zone II reduces and the maximum concentration of component 1 (weakly adsorbed compound) increases. At high concentrations, zone II completely disappears, and the breakthrough curve for component 1 goes through a maximum and drops rapidly (Fig. 8). At very low concentrations of component 2, the effect of component 2 on the adsorption behavior of component 1 is negligible, and the displacement of component 1 does not occur. No maximum is therefore observed for the breakthrough curve of component 1 (Fig. 9). Under certain conditions, as the concentration of

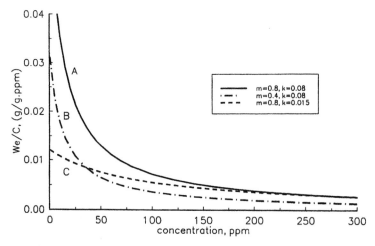

Fig. 6. The effect of concentration on the ratio W_e/C for pure compounds.

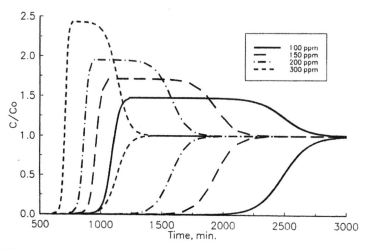

Fig. 7. The effect of concentration of component 2 on breakthrough curves of a binary mixture.

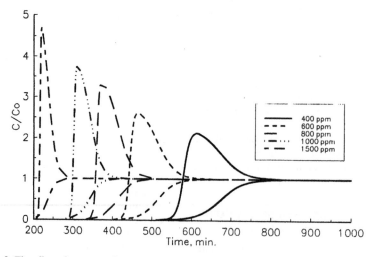

Fig. 8. The effect of concentration of component 2 on breakthrough curves of a binary mixture.

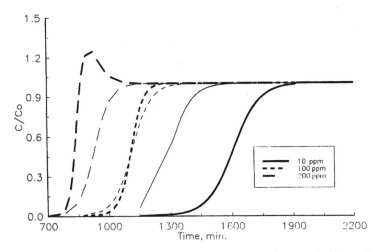

Fig. 9. The effect of concentration of component 1 on breakthrough curves of a binary mixture.

component 2 varies, a very interesting phenomenon occurs. If the isotherms for the two pure compounds intersect (curves B and C in Fig. 2), for certain concentrations of the two compounds, component 1 becomes the more strongly adsorbed component. As was stated earlier, for this to happen, the following relations should hold for the Langmuir constants of the two compounds: $m_2 k_2 < m_1 k_1$ and $m_2 > m_1$. Fig. 9 shows this phenomenon. This figure is plotted for a 100 ppm concentration of component 1 ($m_1 = 0.4$ g g^{-1}, $k_1 = 0.8$ ppm^{-1}) and three different concentrations of component 2 ($m_2 = 0.80$ g g^{-1}, $k_2 = 0.015$ ppm^{-1}), 10, 100 and 200 ppm. For the 10 ppm concentration of component 2, (solid lines) component 1 breaks through first (thin solid line) indicating that component 2 is the more strongly adsorbed component (thick solid line). For a mixture containing 100 ppm component 2 (short dash lines in Fig. 9), the two breakthrough curves intersect indicating a transition from case I (component 2 is the more strongly adsorbed component) to case II (component 1 is the more strongly adsorbed component). For a binary mixture containing 200 ppm component 2 (long dash lines in Fig. 9), component 1 is the more strongly adsorbed component and component 2 breaks through first (thick long dash lines). The curve for component 2 actually goes through a maximum indicating replacement of molecules of component 2 by component 1. This phenomenon can be explained by studying the ratio W_e/C. For a given adsorbent, this ratio determines which component will be adsorbed more strongly. Figure 6 shows the effect of concentration on W_e/C for the binary system presented in Fig. 9. For a given concentration of components 1 and 2, whichever compound has a higher W_e/C will be adsorbed more strongly by the solid adsorbent. This figure clearly shows that for concentrations less than 35 ppm, component 1 ($m = 0.4$ g g^{-1}, $k = 0.08$ ppm^{-1}) has a higher W_e/C, and for concentrations greater than 35 ppm, compo-

nent 2 ($m = 0.8$ g g^{-1}, $k = 0.015$ ppm^{-1}) is the more strongly adsorbed compound. For case I ($C_1 = 100$ ppm, $C_2 = 10$ ppm) and case II ($C_1 = 100$ ppm, $C_2 = 100$ ppm) in Fig. 9, component 2 has a higher W_e/C, and therefore is the more strongly adsorbed component. For case III ($C_1 = 100$ ppm, $C_2 = 200$ ppm), component 1 has a slightly larger W_e/C and is adsorbed more strongly.

The effect of concentration of the less strongly adsorbed component on the breakthrough time is shown in Fig. 10. As the concentration of component 1 increases, breakthrough times of the two components decrease (breakthrough curves shift to the left), the size of zone II increases, and the maximum concentration of component 1 decreases. At very high concentrations, the maximum in the breakthrough curve completely disappears.

4.2 Flow rate

The effect of flow rate on the breakthrough time of pure compounds in an activated carbon bed has been studied by several investigators [14]. For a given concentration of the compound, the breakthrough time is almost inversely proportional to the flow rate. This relationship between breakthrough time and flow rate is also evident from the Wheeler correlation (eqn (8)). No experimental data are available for the effect of flow rate on the breakthrough times of a binary mixture. The theoretical model, however, can be used to explore this relationship. The results of calculations are given in Figs. 11 and 12. As the flow rate increases, breakthrough times of both components decrease, and the breakthrough curves, shift to the left. The flow rate also has an effect on the size of zone II and the value of maximum concentration of the weakly adsorbed component (C_1^{max}). At higher flow rates, the length of zone II decreases, indicating that the two breakthrough curves approach each other. Figure 13 shows the breakthrough time of the two components and the difference between the two

196

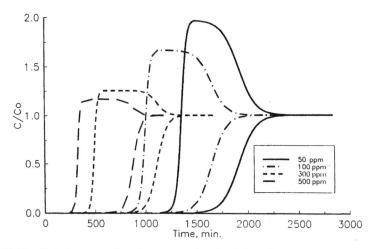

Fig. 10. The effect of concentration of component 1 on breakthrough curves of a binary mixture.

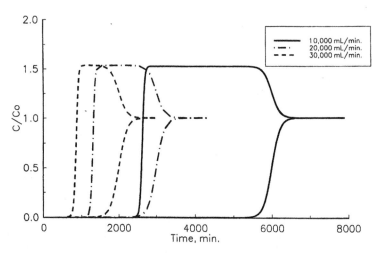

Fig. 11. The effect of flow rate on breakthrough curves of a binary mixture.

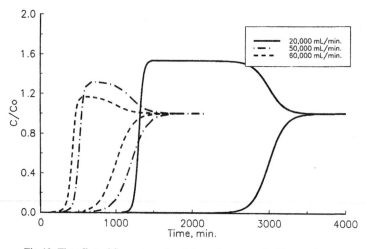

Fig. 12. The effect of flow rate on breakthrough curves of a binary mixture.

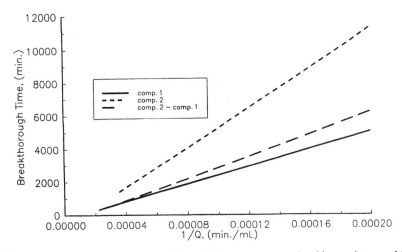

Fig. 13. The relationship between breakthrough times of the two components in a binary mixture and flow rate.

breakthrough times vs reciprocal of flow rate. All three parameters are inversely proportional to flow rate. For the binary system given in Figs 11 and 12, the effect of flow rate on the maximum concentration of component 1 is almost insignificant for flow rates below 30 000 ml min^{-1}. At higher flow rates, however, C_1^{max} decreases as the air flow rate through the adsorbent bed increases.

4.3 The effect of Langmuir constants

The effect of adsorption capacity isotherms of pure compounds on the breakthrough times of a binary system can be investigated by studying the effect of the Langmuir constants on the shape of the break-through curves. The theoretical model was used to develop breakthrough curves for different values of the Langmuir constants (m and k), and the results are given in Figs 14–17. As parameter m_2 increases, the capacity of the solid adsorbent for component 2 increases, and consequently breakthrough time of this component increases too. The parameter also increases the breakthrough time of component 1 slightly (Fig. 14) with a net result of having a larger zone II. The maximum concentration of component 1 (C_1^{max}) increases and then drops as m_2 increases.

The adsorption capacity of pure component 1 is directly proportional to the parameter m_1. One can therefore expect to observe longer breakthrough times for component 1 as m_1 increases. This is in fact the case, as it is shown in Fig. 15. The breakthrough time of component 2 on the other hand decreases with a net result of reducing the size of zone II. At large values of m_1, zone II completely disappears. The maximum concentration of component 1 (C_1^{max}) increases, goes through a maximum and drops as m_1 increases.

The effect of Langmuir constant k on the break-through curves is shown in Figs 16 and 17. The value of k_2 (the parameter for the more strongly adsorbed component 2) has a direct effect on the breakthrough time of component 2 and the maximum concentration of component 1 (C_1^{max}), and has no or little effect on

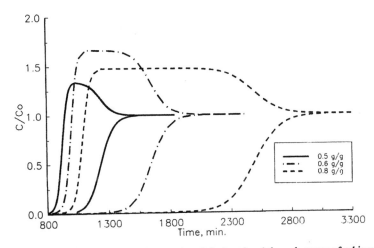

Fig. 14. The effect of Langmuir constant of component 2 (m_2) on breakthrough curves of a binary mixture.

198

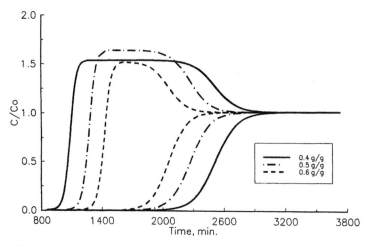

Fig. 15. The effect of Langmuir constant of component 1 (m_1) on breakthrough curves of a binary mixture.

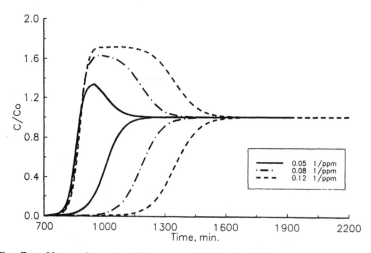

Fig. 16. The effect of Langmuir constant of component 2 (k_2) on breakthrough curves of a binary mixture.

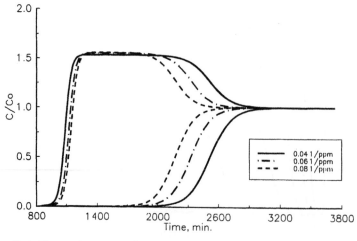

Fig. 17. The effect of Langmuir constant of component 1 (k_1) on breakthrough curves of a binary mixture.

the breakthrough time of the first component. C_1^{max} and breakthrough time of component 2 increase as k_2 increases. The effect of k_1 (the Langmuir constant for the weakly adsorbed component) is to increase the breakthrough time of component 1, and decrease the breakthrough time of the second component. The length of zone II reduces, and C_1^{max} increases, goes through a maximum and drops as k_1 increases.

5. CONCLUSION

The effect of different parameters on the performance of activated carbons in an environment containing two contaminants was determined. Breakthrough time of the components depends on concentrations, adsorption capacities and flow rate. Due to the difference in the adsorption capacity and concentration of the two compounds, one of them is more strongly adsorbed by the carbon bed, and has a longer breakthrough time. The component which has a higher ratio of adsorption capacity to concentration is the more strongly adsorbed component. The effect of concentration of the more strongly adsorbed component is to reduce the breakthrough times of the compounds, and to increase the rate of replacement of the weakly adsorbed component (and thus an increase in the maximum concentration).

The effect of increasing the concentration of the weakly adsorbed compound is to decrease its rate of replacement (and thus a decrease in its maximum concentration) and reduce the breakthrough times. Under certain conditions for the isotherms, varying the concentration of the compounds can cause the more strongly adsorbed component to become less strongly adsorbed.

The effect of the adsorption isotherm of pure compounds on the breakthrough curves was evaluated by varying Langmuir constants for each compound. An increase in the adsorption capacity of the more strongly adsorbed component causes an increase in the breakthrough time of both compounds. The effect of the adsorption capacity of the weakly adsorbed compound on the other hand is to increase its breakthrough time and to reduce the breakthrough time of the other compound.

The effect of flow rate on breakthrough times of the two components of a binary mixture is similar to that of pure compounds. Both breakthrough times vary linearly with the reciprocal of flow rate.

REFERENCES

1. Ruch, W. E., Nelson, G. O., Lindeken, C. L., Johnson, R. G. and Hodgkins, D. J., *Am. Ind. Hyg. Assoc. J.*, 1972, **33**, 105.
2. Nelson, G. O. and Hodgkins, D. J., *Am. Ind. Hyg. Assoc. J.*, 1972, **33**, 110.
3. Nelson, G. O., Johnson, R. E., Lindeken, D. L. and Taylor, R. D., *Am. Ind. Hyg. Assoc. J.*, 1972, **33**, 745.
4. Nelson, G. O. and Harder, C. A., *Am. Ind. Hyg. Assoc. J.*, 1972, **33**, 797.
5. Nelson, G. O. and Harder, C. A., *Am. Ind. Hyg. Assoc. J.*, 1974, **37**, 205.
6. Nelson, G. O. and Harder, C. A., *Am. Ind. Hyg. Assoc. J.*, 1976, **37**, 205.
7. Guenier, J. P. and Muller, J., *Ann. Occup. Hyg.*, 1984, **28**, 61.
8. Jones, L. A., Tewari, Y. B. and SaSone, E. B., *Carbon*, 1979, **17**, 345.
9. Moyer, E. S., *Am. Ind. Hyg. Assoc. J.*, 1978, **48**, 791.
10. Wood, G. and Moyer, E. S., *Am. Ind. Hyg. Assoc. J.*, 1989, **50**, 400.
11. Cohen, H. J. and Garrison, R. P., *Am. Ind. Hyg. Assoc. J.*, 1989, **50**, 486.
12. Wood, G. O. and Moyer, E. S., *Am. Ind. Hyg. Assoc. J.*, 1991, **52**, 235.
13. Vahdat, N., Searengen, P. M., Johnson, J. S., Priante, S., Mathews, K. and Neihart, A., *Am. Ind. Hyg. Assoc. J.*, 1995, **56**, 32.
14. Wood, G. O., *Carbon*, 1992, **30**, 593.
15. Jonas, L. A., Rehmann, *Carbon*, 1972, **10**, 657.
16. Wood, G. O. and Stampfer, J. F., *Carbon*, 1993, **31**, 195.
17. Wood, G. O., *Am. Ind. Hyg. Assoc. J.*, 1994, **55**, 11.
18. Swearengen, P. M. and Weaver, S. C., *Ind. Hyg. Assoc. J.*, 1988, **49**, 70.
19. Siddiqi, K. S. and Thomas, W. J., *Carbon*, 1982, **20**, 473.
20. Yoon, Y. H., Nelson, J. H., Lara, J., Kamal, C. and Fregeau, D., *Am. Ind. Hyg. Assoc. J.*, 1991, **52**, 65.
21. Yoon, Y. H., Nelson, J. H., Lara, J., Kamal, C. and Fregeau, D., *Am. Ind. Hyg. Assoc. J.*, 1992, **53**, 493.
22. Vahdat, N., Swearengen, P. M. and Johnson, J., *Am. Ind. Hyg. Assoc. J.*, 1994, **55**, 909.
23. Myers, A. L. and Prausintz, J. M., *AIChE J.*, 1965, **11**, 121.
24. Langmuir, I., *Am. Chem. Soc.*, 1916, **38**, 1195.
25. Noll, K. E., Wang, D. and Shen, T., *Carbon*, 1989, **27**, 239.
26. Urano, K., Omori, S. and Yamamoto, E., *Environ. Sci. Technol.*, 1982, **16**, 10.
27. Reucroft, R. J., Simpson, W. H. and Jonas, L. A., *J. Phys. Chem.*, 1971, **75**, 3526.
28. Zwiebel, I., Myers, F. R. and Neusch, D. A., *Carbon*, 1987, **25**, 85.

Activated Carbon Compendium
H. Marsh (Editor)

APPLICATION OF THE MONO/MULTILAYER AND ADSORPTION POTENTIAL THEORIES TO COAL METHANE ADSORPTION ISOTHERMS AT ELEVATED TEMPERATURE AND PRESSURE

C. R. CLARKSON,[a,*] R. M. BUSTIN[a] and J. H. LEVY[b]

[a]Department of Earth and Ocean Sciences, University of British Columbia, 6339 Stores Road, Vancouver, BC, Canada V6T 1Z4
[b]CSIRO Division of Coal and Energy, Menai, NSW, Australia

(*Received* 27 *November* 1996; *accepted in revised form* 12 *June* 1997)

Abstract—Accurate estimates of gas-in-place and prediction of gas production from coal reservoirs require reasonable estimates of gas contents. Equations based on pore volume filling/potential theory provide a better fit than the Langmuir equation to both high-pressure (up to 10 MPa), high-temperature ($>1.5T_c$) methane isotherm data, and low-pressure (<0.127 MPa) carbon dioxide isotherm data for 13 Australian coals. The assumption of an energetically homogeneous surface as proposed by Langmuir theory is not true for coal. Application of potential theory to the methane–coal system results in temperature-invariant methane characteristic curves, obtained with the assumption of liquid molar volume of the adsorbate and extrapolated vapour pressures. Temperature-invariant characteristic curves are obtained for carbon dioxide, although further testing is required. The application of isotherms equations based upon pore volume filling/potential theory, in particular the Dubinin–Astakhov equation, have general validity in their application to high-pressure supercritical methane–coal systems as well as providing a better fit to isotherm data. © 1997 Elsevier Science Ltd

Key Words—A. Coal, C. adsorption.

1. INTRODUCTION

Recent interest in both recovery of natural gas from coal seams and in outburst hazards related to coal mining has led to extensive study of gas sorption in coal. The accurate prediction of coal gas capacities are important in gas reserve estimates and for input to production simulators. To simulate reservoir conditions, laboratory sorption isotherm data are generally collected at elevated temperature, usually between 0 and 50°C, and elevated pressure (up to 100 MPa). Methane gas sorption isotherms in coal, commonly Type I [1] in nature, are most often modeled using the Langmuir isotherm [2] and less frequently the Freundlich [3] or linear [4] isotherms. A study performed by Hall *et al.* compared the Langmuir model with various two-dimensional equations of state, the ideal adsorbed solution model, and loading ratio correlations for the adsorption of various gases and their mixtures on wet coal [5]. Only limited attention has been focused upon the application of adsorption potential theories to the description of methane adsorption isotherms collected for coals at elevated temperature and pressure [2]. Both the Dubinin–Radushkevich (D–R) and Dubinin–Astakhov (D–A) equations have been used to model Type I isotherms [1]. These isotherm equations are based upon the potential theory developed by Polanyi [6].

Ruppel *et al.* [2] applied the Langmuir and Polanyi adsorption models to methane adsorption isotherms collected for coals at temperatures ranging from 0 to 50°C and pressures from 10 to 150 atm (1–15 MPa). The Ruppel *et al.* study is the first detailing the application of adsorption-potential theory to methane sorption on dry coal at elevated temperature and pressure, although the model had previously been applied to other porous media under various conditions [7–10]. Ruppel *et al.* [2] found that the coal–methane adsorption system was well described by the Langmuir model, but the Polanyi theory did not accurately describe the system for all coals. It should be noted that thermal expansion of the adsorbate was not accounted for by Ruppel *et al.*, which, as shown in the case of activated carbons, may lead to the failure of one of the fundamental postulates of potential theory [11].

More recent studies [11,12] have applied Dubinin's volume filling theory, an adaptation of the original potential theory developed by Polanyi, to the high-pressure adsorption of various gases on activated carbon, in particular methane, above the critical point. These studies [11,12] outline the difficulties of applying the Dubinin postulates to supercritical fluids, such as the attainment of saturation vapour pressures and adsorbate densities. Agarwal and Schwartz [11] and Yang [13] provide a summary of the different approaches used to obtain these parameters. The desirability of using potential theory for the

*Corresponding author. E-mail: cclarkso@eos.ubc.ca

Reprinted from *Carbon* 35 (**12**), 1689-1705 (1997)

description of supercritical gas adsorption is that a single characteristic curve may be used to describe gas adsorption at a variety of temperatures.

The adsorption of methane in coal is generally believed to be due to physical adsorption. This is demonstrated by the small heats of adsorption [1,14] and that methane isotherms are reversible [15,16]. It is therefore reasonable to apply theories based on physical adsorption to the problem of methane adsorption in coal. In particular, Dubinin–Polanyi potential theory may be applied to the current problem, if properly validated.

The purpose of the current study is to apply the Brunauer, Emmett and Teller (BET), Dubinin–Radushkevich (D–R) and Dubinin–Astakhov (D–A) equations to methane–wet coal isotherm data at elevated temperature and pressure, above the critical point for methane ($T > 1.5 T_c$), and compare the results to isotherms obtained by the more commonly applied Langmuir equation. The BET, D–R and D–A equations contain either two or three parameters that are easily determined from experimental isotherm data. The goal of this research is to determine the optimum equation to be applied to adsorption isotherm data. In addition, the general validity of the monolayer and pore volume filling theories is tested for methane–coal systems. The conceptual model of pore-filling is different for these theories, and therefore the current study will have important implications for the modelling of equilibrium isotherm data in coal.

1.1 Background

1.1.1 Langmuir (monolayer) and BET (multilayer) theory.
The most commonly applied adsorption isotherm model for coal at elevated temperature and pressure is the Langmuir model [17], from which the Langmuir equation is obtained. The Langmuir equation is written in the following form for plotting purposes:

$$\frac{P}{V} = \frac{1}{B V_m} + \frac{P}{V_m} \qquad (1)$$

where P is the equilibrium gas or vapour pressure; V is the volume of gas adsorbed, commonly reported at standard temperature and pressure (STP), per unit mass of coal; V_m is the Langmuir monolayer volume; and B is an empirical constant.

The Langmuir model is based upon the assumption that a state of dynamic equilibrium exists (at constant T and P) between adsorbed and non-adsorbed species and that adsorption is restricted to a single monolayer [1]. In addition, the adsorbent surface is assumed to be energetically homogeneous with respect to adsorption. In many cases, a plot of P/V versus P yields a straight line whose slope yields V_m. The Langmuir model has frequently been applied to the description of Type I isotherms obtained for microporous solids such as activated carbons. Several studies of methane adsorption on coal have shown that the Langmuir

equation fits well over the range of temperatures and pressures applied [2,14,15,18].

The BET model is an extension of the Langmuir model that accounts for the formation of multilayers [19]. The model was developed for the interpretation of Type II isotherms and the reversible part of Type IV isotherms. The BET equation has the following form:

$$\frac{1}{V(P_o/P - 1)} = \frac{1}{V_m C} + \frac{C-1}{V_m C} \frac{P}{P_o} \qquad (2)$$

where P/P_o is the relative pressure, and C is a constant related to the net heat of adsorption. A plot of the left-hand side of eqn (2) versus relative pressure should yield a straight line in the relative pressure range $0.05 < P/P_o < 0.35$ [20]. The application of the BET equation to supercritical fluid adsorption cannot be justified physically as multilayer formation is considered unlikely [14].

1.1.2 Dubinin theory of adsorption of vapours in micropores.
Dubinin theory has commonly been applied to the description of Type I isotherms [21]. A fundamental difference between the Dubinin and Langmuir theories of adsorption is in the postulated mechanism of pore filling. In the Langmuir theory, the sorbed phase is assumed to occupy a monolayer on the adsorbent surface, which is in turn assumed to be homogeneous. Dubinin [21,22] theory, however, assumes that, in micropores, the adsorbate fills the adsorption space via the mechanism of volume filling and hence does not form discrete monolayers in the pores. Dubinin [21] showed that, for several vapours, the ratio of limiting adsorption values on two varieties of zeolite crystals is essentially constant and equal to the ratio of void volumes calculated from X-ray data. The ratio was, however, not equal to the geometric surface area of the zeolites. This observation was given as proof of the volume filling mechanism of micropores with radii less than 0.6–0.7 nm. Dubinin [22] likens the process of adsorption in micropores to the process of solution.

Two equations developed by Dubinin and his coworkers are the Dubinin–Radushkevich (D–R) equation [22,13,23]:

$$W = W_o \exp \left\{ -\left(\frac{RT}{\beta E} \ln \frac{P_o}{P} \right)^2 \right\} \qquad (3)$$

and the Dubinin–Astakhov (D–A) equation:

$$W = W_o \exp \left\{ -\left(\frac{RT}{\beta E} \ln \frac{P_o}{P} \right)^n \right\} \qquad (4)$$

where W is the volume of adsorbate adsorbed at equilibrium, W_o is the micropore volume, β is a sorbate affinity coefficient, E is the characteristic energy, R is the universal gas constant, T is temperature, P_o is the saturation vapour pressure for the adsorbate, P is the equilibrium vapour pressure and n is a small integer (1–4) and is related to the distribution of pore sizes. These equations have been applied to a variety of microporous solid–adsorbate systems, including activated carbons. The D–A equation is a general form of the D–R equation in which the coefficient n may be optimized. For plotting purposes, eqns (3) and (4) may be written in the following form:

$$\log W = \log W_o - D(\log P_o/P)^n \qquad (5)$$

where n is equal to 2 for the D–R equation, and is optimized for the D–A equation. A plot of $\log W$ versus $(\log P_o/P)^n$ should yield a straight line. P and P_o may be replaced by f and f_o, the equilibrium and saturation fugacities, to account for non-ideality.

The Dubinin equations are valid for a particular adsorbate–adsorbent system only if certain fundamental postulates of the Dubinin theory are adhered to. An important postulate is that characteristic curves, which are plots of the degree of filling $\theta(W/W_o)$ versus the parameter $A = RT \ln P_o/P$, defined as the differential molar work of adsorption by Dubinin [21], are invariant with temperature for a particular adsorbate–adsorbent system. This may be expressed analytically as $(\partial A/\partial T)_\theta = 0$. Temperature invariance holds if adsorption forces are temperature-independent dispersion forces. Dubinin demonstrated this to be the case for a variety of systems [21]. Bering et al. [24] give the thermodynamic limits for which temperature invariance holds. Dubinin [22] states that temperature invariance of the characteristic curve is not a necessary requirement for all microporous systems, but simply that for many systems it appears to be true. Deviations of the characteristic curve from temperature independence reflect the temperature dependence of the work of adsorption.

The Dubinin eqns (3) and (4) were mainly developed for the adsorption of vapours below the critical point but may also be modified for gases above the critical point. Several authors [2,7–12,25,26] have extrapolated the Polanyi adsorption-potential theory, upon which the Dubinin equations are based, to the supercritical region for methane in various adsorption systems. Recent work has focused upon the applica-

tion of this theory to the adsorption of gases above their critical point [11,12] upon activated carbons. Two important problems associated with the application of potential theories are: 1) the determination of a suitable molar volume of the adsorbate at a given temperature and a pseudo-saturation pressure value; and 2) the form of the temperature-invariant characteristic curve to be utilized. Several approaches to these problems have been used [11,13].

2. METHODS

2.1 Sample collection and preparation

Seven lithotypes of Upper Permian coal were collected from the Appin mine (Bulli seam) and five from the Wongawilli mine (Wongawilli seam) in New South Wales, Australia [27]. In addition, one sample (GHA1-09) was obtained from the Bowen Basin. Each sample was prepared for petrographic (maceral and mineral) analysis and proximate analysis according to procedures discussed in Bustin et al. [28]. High-pressure methane isotherm analysis sample preparation was performed according to procedures described in Mavor et al. [15]. A split of each sample was also taken for low-pressure carbon dioxide analysis.

2.2 Isotherm analysis

High-pressure methane adsorption isotherms at 30°C were collected using a volumetric apparatus constructed by the CSIRO of Lucas Heights, Australia. In addition, one sample (GHA1-09) was run at 25, 30 and 50°C on a similar apparatus at The University of British Columbia, Earth and Ocean Sciences Department. The instrument is described in Levy et al. [29]. Instrument design is based on the volumetric adsorption apparatus described by Mavor et al. [15]. Samples were equilibrated with moisture prior to isotherm analysis using the following procedure: samples were placed in a vacuum type desiccator containing a saturated solution of K_2SO_4; the desiccator was evacuated and placed in an oven set at 30°C; at least 48 hours were allowed for equilib-

Table 1. Maceral composition (vol%, mmf) and proximate analysis of samples studied

Sample	Telocollinite	Desmocollinite	Semifusinite	Fusinite	Inertodetrinite	Moisture (%)	Volatiles (%)	Fixed C (%)	Ash (%)
B1	9	24	34	33	–	0.9	20.3	68.6	10.2
B2	11	37	24	28	–	0.8	22.0	67.6	9.6
B3	10	39	37	14	–	0.7	21.8	68.9	8.6
B4	16	39	24	22	–	0.8	25.1	66.6	7.5
B5	30	36	20	15	–	0.7	25.4	64.1	9.8
B6	39	37	12	11	–	0.6	26.8	69.2	3.4
B7	90	0	1	9	–	0.7	21.8	67.7	9.8
W1	7	40	41	12	–	0.8	18.3	58.6	22.3
W2	13	54	18	15	–	0.6	24.6	59.4	15.4
W3	41	38	13	9	–	0.8	23.8	43.0	32.4
W4	62	31	4	3	–	1.0	21.7	58.0	19.3
W5	79	16	3	2	–	0.9	25.0	63.3	10.8
GHA1-09	13	48	24	1	14	1.4	24.1	67.0	7.5

204

rium prior to isotherm analysis. 10 or 11 pressure points were collected during isotherm analysis up to a pressure of about 10 MPa (absolute) for each sample. Equilibrium at each point is assumed to have been achieved if the pressure reading is constant ($\Delta P = 0.000$ MPa) over a 40 minute interval. Volumes of gas adsorbed are calculated using the Real Gas Law, and the void volume of the system, determined through helium expansion, is corrected for gas adsorbed at each pressure step, assuming liquid density of the adsorbate. Precision of isotherm runs is about ±1% relative.

Low-pressure (<0.127 MPa) carbon dioxide isotherms at 0°C were performed on dry (evacuated) coal samples on a Micromeritics ASAP 2010® automated volumetric gas sorption apparatus. In addition, the GHA1-09 sample was run at 25°C. Coal samples were evacuated at 100°C overnight to a pressure of <0.25 Pa prior to analysis. No nonideality correction was made for carbon dioxide at these low pressures. A saturation vapour pressure of 3.48 MPa was used for carbon dioxide at 273 K and 6.42 MPa at 298 K. Precision of isotherm runs for coal is about ±1% relative. The instrument is periodi-

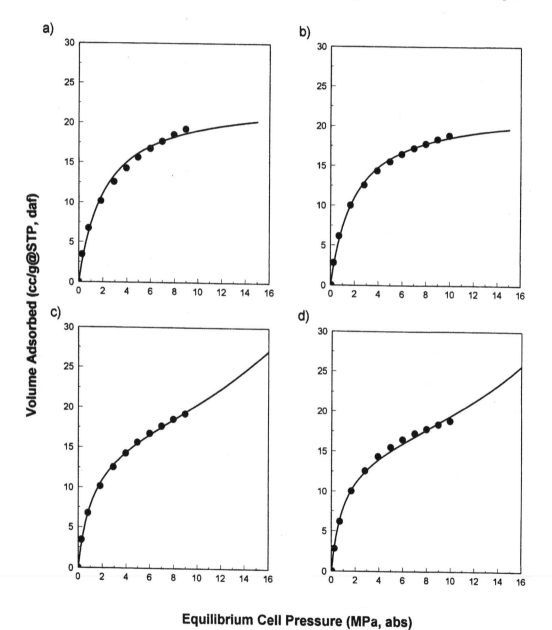

Fig. 1. Langmuir (a,b) and BET (c,d) curve fits to methane isotherm data for samples W1 (a,c) and B4 (b,d). Solid line is curve fit.

cally calibrated against a zeolite standard to check for systematic error.

2.3 *Langmuir, BET, D–R and D–A regression analysis*

The Langmuir isotherm was fit to the methane and carbon dioxide experimental data through the following procedure:

(1) A linear regression was performed for P/V versus P plots (referred to as Langmuir plots) and the Langmuir constants (B and V_m) were calculated from the slope and intercept.

(2) The calculated Langmuir sorbed volumes were obtained from the Langmuir equation in the usual form: $V = (V_m BP)/(1 + BP)$.

(3) The Langmuir isotherm, plotted for 0.1 MPa pressure increments, was superimposed upon the experimental data.

(4) A similar procedure was followed for the BET analysis except a linear regression was performed upon $1/V(P_o/P - 1)$ versus relative pressure plots (BET plots).

The saturation vapour pressure (or pseudo-saturation pressure) for the high-pressure/temperature methane analysis was obtained from the extrapolation of the log of vapour pressure, obtained from the CRC handbook [30], versus reciprocal of absolute temperature to the temperature of actual analysis. A similar procedure was used by Grant et al. [9]. This approach is compared against other methods in a later section. The values obtained from this extrapolation method agree most closely with values obtained from the use of the reduced Kirchoff equation as utilized by Kapoor et al. [23].

Dubinin regression analysis was performed in the following fashion:

(1) $\log V$ is plotted against $(\log P_o/P)^n$ (D–R or D–A transformed plots) and a least squares fit performed. For the D–R equation the value of n is equal to 2, but for the D–A equation the value of n is optimized by recalculating the linear regression until the standard error of the Y-intercept is minimized. The value of n is optimized to within 10^{-4}.

(2) The micropore capacity (V_o) is obtained from the Y-intercept.

(3) The calculated volumes adsorbed are obtained from eqn (4). The Dubinin isotherms are then plotted for 0.1 MPa increments and superimposed upon the experimental data.

Although conventional linearized unweighted regression models [2,15,31,32] were used to obtain fit parameters for the four isotherm equations, more refined and statistically rigorous regression models were also considered. Non-linear and non-linear weighted ordinary least squares (OLS) V on P, and non-linear weighted OLS P on V regression models were also applied to representative samples. These regression models more rigorously accommodate the proportional measurement error observed in pressure and volume adsorbed measurements. They have unbiased V residuals (see below), as opposed to the linearized regression model, which biases the V residual plots during transformation. Furthermore, comparison of the V on P and P on V non-linear weighted regression results allows determination of whether a regression model accommodating errors in both P and V variables is required.

Results from the different regressions were insig-

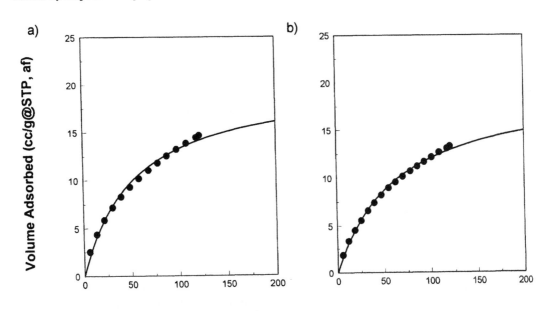

Fig. 2. Langmuir curve fits to carbon dioxide isotherm data for samples W1 (a) and B4 (b). Solid line is curve fit.

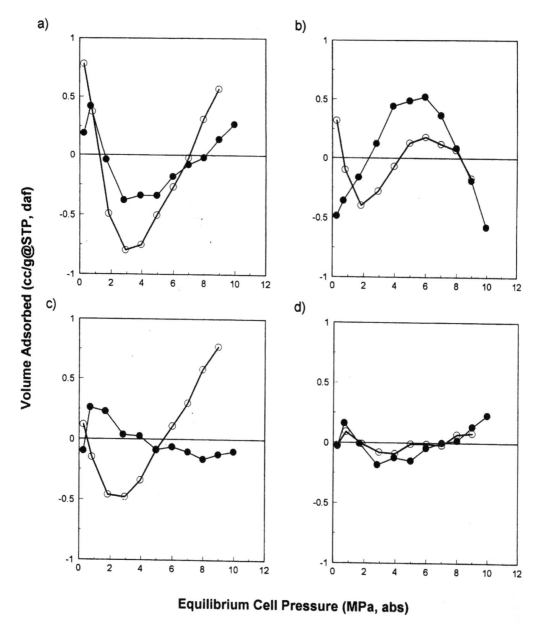

Fig. 3. Plots of residuals for Langmuir (a), BET (b), D–R (c) and D–A (d) curve fits. Samples are W1 (open circles) and B4 (solid circles).

nificantly different. This is a consequence of the relatively high degree of fit of all isotherm regression models. The average relative error (see below) averaged over all samples is less than 4.2, 2.1, 1.6 and 0.4% for the Langmuir, BET, D–R and D–A equations, respectively, regardless of the regression model applied. Results did not reorder the relative quality of fit for the four isotherm models (discussed below) and demonstrate that an "error in both variables" regression model is not required. Thus, although the non-linear weighted V–P regression model is the most consistent with the nature of the data and their errors, the conventionally applied

linearized regression results are presented below in order to allow comparison with previous isotherm results (e.g. refs. [2,15]).

3. RESULTS

3.1 Petrographic and proximate analysis

The results of petrographic (maceral) and proximate analysis are presented in Table 1. Fixed carbon content of the Bulli seam samples ranges from 71 to 77% (dry, ash-free basis) whereas the Wongawilli seam values range from 64 to 76% [27]. Samples are composed mainly of the vitrinite macerals telocollinite and desmocollinite, and inertinite macerals fusinite and semifusinite as well as mineral matter. Liptinite is rare.

3.2 *Langmuir correlations*

Correlation coefficients (r^2) calculated from linear regression analysis of the Langmuir plots (P/V vs P) are greater than 99% for all coals, and range from 99.06% (W1) to 99.85% (B4).

The calculated Langmuir isotherms (solid line) for the coals with the worst (W1) and best (B4) Langmuir correlations are shown in Fig. 1 along with the experimental high-pressure methane isotherms.

The adsorbed volumes are presented on a dry, ash-free basis (DAF) and are corrected to standard temperature and pressure. The Langmuir isotherm calculated for W1 and B4 underestimates the volumes adsorbed at low (<2 MPa) and high pressure (>7 MPa) and overestimates the volume adsorbed in the mid-region of the isotherm (2–7 MPa). Although data for only two samples is shown here, the Langmuir isotherm breaks down in the same pressure region for all samples analyzed.

Langmuir isotherms are plotted for the low-pressure carbon dioxide analyses at 273 K (Fig. 2). The Langmuir plot r^2 values (not shown) are around

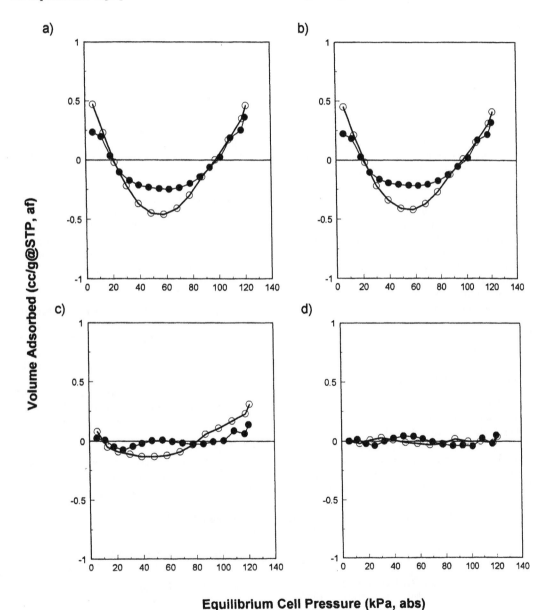

Fig. 4. Plots of residuals for Langmuir (a), BET (b), D–R (c) and D–A (d) curve fits. Samples are W1 (open circles) and B4 (solid circles). Carbon dioxide is the adsorbate used.

Table 2. Relative error calculations for isotherm fits

Average relative error					V_o (monolayer vol./micropore cap.)			
Sample	Langmuir (%)	BET (%)	D–R (%)	D–A (%)	Langmuir ($cm^{-3}\,g^{-1}$, DAF)	BET ($cm^{-3}\,g^{-1}$, DAF)	D–R ($cm^{-3}\,g^{-1}$, DAF)	D–A ($cm^{-3}\,g^{-1}$, DAF)
B1	3.05	1.82	0.67	0.67	21.4	15.1	20.8	20.7
B2	3.99	1.65	1.01	0.15	21.7	15.5	21.1	22.2
B3	3.93	2.65	1.14	0.36	21.7	15.0	20.9	21.9
B4	2.32	3.83	1.29	0.86	22.3	15.3	21.9	20.8
B5	4.18	1.48	1.06	0.47	22.3	15.9	21.7	22.6
B6	4.01	2.57	1.10	0.25	22.7	15.8	22.0	23.1
B7	4.29	1.63	1.18	0.62	21.4	15.2	20.7	21.9
W1	5.44	2.12	2.65	0.41	23.1	16.2	21.6	25.0
W2	4.86	1.95	2.49	0.37	24.8	17.3	23.1	26.3
W3	5.02	2.21	2.71	0.25	22.1	15.4	20.5	23.9
W4	5.31	1.78	2.31	0.77	23.9	16.8	22.5	25.4
W5	4.82	1.94	2.28	0.30	24.7	17.3	23.2	26.2
GHA1-09	2.71	2.19	0.86	0.31	23.2	15.6	21.7	22.6
Average (%)	4.15	2.14	1.60	0.45				

Average relative error (%) $= (100/N)\mathrm{Sum}(\mathrm{abs}(V_{cal} - V_{exp})/V_{exp})$.

99%, and the correlation for B4 (99.17%) is better than for W1 (98.50%).

A plot of the residuals, the difference between experimentally-determined and calculated adsorbed volumes of methane for the Langmuir fit, is shown in Fig. 3 for samples B4 and W1. The same trend is apparent for the low-pressure carbon dioxide analyses (Fig. 4), despite the difference in experimental conditions. It is important to note, however, that these plots may not be compared directly with the methane results as the isotherms for methane and carbon dioxide were run over much different relative pressure ranges (~0.001–0.3 for methane, and 0.002–0.035 for carbon dioxide).

In order to quantify the curve-fit for all samples, the average relative errors (RE) between the calculated and experimental adsorbed volumes of methane were determined (Table 2). The mean average relative errors for all samples tested is about 4.2% for the Langmuir correlation.

3.3 BET correlations

BET isotherms for samples W1 and B4 are shown in Fig. 1. BET plot (not shown) r^2 values are greater than 99% (99.71–99.92%), and are on average greater than the Langmuir values. The BET isotherm better fits the experimental data for sample W1 than the Langmuir isotherm (2.1% RE versus 5.0%), but not

for B4 (3.8 versus 2.3%). In general, however, the BET isotherm provides a better fit to the high-pressure methane experimental data (Table 2).

A plot of the BET residuals for samples W1 and B4 is shown in Fig. 3. The opposite trend is observed than for the Langmuir residuals: the BET isotherm underestimates the volumes adsorbed in the middle region of the isotherm and overestimates at the low and high-pressure ends. This trend is obeyed for all samples studied. The same plot for carbon dioxide analyses (Fig. 4) illustrates that the BET residual trend is very similar to the Langmuir trend.

3.4 D–R correlations

D–R plots for the Australian coals are linear (not shown) and correlations are greater than 99% (99.66–99.98%) for all samples. The mean average relative error for the D–R fit for all samples is slightly lower than the BET fit (1.6 versus 2.1%). The D–R isotherm fit (Fig. 5) is better for sample B4 than for W1 and the residual plot trends (Fig. 3) are similar to that for the Langmuir fit. Residual plots are also shown for the carbon dioxide analyses (Fig. 4).

3.5 D–A correlations

The D–A equation yielded the highest correlations for the experimental data (99.96–100%); the mean average relative error for all samples is around 0.5%.

Table 3. Summary of the various methods used for the attainment of pseudo-saturation pressure and molar volume of the adsorbate

Pseudo-saturation pressure	Molar volume of adsorbate
$P_s = (T/T_c)^2$ [20]	$V = V_s(T_b)^a$
From reduced Kirchoff [22]	$V = V_s(T_b) \exp(\Omega(T - T_b))^b$ [10,21,25]
Extrapolated log vp versus $1/T$ plot[c]	

[a] $V_s(T_b) =$ liquid molar volume of the adsorbate at normal boiling point.
[b] $\Omega =$ thermal expansion coefficient of the adsorbate, taken as 0.0016 [10].
[c] $vp =$ vapour pressure.

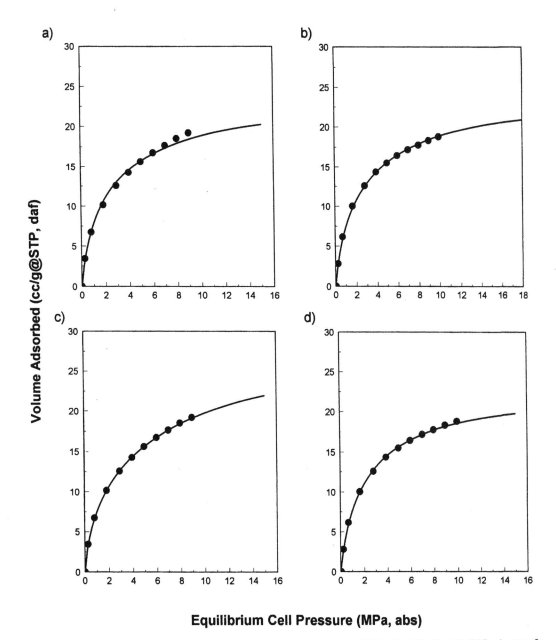

Fig. 5. D–R (a,b) and D–A (c,d) curve fits to methane isotherm data for samples W1 (a,c) and B4 (b,d). Solid line is curve fit.

Residual plots for samples W1 and B4 show no distinct trend (Fig. 3), which indicates an excellent fit to the data.

4. DISCUSSION

In general, for the coals studied, the isotherm equations based upon adsorption potential theory (D–R and D–A) yield a better curve-fit to the experimental data than those based upon the mono/multi-layer pore filling models (Langmuir and BET). Three equations are two-parameter models (Langmuir, BET and D–R) while the fourth is a three-parameter model (D–A). Both the BET and D–R equations yield better fits to the data than the Langmuir equation for all coals, except sample B4. Although all the isotherm equations applied yield a reasonable approximation to the experimental data, the validity of the underlying theories requires testing.

4.1 Langmuir theory

Criteria for Langmuir-type adsorption may be summarized as follows:
(1) The adsorption surface is energetically homogeneous.

210

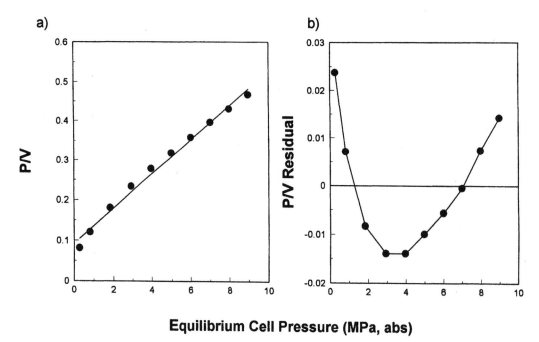

Fig. 6. Langmuir plot (a) and plot of residuals (b) for W1.

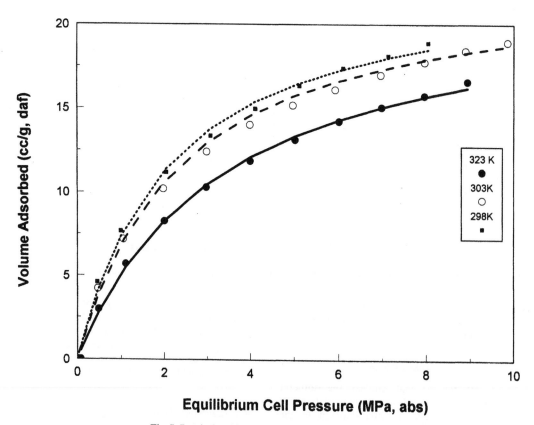

Fig. 7. Recalculated Langmuir isotherms for GHA1-09.

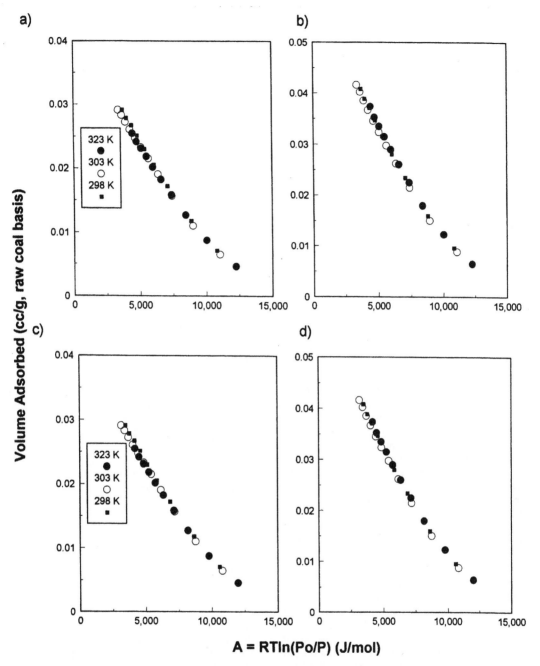

Fig. 8. Methane characteristic curves for sample GHA1-09. P_o was calculated using extrapolated vapour pressures (a,b) and the Kirchoff equation (c,d). Thermal expansion of the adsorbate is accounted for in (b) and (d).

(2) The monolayer volume, and hence the slope of the Langmuir plot, is temperature invariant.

(3) The fitting parameter B decreases exponentially with an increase in temperature.

Brunauer [33] showed that the Langmuir plot may not have a constant slope for the complete range of relative pressures, but may be subdivided into linear subsegments. This behaviour was attributed to surface heterogeneity. Several studies have shown that the monolayer amounts do vary with temperature, violating (2), but, as stated by Koresh [34], these experiments were performed at the same pressure range at different temperatures, and hence at higher temperatures, the isotherms represent less surface coverage. The consequent lower monolayer amounts may therefore be attributed to surface heterogeneity, as stated above. In this study, as with Koresh [34], criteria 1 and 2 will be considered as one.

212

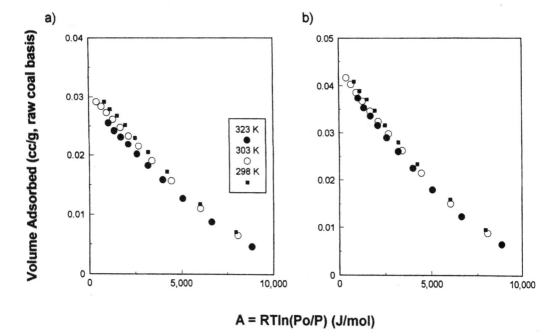

A = RTln(Po/P) (J/mol)

Fig. 9. Methane characteristic curves for sample GHA1-09. P_o was calculated using the Dubinin method. Thermal expansion of adsorbate is accounted for in (b).

Figure 6 shows the Langmuir plot (303 K isotherm data) for sample W1, along with a plot of the residuals. A line fit to the low pressure data (< 3 MPa) would yield a larger slope of the Langmuir plot compared to that at higher pressures, resulting in a smaller predicted monolayer volume at the lower pressures. In addition, the B constant would be larger. If the first three points of the Langmuir plot are used to calculate the monolayer volume, a value of 15.8 $cm^{-3} g^{-1}$ (DAF) is obtained, which is about 32% smaller than the value obtained from the entire range of pressure (23.1 $cm^{-3} g^{-1}$). Note that although sample W1 is an extreme case of the failure of the Langmuir model in this regard, all the samples studied show a similar trend to varying degrees.

Ruppel *et al.* [2] found that correcting for the change in dead space volume with gas adsorption actually introduced curvature into the Langmuir plot. In the current study, Langmuir plots were generated for uncorrected (no dead space correction) adsorbed volumes, and the curvature in the plots still existed (not shown). Correction for dead space error is therefore not the cause of curvature in the Langmuir plots shown.

According to 3, a plot of the parameter B versus temperature should be exponential in form [2] and this was demonstrated for sample GHA1-09 (not shown). Langmuir isotherms were recalculated by fitting a linear relationship between the (natural) logarithm of the experimentally-derived B parameter and the reciprocal temperature. These isotherms are shown in Fig. 7 along with the original experimentally determined isotherms. The fit to the original data is

good with a relative error of around 2–3%. The Langmuir volumes determined from the Langmuir fit to each of the three isotherms separately were used in the recalculated Langmuir isotherms.

4.2 Dubinin theory

The thermal equation for adsorption may be written in the following form [22]:

$$\theta = F(A/E, n)$$

where θ is the degree of filling of the adsorption space; A is as defined previously (for vapours); E is the characteristic energy, which is equal to A at a particular value of θ; and n is a constant parameter. The function F is a distribution function of θ with respect to A; in the case of the D–A equation, the distribution adopted is that due to Weibull [35]. If E and n are temperature invariant parameters of this distribution, then the characteristic curve defined by the relationship $A = E\phi(\theta, n)$ should also be temperature invariant.

Methane adsorption characteristic curves for sample GHA1-09 at 298, 303 and 323 K were calculated to determine the effect of using several techniques for the determination of saturation pressure, adsorbed phase volume, differential molar work of adsorption (A), as well as for the correction for gas non-ideality. Table 3 outlines the various techniques utilized to obtain these parameters.

Figure 8 shows the characteristic plots for the case of no correction for gas non-ideality (pressure instead of fugacity), the assumption of liquid molar volume of the adsorbate at boiling point (Fig. 8(a)), and the

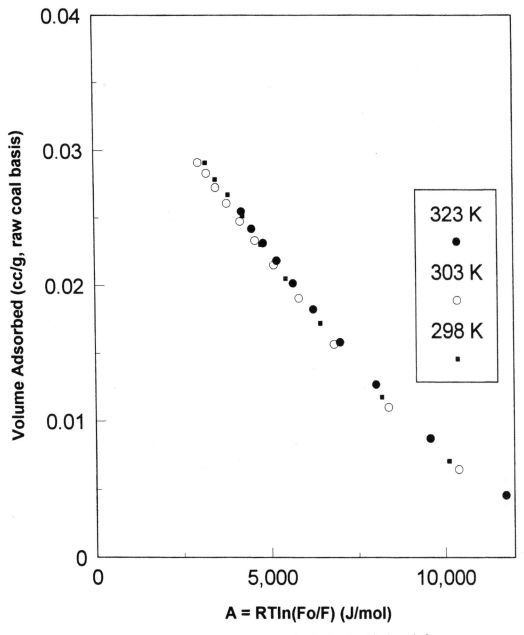

Fig. 10. Methane characteristic curves for sample GHA1-09 using fugacities instead of pressure.

calculated adsorbate molar volume using Dubinin's technique (Fig. 8(b)), which corrects for the thermal expansion of the adsorbate. For plots 8(a) and 8(b), the pseudo-saturation pressure was calculated using the extrapolation procedure discussed above. For the two plots, the maximum deviations from the characteristic curve are around ±2%. Deviations are defined as the maximum deviation of adsorbed volume at a particular value of A divided by the maximum adsorption volume (×100) [11]. The correction for the thermal expansion of the adsorbate does not appear to have a significant effect upon the characteristic curve.

Characteristic curves calculated using the reduced Kirchoff equation to determine pseudo (extrapolated) vapour pressures are shown in Fig. 8(c) and (d). The maximum deviation for these plots is also about ±2%. The use of the Kirchoff extrapolation for pseudo-vapour pressures in the calculation of the characteristic curves therefore yields similar results to those calculated using the extrapolation described above.

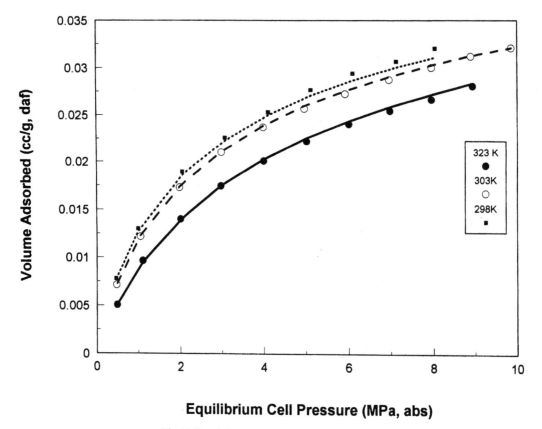

Fig. 11. Recalculated D–A isotherms for GHA1-09.

Characteristic curves using the Dubinin technique for pseudo-vapour pressure extrapolation however display larger deviations than the other two vapour pressure extrapolation techniques (Fig. 9). The maximum deviation is over 4%. Correction for the thermal expansion of the adsorbate does appear to decrease the deviation from temperature invariance of the characteristic curve, however.

Correcting for non-ideality of the adsorbate vapour does not have a significant effect upon the characteristic curve. A characteristic curve calculated using the extrapolated vapour pressure technique and the assumption of a constant liquid density is shown in Fig. 10. Fugacities of the free-gas in equilibrium with the adsorbate are obtained by calculation of the fugacity coefficient using a virial series with Redlich–Kwong constants. This procedure is outlined in Noggle [36].

A plot of the experimental isotherms and the D–A fit calculated from the characteristic curve is shown in Fig. 11. The curve fit was performed by plotting $\log W$ vs A^n, where W is equal to the product of the number of moles of gas adsorbed and the liquid molar volume of methane at normal boiling point, A is the calculated differential molar work of adsorption and n is the optimized Astakhov coefficient

(1.9). The curve fit appears to be reasonable, with an average relative error of less than 2%.

The characteristic curve for sample GHA1-09, plotted using the low pressure carbon dioxide data at 298 and 273 K and no correction for adsorbate density variation with temperature shows a deviation of up to 9%. Correction for adsorbate density, using the values of Toda et al. [35] at 273 and 298 K decreases the deviation to $\pm 2\%$ (Fig. 12). These results are in agreement with Toda et al. [37] who found that carbon dioxide adsorption at a variety of temperatures could be expressed by a single Dubinin–Polanyi plot.

There is some evidence that adsorption of carbon dioxide on coal may not be strictly physical adsorption. A study by Greaves et al. [38] demonstrated that high pressure sorption (up to about 7 MPa) of methane and carbon dioxide mixtures on dry coal exhibited hysteresis between the adsorption and desorption branches of the isotherm which became more pronounced with percentage of carbon dioxide used in the mix. The hysteresis was interpreted by the authors to be due to the retention of carbon dioxide preferentially over methane upon desorption. This experimental behaviour indicates that sorption of carbon dioxide in coal is not strictly due to

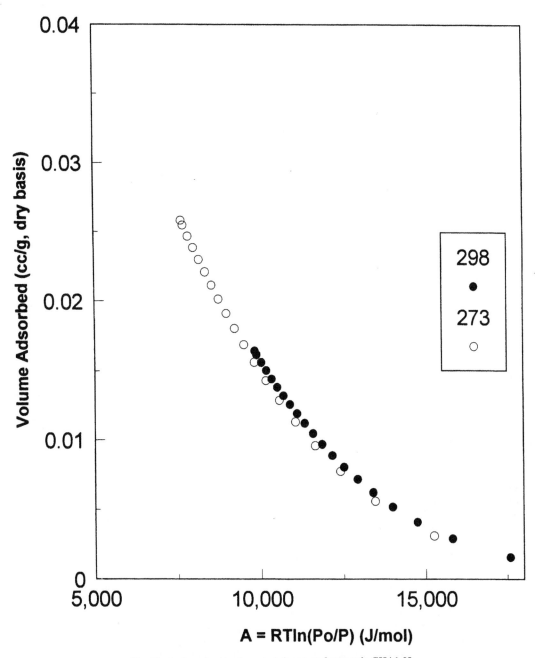

Fig. 12. Carbon dioxide characteristic curves for sample GHA1-09.

physical sorption. This phenomenon is currently being investigated.

In summary, high-pressure methane and low-pressure carbon dioxide adsorption characteristic curves are temperature-independent in the range of temperatures studied. The assumption of liquid density of the adsorbate for methane is sound and temperature dependence of this parameter need not be accounted for to attain temperature-invariant characteristic curves in the range of temperatures utilized here. For carbon dioxide, thermal expansion of the adsorbate

must be accounted for, however. Of the several methods used to obtain pseudo-saturation pressures for methane, the extrapolation of the vapour pressure curve and the reduced Kirchoff equation are the best for calculating characteristic curves. The Dubinin technique for obtaining saturation pressures does not yield temperature-invariant characteristic curves regardless of the value used for adsorbate density. A recent study by Amankwah and Schwartz [12] has shown that a modified Dubinin equation for the attainment of pseudo-saturation fugacity yields better results than the original Dubinin equation for high-pressure methane and hydrogen adsorption. The equation, however, introduces yet another parameter into the D–A equation that needs to be optimized.

5. CONCLUSIONS

A number of classical isotherm equations have been applied to model the adsorption of supercritical methane on coals at high pressures (up to 10 MPa) in an attempt to determine which yield the best curve-fit to the experimental data and which are the most physically realistic models. In addition carbon dioxide adsorption at low pressures on these same coals was analyzed by these equations.

The three-parameter Dubinin–Astakhov equation yields the best curve-fit to the high-pressure (>0.101 MPa) methane experimental data, but both the two-parameter Dubinin–Radushkevich and BET equations are better than the Langmuir equation. The same is true for the adsorption of carbon dioxide at low pressure (<0.127 MPa).

The validity of the monolayer and adsorption potential theories was also tested. The temperature dependence of the Langmuir B parameter is exponential in accordance with theory, although only three estimates of the parameter were obtained. The calculated isotherms from this relationship were in reasonable agreement with the experimental data. The assumption of an energetically homogeneous surface of adsorption is not strictly true, thus one of the postulates of Langmuir theory is violated for methane adsorption on coal.

In order to test the validity of potential theory, methane characteristic curves were plotted for a coal run at several temperatures. The curves are coincident if either the reduced Kirchoff equation or extrapolated vapour pressures are used to obtain the pseudo-saturation pressure of the adsorbate. The adsorbate is assumed to have the same density as the liquid adsorptive at the normal boiling point. Correction for thermal expansion of the adsorbate in the limited range of temperatures used here does not appear to be necessary, although the correction may be appropriate for a wider temperature range.

In the pressure and temperature range studied here, the adsorption potential theory for methane adsorption appears to be valid. For higher pressures, application of the BET isotherm equation may be more appropriate, but this requires further testing.

Low-pressure characteristic curves were also plotted for carbon dioxide at two different temperatures for the same coal. The characteristic curve for this adsorbate appears to be temperature-invariant only for the case where thermal expansion is corrected for.

Future studies will involve testing the validity of potential theory for high-pressure methane and carbon dioxide adsorption on a variety of coals at a wider range of temperatures than those applied here. In addition mixed gas analyses will be performed in order to test the validity of potential theory for these gases.

Acknowledgements—Funding for this study was provided by NSERC grant A-7337 to R. M. Bustin. The authors would like to thank Dr Cliff Stanley for his help with statistical analysis.

REFERENCES

1. Gregg, S. J. and Sing, K. S. W., *Adsorption, Surface Area and Porosity*, 2nd edn. Academic Press, New York, 1982.
2. Ruppel, T. C., Grein, C. T. and Beinstock, D., *Fuel*, 1974, **53**, 152.
3. Kim, A. G., U.S. Bureau of Mines Report of Investigations 8245, 1977.
4. Smith, D. M. and Williams, F. L., *SPEJ* (October), 1984, 529.
5. Hall, F. E., Zhou, C., Gasem, K. A. M. and Robinson Jr, R. L., *SPE* 29194, paper presented at the 1994 Eastern Regional Conference and Exhibition, Charleston, WV, November 8–10, 1994, p. 329.
6. Polanyi, M., *Science*, 1963, **141**, 1010.
7. Lewis, W. K., Gilliland, E. R., Chertow, B. and Cadogen, W. P., *Ind. Eng. Chem.*, 1950, **42**, 1319.
8. Maslan, F. D., Altman, M. and Alberth, E. R., *J. Phys. Chem.*, 1953, **57**, 106.
9. Grant, R. J., Manes, M. and Smith, S. B., *AIChEJ.*, 1964, **8**, 403.
10. Cook, W. H. and Basmadjian, D., *Can. J. Chem. Eng.*, 1964, **42**, 146.
11. Agarwal, R. K. and Schwartz, J. A., *Carbon*, 1988, **26**, 873.
12. Amankwah, K. A. G. and Schwartz, J. A., *Carbon*, 1995, **33**, 1313.
13. Yang, J. T., *Gas Separation by Adsorption Processes*. Butterworth Publishers, London, 1987.
14. Yang, R. T. and Saunders, J. T., *Fuel*, 1985, **64**, 616.
15. Mavor, M. J., Owen, L. B. and Pratt, T. J., *SPE* 20728, paper presented at the SPE 65th Annual Technical Conference and Exhibition, New Orleans, LA, September 23–26, 1990b, p. 157.
16. Harpalani, S. and Pariti, U. M., in *Proceedings of the 1993 International Coalbed Methane Symposium*. The University of Alabama/Tuscaloosa, May 17–21, 1993, p. 151.
17. Langmuir, I., *J. Am. Chem. Soc.*, 1918, **40**, 1361.
18. Bell, G. J. and Rakop, K. C., *SPE* 15454, paper presented at the SPE 61st Annual Technical Conference and Exhibition, New Orleans, LA, October 5–8, 1986, p. 1.
19. Brunauer, S., Emmett, P. H. and Teller, E., *J. Am. Chem. Soc.*, 1938, **60**, 309.
20. Lowell, S. and Shield, J. E., *Powder Surface Area and Porosity*, 2nd edn. Chapman and Hall, London, 1984.
21. Dubinin, M. M., in *Chemistry and Physics of Carbon*, Vol. 2, ed. P. L. Walker, Jr. Edward Arnold, Ltd., New York, 1966.
22. Dubinin, M. M., in *Progress in Surface and Membrane Science*, Vol. 9, Ch. 1, ed. D. A. Cadenhead *et al.* Academic Press, New York, 1975.
23. Kapoor, A., Ritter, J. A. and Yang, R. T., *Langmuir*, 1989, **5**, 1118.
24. Bering, B. P., Dubinin, M. M. and Serpinsky, V. V., *J. Colloid Int. Sci.*, 1966, **21**, 378.
25. Wakasugi, Y., Ozawa, S. and Ogino, Y., *J. Colloid Int. Sci.*, 1976, **79**, 399.
26. Ozawa, S., Kusumi, S. and Ogino, Y., *J. Colloid Int. Sci.*, 1981, **56**, 83.
27. Bustin, R. M., Clarkson, C. R. and Levy, J., paper presented at the Twenty Ninth Newcastle Symposium

on "Advances in the Study of the Sydney Basin". Newcastle, NSW, Australia, April 6–9, 1995.

28. Bustin, R. M., Cameron, A. R., Grieve, D. A. and Kalkreuth, W. D., *Coal Petrology: Its Principles, Methods and Applications*, 2nd edn. Geological Association of Canada Short Course Notes 3, 1985.

29. Levy, J. H., Killingley, J. S., Day, S. J. and Liepa, I., in *Proceedings of the Symposium on Coalbed Methane Research and Development in Australia*. Townsville, Queensland, Australia, 1992.

30. *Handbook of Chemistry and Physics*, section 6-82, 71st edn., editor-in-chief David R. Lide. CRC Press, Boston, USA 1990–1991.

31. Yee, D., Seidle, J. P. and Hanson, W. B., in *Hydrocarbons from Coal*, AAPG Studies in Geology #38, Ch. 9, ed. B. E. Law and D. D. Rice. The American Association of Petroleum Geologists, Tulsa, Oklahoma, USA 1993, p. 203.

32. Marsh, H. and Siemieniewska, T., *Fuel*, 1965, **44,** 355.

33. Brunauer, S., *The Adsorption of Gases and Vapors*, Vol. 1. Princeton University Press, Princeton, 1943.

34. Koresh, J., *J. Colloid Int. Sci.*, 1982, **88,** 398.

35. Dubinin, M. M. and Astakhov, V. A., in *Advances in Chemistry Series*, No. 102. American Chemical Society Publications, Washington D.C. 1971, p. 102.

36. Noggle, J. H., *Physical Chemistry*, 2nd edn. Harper Collins Publishers, New York, NY 1989.

37. Toda, Y., Hatami, M., Toyoda, S., Yoshida, Y. and Honda, H., *Fuel*, 1971, **50,** 187.

38. Greaves, K. H., Owen, L. B. and McLennan, J. D., in *Proceedings of the 1993 International Coalbed Methane Symposium*. The University of Alabama/Tuscaloosa, May 17–21, 1993, p. 151.

Activated Carbon Compendium
H. Marsh (Editor)

Applicability of adsorption equations to argon, nitrogen and volatile organic compound adsorption onto activated carbon

Paul D. Paulsen, Brian C. Moore, Fred S. Cannon*

Department of Civil and Environmental Engineering, The Pennsylvania State University, University Park, PA 16802, USA

Received 10 April 1998; accepted 15 February 1999

Abstract

This research evaluates adsorption equations for argon, nitrogen, and volatile organic compound adsorption onto several commercially available activated carbons that represented a broad range of pore volume character, from predominantly microporous to predominantly mesoporous. For all these carbons, both the recently introduced (Paulson PD, Cannon FS, submitted to *Carbon*) Modified Freundlich equation and the Dubinin–Astakhov equation accurately characterized adsorption behavior in the relative pressure range of 1.0×10^{-5} to 0.1 for argon adsorption onto six different activated carbons and for nitrogen adsorption onto two different activated carbons. In particular, the Modified Freundlich equation offered a slightly better fit to the experimental data for mesoporous carbons in the full relative pressure range, and to all activated carbons in the relative pressure range of $10^{-2.5}$ to 0.1.

Micromeritics' density functional analysis software further used this argon and nitrogen adsorption data to characterize these activated carbons by their pore volume distribution. The Modified Freundlich's v_{0MF} term correlated best to the cumulative pore volume up to 60 Å and the Dubinin–Astakhov's v_{0DA} term correlated best to the cumulative pore volume up to 30 Å as determined via density functional theory. In addition, the research herein identified a consistency in the adsorption behavior of methylisobutylketone, *m*-xylene, and argon at their respective boiling points. © 1999 Elsevier Science Ltd. All rights reserved.

Keywords: A. Activated carbon; C. Adsorption; D. Adsorption properties; Microporosity

1. Introduction

An activated carbon system has been operating at the Marine Corps Maintenance Center in Barstow, CA. This activated carbon system, coupled with other unit processes, has been designed to remove and destroy volatile organic compounds (VOCs) resulting from spray coating operations. The applied goal for the research herein and related activity has been to better understand the properties of activated carbon in order to improve the performance of this system.

Activated carbons exhibit a high degree of porosity and an extensive internal surface area. The adsorption ability of activated carbon is most often experimentally characterized with an adsorption isotherm. Herein, the term 'adsorption isotherm' will be used to describe a collection of equilibrium measurements of the adsorption of a gas or vapor under isothermal conditions [2].

---

*Corresponding author.

The ability of an equation to model adsorption on porous materials is frequently improved as the number of independent variables is increased. However, the complexity involved in applying an equation to experimental adsorption data also increases when more coefficients are involved. Thus, the utility of a model is a function both of its accuracy and simplicity. For the comparisons herein, we have limited the isotherm equations to those with three independent coefficients.

Several of the more common techniques for modeling gas or vapor adsorption onto activated carbon are based on the Dubinin theory of micropore filling [3–6]. The two parameter Dubinin–Radushkevich (DR) equation [3,7,8] has been suggested to describe adsorption on homogeneous microporous solids [9–11].

Dubinin–Raduschkevich [3,7,8]: $\ln(v) = \ln(v_{0DA})$

$$- \left[\frac{RT}{\beta E_0} \ln\left(\frac{p_{SAT}}{p} \right) \right]^2 \qquad (1)$$

In Eq. (1), the coefficient v_{0DA} represents the extrapolated volume adsorbed when the system is at saturation, which corresponds to the volume inside the adsorbent's pores up to a threshold pore size. E_0 is the characteristic adsorption energy for a standard vapor (typically benzene), and β is a similarity coefficient [5,6]. The term v represents the volume adsorbed at a pressure, p.

Over the past 50 years, the DR equation has been widely used for the analysis of vapor adsorption on microporous adsorbents with type I isotherms [12–14]. However, for many activated carbons, the DR has not adequately described experimental adsorption isotherm data, especially in cases of high burn-off [2,15,16].

In these cases, a modification of the DR equation has been proposed, whereby the exponent 2 in the DR equation has been generalized with the variable n [5,17,18]. The resulting three coefficient equation has been referred to as the Dubinin–Astakhov (DA) equation.

Dubinin–Astakhov [5,17,18]: $\ln(v) = \ln(v_{0DA})$

$$+ k_{DA} \left[\ln\left(\frac{p_{SAT}}{p}\right) \right]^n \tag{2}$$

In Eq. (2), the coefficient k_{DA} is equal to $-(RT/\beta E_0)^n$; where the terms β and E_0 are the same as defined for Eq. (1). The improved accuracy of the DA equation, as compared with the DR equation, is due to the additional coefficient n. Typically, n ranges from 1 to 3, and it has been associated with the type of microporous structure [19]. To illustrate, isotherms of non-porous or mesoporous materials have been characterized by n values near 1 [20], while molecular sieve carbons have typically been characterized by n values close to 3 [19–22]. Expressed in another way, a small n value denotes a heterogeneous micropore structure with a broad size distribution of micropores, while a large n value suggests a homogeneous micropore structure with a narrow size distribution of micropores [15]. Bansal et al. [23] have fit adsorption data for strongly activated and heterogeneous carbons with n values between 1.5 and 2. For the activated carbons studied herein, the authors have found n to be between 1.6 and 2.4. Although the DA equation has enjoyed wide use, some researchers have questioned the applicability of the DA equation to carbons with wide distributions of micropores [2,15].

It is worth mentioning that some literature has discussed that an improved fit of the gas adsorption data can be achieved by using a series of two or more DR/DA equations to describe a single experimental adsorption isotherm (for instance, Marcilla-Gomis et al. [16] and Dubinin [6]). In short, each equation in the series is intended to correspond to adsorption in different ranges of microporosity [6,16]. Certainly, the added complexity of this type of technique will predict adsorption isotherm data more accurately than an isotherm produced via a single DA/DR equation. However, due to the number of parame-

ters in these equations (i.e. greater than 3), it was beyond the scope of this paper to evaluate them in more detail.

In a more complex generalization of the DR equation, the three-coefficient Dubinin–Stoeckli (DS) equation [5,6] has been proposed for describing adsorption in activated carbons with non-homogeneous micropore structure [6].

Dubinin–Stoeckli [5,6]: $v = \dfrac{v_0^0}{2\sqrt{1 + 2m\delta^2 A^2}} \exp$

$$\times \left[-\frac{mx_0^2 A^2}{1 + 2m\delta^2 A^2} \right] \left[1 + erf \frac{x_0}{\delta\sqrt{2}\sqrt{1 + 2m\delta^2 A^2}} \right] \tag{3}$$

In Eq. (3), the parameter A is the differential molar work of adsorption, defined as: $A = RT \ln(p_s/p)$ [5,6]. The three independent coefficients of this equation are v_0^0, the total volume of micropores; x_0, micropore half-width for the distribution curve maximum; and δ, the dispersion of pore sizes [5,6]. For adsorbents with homogeneous microporous structures, the DS equation reduces to the DR equation [6]. To illustrate the applicability of the DS equation, Dubinin [6] compared calculated adsorption values with experimental values. For the three activated carbons he studied, the DS method predicted values that were in good agreement with benzene adsorption data over wide relative pressure ranges.

In a companion paper [1], Paulsen and Cannon introduced a new algebraic equation called the 'Modified Freundlich Isotherm' (MF equation) to describe the physical adsorption of vapors in microporous adsorbents at equilibrium. Using a simpler temperature dependence for its equation coefficients, the MF equation compared favorably to the widely used DA equation when modeling the multi-temperature adsorption of methylisobutylketone (MIBK).

The MF equation has been developed by generalizing the Freundlich [24] isotherm equation through addition of a third term, $(p/p_{SAT})^{-\lambda}$, which contains the additional coefficient, λ.

Freundlich [24]: $\ln(v) = \ln(v_{0F}) + k_F \ln\left(\dfrac{p}{p_{SAT}}\right) \tag{4}$

Modified Freundlich [1]: $\ln(v) = \ln(v_{0MF})$

$$+ k_{MF}\left(\frac{p}{p_{SAT}}\right)^{-\lambda} \ln\left(\frac{p}{p_{SAT}}\right) \tag{5}$$

Having introduced this new equation and evaluated it for one adsorbent, this paper's first objective was to evaluate how well the MF equation performed for argon and nitrogen adsorption onto a variety of activated carbons, as compared with the DA equation. The second objective of this paper was to correlate the optimal fitting coefficients for the MF and DA equations to the activated carbon's characteristics of source material and pore size distribution. The third objective was to examine the influence of dataset size and distribution on the accuracy of these two equations. Our final objective was to determine whether the gas-phase adsorption behavior of different adsorbates could be normalized with respect to one another.

2. Experimental

2.1. Materials

In order to evaluate each equation's ability to model adsorption behavior, we examined several different types of activated carbons that covered a variety of carbon source materials and pore volume distributions that ranged from the predominantly microporous to the predominantly mesoporous. The names, characteristics and uses of these carbons were as follows.

- The Coconut BS was a virgin, coconut shell based granular activated carbon (GAC) that was manufactured by the Barnebey-Sutcliffe corporation. This GAC has been typically used for air phase adsorption applications including use in the aforementioned Barstow, CA air pollution control system. In addition, the Coconut BS evaluated herein was also studied in the companion paper [1].
- Coconut SA was a virgin, coconut shell based, GAC that was manufactured in Southeast Asia. This GAC type has been employed in air pollution control systems that are similar to the one in Barstow, CA.
- Agro A was a virgin, peat based, powdered activated carbon that was manufactured by Agrosorb of Russia. Agro A was manufactured for the cleanup of chemical spills.
- F-200 and F-400 were virgin, bituminous coal based, GACs that were manufactured by the Calgon Carbon Corporation. The F-400 that was used for argon analyses (F-400 A) had been placed in-situ, water fluidized once, cored, and sampled from the City of Cincinnati treatment plant. The F-400 for the nitrogen tests (F-400 B) was an unfluidized virgin GAC from a different lot.
- The Hydro 4000 was a virgin, lignite coal based, GAC manufactured by Norit Americas Incorporated. This GAC brand, known as Hydrodarco 4000, has commonly been employed in water treatment.
- The F-400 Reac. was a Filtrasorb 400 GAC that had undergone six cycles of water treatment service and thermal reactivation at the Richard Miller water treatment plant in Cincinnati, Ohio.

2.2. Methods

2.2.1. Activated carbon sample preparation

Prior to adsorption experiments, the activated carbon samples were heated for 1 day in an oven at 383 K to remove most of the moisture. Next, a test analysis tube that contained the activated carbon sample was attached to the degas port of a Micromeritics Accelerated Surface Area and Porisimetry (ASAP) 2000 instrument, where a degassing process removed yet further moisture and weakly bound volatiles that could interfere with the subsequent analysis process. The degassing process heated samples to 383 K while evacuating the tube to a pressure of 5×10^{-3} Torr (7×10^{-4} kPa).

2.2.2. Argon and nitrogen adsorption analyses

In a separate analysis port, the ASAP 2000 instrument collected adsorption data through repeated cycles of incremental addition of adsorbate to the tube, followed by equilibration of the pressure inside the tube, with procedures similar to those of Krupa and Cannon [25]. Adsorption isotherms were determined in the relative pressure range of 10^{-6} to 0.99 for argon at 87.3 K and nitrogen at 77.2 K, which were their respective boiling points at atmospheric pressure. These adsorption isotherms included 60–133 data points, and we intentionally collected more data at lower relative pressures in order to accommodate the demands of characterizing the micropore volume distributions with the density functional theory (DFT). The DFT software that was provided by Micromeritics contained a template of model adsorption isotherms that were calculated for 92 different pore widths. Specifically, this DFT software package applied this template of model adsorption isotherms to the experimental gas-adsorption isotherms for determination of a pore size distribution [25–27]. When considering any gas adsorption technique, the accuracy of such pore volume data for pores larger than 500 Å has decreased quantitative value [28]. For this reason, we have only evaluated pore volume data for pores smaller than 500 Å.

The saturation pressure of each adsorbate initially equaled atmospheric pressure. As these liquids accumulated dissolved gases and ice with time, their corresponding saturation pressures rose above ambient pressure by as much as 15–20 Torr (2.0–2.7 kPa) in prolonged runs (i.e. by 2.5% of the full atmospheric pressure range). The instrument included a saturation pressure tube that was also immersed in the bath, and the instrument corrected for thermal transpiration effects. Moreover, the authors' interpretation of volume data only up to a 500 Å pore width was done in order to avoid the confounding influences of evaluating relative pressures that approached the saturation pressure.

For duplicate argon adsorption experiments of the same type and lot of activated carbon, the cumulative pore volume measurements at 500 Å differed by a maximum of ±2.0%. For duplicate nitrogen adsorption experiments, the cumulative pore volume measurements differed by a maximum of ±1.0%.

2.2.3. Thermogravimetric analyses

Thermogravimetric analyses (TGA) were conducted with an ATI-Cahn TG-131 (Cerritos, CA). The methods and standard experimental error for the TGA data have been previously described [1]. We note that the argon adsorption experiments could inherently achieve tighter experimental accuracy than could the MIBK or m-xylene TGA experiments. Moreover, the argon adsorption experiments were comprised of multiple data points with progressively higher relative pressures for a single carbon sample, whereas the TGA experiments engaged a number of carbon samples for a given isotherm. Thus, the argon adsorption experiments could achieve tighter experimental precision relative to within-sample error.

2.2.4. Volumetric data presentation

Due to the way in which the data was collected in the companion paper [1], Eqs. (2), (4) and (5) were previously written in a gravimetric format (i.e. in terms of mass adsorbed per gram GAC, denoted as q and q_0). In this paper, it was more appropriate and convenient to represent adsorption data in terms of the equivalent liquid volume adsorbed (volume adsorbed per gram GAC, denoted as v and v_0). For the presentation of data herein, the adsorbate concentration was presented as if the adsorbate existed at its free liquid density. For argon at 87.3 K $\rho = 1.40$ g/ml, for nitrogen at 77.3 K $\rho = 0.808$ g/ml, for MIBK $\rho = 0.798$ g/ml was used for all temperatures, and for m-xylene $\rho = 0.864$ g/ml was used for all temperatures [29].

3. Results and discussion

3.1. Pore volume distributions

Argon adsorption data were collected for six different activated carbon types, while nitrogen adsorption data were collected for two different carbon types. Five of the pore volume distributions that were determined from argon adsorption data are shown in Fig. 1, while Fig. 2 shows typical incremental DFT pore volume data for two activated carbons. These pore volume distributions demonstrate the wide variety of samples that have been used to compare our adsorption equilibrium equations. The total micropore volumes, mesopore volumes, and surface areas of all the activated carbons studied are listed in order of descending microporosity in Table 1. Table 1 also shows the cumulative volume in pores with widths less than 30 Å and less than 60 Å.

We note that the cumulative pore size distributions all showed a plateau in the 8–10 Å region (Fig. 1). This result was thought to be an artifact of the ASAP 2000 software [25], and we have observed this in almost all of the activated carbons that we have evaluated (both herein and on other occasions). We are not aware of any intrinsic property of all these activated carbons that would cause

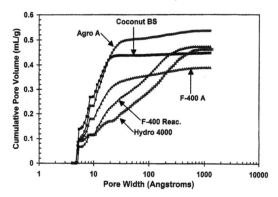

Fig. 1. Cumulative pore volume distributions for five activated carbons as calculated by Micromeritics 2000 density functional theory software.

this. Moreover, none of our data compilation over several years has led us to believe that this artifact has inconsistently skewed the proper interpretations for one activated carbon over another relative to determining such summary information as that shown in Table 1.

3.2. Comparison of isotherm equations

In order to compare how well the MF, DA, and DS equations predicted equilibrium behavior, argon adsorption equilibrium data were fitted to these equations over the same relative pressure range of 10^{-5} to 0.1. For meaningful interpretation, our results indicated that both the MF or DA equations provided appropriate fits over the relative pressure range of 10^{-5} to 10^{-1} for these sorbate/activated carbon systems. The authors would caution against extending these equations above a relative pressure of 0.1 to where adsorption was more influenced by filling of the mesopores. The adsorption data in the relative pressure range above 0.1 did not agree well with the fits developed

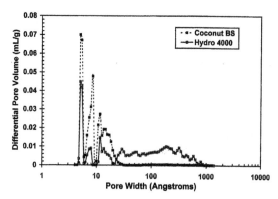

Fig. 2. Differential pore volume distributions for Coconut BS and Hydro 4000 as calculated by Micromeritics 2000 density functional theory software.

Table 1
Pore volume and surface area distributions for activated carbons in the micropore and mesopore regions as computed by Micromeritics 2000 density functional theory software

GAC type: (Adsorbate type)	Micropores (Less than 20 Å)			Mesopores (20 to 500 Å)		Pores <30 Å Pore volume ml/g	Pores <60 Å Pore volume ml/g
	Pore volume ml/g	Volume fraction in micropores	Surface area m^2/g	Pore volume ml/g	Surface area m^2/g		
Using argon							
Coconut BS	0.427	96.0%	1097	0.018	10	0.438	0.438
Coconut SA	0.390	90.5%	1058	0.041	20	0.406	0.415
Agro A	0.435	81.9%	1024	0.096	54	0.487	0.503
F-400 A	0.303	78.7%	763	0.082	36	0.330	0.349
F-400 Reac.	0.226	48.1%	534	0.244	76	0.261	0.312
Hydro 4000	0.170	38.2%	494	0.275	70	0.194	0.249
Using nitrogen							
F-200	0.248	79.5%	602	0.064	30	0.269	0.288
F-400 B	0.299	71.0%	652	0.122	67	0.356	0.386

from the adsorption data from the lower relative pressure range listed above. The MF and DA fits shown in the figures were extended to the relative pressure of 0.2 to demonstrate how these fits began to deviate from the adsorption data in the higher relative pressure regions.

The subtle difference between the curve fits produced by the DA [4,17,18], DS [5,6], and MF [1] equations are illustrated in Fig. 3. The two activated carbons shown in

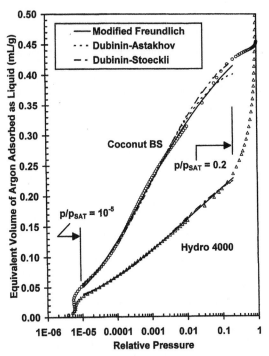

Fig. 3. Comparison of three isotherm equation types for argon adsorption onto Coconut BS and Hydro 4000.

Fig. 3 represent the most microporous (Coconut BS) and the most mesoporous (Hydro 4000) activated carbons that we studied. Furthermore, these activated carbons had the highest and lowest DFT surface areas, respectively. For the carbons we evaluated including those shown in Fig. 3, it seemed that the DS offered no clear advantage over the less complex DA and MF equations in this relative pressure range. The authors note that others have most often applied the DS equation to benzene adsorption data.

In order to develop the best MF and DA equation fits, the parameters n (DA) and λ (MF) of Eqs. (2) and (5) were varied to minimize the deviation between the equation and the equilibrium data for each adsorbent. At each different n and λ value, new optimal v_0 and k coefficients were calculated. As defined by the minimum coefficient of variation over the full pressure range, the optimum λ and n values were 0.14 and 2.4, respectively, for the Coconut BS. This minimization is illustrated in Fig. 4. The coefficients of variation herein were calculated using the following equation [30]:

$$CV = 100 \times \sqrt{\frac{\sum_{i=1}^{m} (v_{i,\text{Model}} - v_{i,\text{Data}})^2}{m}} \bigg/ \frac{\sum_{i=1}^{m} v_{i,\text{Data}}}{m} \quad (6)$$

In Eq. (6), CV is the coefficient of variation in units of percent, $v_{i,\text{Data}}$ is the equivalent volume adsorbed in ml/g for point i at its given relative pressure, $v_{i,\text{Model}}$ is the corresponding equivalent volume adsorbed as calculated by the isotherm equation at this same relative pressure, and m is the number of data points inside the relative pressure range of the calculation. The denominator of Eq. (6) represents the average value of v_{Data} over the relative pressure range of the calculation.

Fig. 5 presents some typical data for argon adsorption onto four other GACs, as well as the MF modeling of this data in the range of $p/p_{\text{SAT}} = 10^{-5}$ to 0.1. The MF and DA

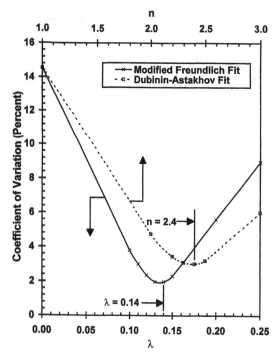

Fig. 4. Optimizing n and λ coefficients for Dubinin–Astakhov and Modified Freundlich equations for Coconut BS, using full set of argon adsorption data.

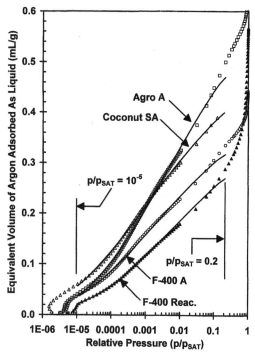

Fig. 5. Argon adsorption data for Agro A, Coconut SA, F-400 A, and F-400 Reac. with Modified Freundlich equation fits.

have also been compared with the full sets of experimental data for all the carbons used in this study, and their coefficients were optimized in the same manner as has been presented in Fig. 4.

Both the best fit coefficients of v_{0MF}, k_{MF}, and λ for the MF equation and the best fit coefficients of v_{0DA}, k_{DA}, and n for the DA equation are listed in Table 2. The bold values in Table 2 represent the coefficients obtained using the full datasets. For the argon data, these values indicated that increased mesopore volume corresponded to decreased λ and n values. For example, in the most mesoporous carbon, the Hydro 4000, the optimal λ and n values (0.07 and 1.7, respectively) were the lowest values among the carbons studied when argon was used as an adsorbate. Likewise, the λ and n values for the most microporous carbon, the Coconut BS, were the highest values (0.14 and 2.4, respectively). With regard to the adsorbate type, we note that the optimal λ and n coefficients were higher for the argon adsorption data when compared to the nitrogen adsorption data. We note again that nitrogen adsorption was studied for only two activated carbons. The two most microporous carbons, which were both coconut based, had the smallest (in magnitude) k_{DA} and k_{MF} values.

For each carbon, the authors optimized one single MF equation and one single DA equation for the full relative pressure range of 10^{-5} to 0.1. Using Eq. (6), we then computed the coefficient of variation (CV) that depicted how these full-range fitted values differed from the actual data over three relative pressure ranges: (1) 10^{-5} to 10^{-4}, (2) 10^{-4} to $10^{-2.5}$, and (3) $10^{-2.5}$ to 10^{-1}. For the argon data, the CV values varied between 0.74 and 4.75%, as shown in Table 3. These small values indicated a strong agreement between the MF and DA equations and the isotherm data. To put these CV values into perspective, the standard deviations (the numerator of Eq. (6)) for these fits ranged from 0.00075 to 0.012 ml/g.

For the less microporous activated carbons, the MF typically gave a slightly better fit (i.e. lower CV) in the p/p_{SAT} range of 10^{-5} to 10^{-4}. Conversely, for the more microporous activated carbons, the DA was slightly more accurate at representing the experimental data in the p/p_{SAT} range of 10^{-5} to 10^{-4}. In the p/p_{SAT} range of 10^{-4} to $10^{-2.5}$, the MF equation performed equal to, or better than, the DA equation for all of the carbons and both adsorbates.

Above the relative pressure of $10^{-2.5}$, the MF equation provided a better fit to the experimental adsorption data than did the DA equation for all of the activated carbons and both adsorbates that we studied. The fewest number of data points in these standard argon adsorption experiments occurred between the relative pressures of 0.01 and 0.1 (as

Table 2

Coefficients for Modified Freundlich (MF) and Dubinin–Astakhov (DA) adsorption isotherms or polytherms (Eqs. (2) and (5)) using argon, nitrogen, methylisobutylketone, and m-xylene data in the region from $p/p_{SAT} = 10^{-5}$ to 0.1. Values presented for full dataset, nine-point set, and five-point set

Sample: (Adsorbate type)	Number of data points	Modified Freundlich			Dubinin–Astakhov		
		v_{0MF} ml/g GAC	k_{MF} unitless	λ	v_{0DA} ml/g GAC	k_{DA} unitless	n
Argon at 87.3 K							
Coconut BS	**77**	**0.446**	**0.0381**	**0.14**	**0.408**	**−0.0058**	**2.4**
	9	0.453	0.0422	0.13	0.415	−0.0074	2.3
	5	*0.466*	*0.0474*	*0.12*	*0.428*	*−0.0096*	*2.2*
Coconut SA	**50**	**0.437**	**0.0451**	**0.12**	**0.389**	**−0.0070**	**2.3**
	9	0.430	0.0442	0.12	0.395	−0.0088	2.2
	5	*0.441*	*0.0495*	*0.11*	*0.387*	*−0.0069*	*2.3*
Agro A	**84**	**0.527**	**0.0586**	**0.12**	**0.469**	**−0.0116**	**2.2**
	9	0.535	0.0646	0.11	0.478	−0.0147	2.1
	5	*0.556*	*0.0727*	*0.10*	*0.482*	*−0.0148*	*2.1*
F-400 A	**59**	**0.351**	**0.0506**	**0.12**	**0.317**	**−0.0100**	**2.2**
	9	0.357	0.0562	0.11	0.323	−0.0128	2.1
	5	*0.358*	*0.0561*	*0.11*	*0.326*	*−0.0128*	*2.1*
F-400 Reac.	**51**	**0.307**	**0.0659**	**0.11**	**0.273**	**−0.0149**	**2.1**
	9	0.299	0.0599	0.12	0.267	−0.0120	2.2
	5	*0.305*	*0.0614*	*0.12*	*0.273*	*−0.0124*	*2.2*
Hydro 4000	**34**	**0.262**	**0.0765**	**0.07**	**0.240**	**−0.0291**	**1.7**
	9	0.251	0.0677	0.08	0.231	−0.0227	1.8
	5	*0.253*	*0.0682*	*0.08*	*0.233*	*−0.0230*	*1.8*
Nitrogen at 77.3 K							
F-200	**37**	**0.350**	**0.0489**	**0.07**	**0.338**	**−0.0238**	**1.6**
	9	0.342	0.0435	0.08	0.323	−0.0145	1.8
	5	*0.345*	*0.0442*	*0.08*	*0.326*	*−0.0149*	*1.8*
F-400 B	**44**	**0.445**	**0.0690**	**0.06**	**0.416**	**−0.0300**	**1.6**
	9	0.418	0.0542	0.08	0.390	−0.0181	1.8
	5	*0.425*	*0.0553*	*0.08*	*0.397*	*−0.0187*	*1.8*
Methylisobutylketone adsorption [1]							
Coconut BS at 298 K	51	0.462	0.0134	0.18	0.452	−0.0030	2.4
Coconut BS at 323 K	13	0.444	0.0147	0.18	0.431	−0.0031	2.4
Coconut BS at 348 K	14	0.434	0.0186	0.18	0.419	−0.0041	2.4
Coconut BS at 378 K	48	0.454	0.0276	0.18	0.444	−0.0063	2.4
Coconut BS MIBK polytherm	126	0.459	-0.041 $+1.81\times10^{-4}T$	0.18	0.447	-0.0018 $-5.28\times10^{-9}T^{2.4}$	2.4
m-Xylene adsorption [31]							
Coconut BS at 298 K	29	0.463	0.0121	0.13	0.455	−0.0040	2.0
Coconut BS at 318 K	24	0.472	0.0204	0.13	0.457	−0.0066	2.0
Coconut BS at 338 K	26	0.467	0.0238	0.13	0.452	−0.0077	2.0
Coconut BS at 358 K	25	0.477	0.0318	0.13	0.462	−0.0107	2.0
Coconut BS at 378 K	30	0.464	0.356	0.13	0.451	−0.0123	2.0
Coconut BS m-xylene polytherm	134	0.466	-0.075 $+2.93\times10^{-4}T$	0.13	0.478	-1.54×10^{-7} $-8.92\times10^{-8}T^{2.0}$	2.0

shown in Fig. 6). When we optimized the MF and DA equations for the full set of data, we weighed every point equally throughout the full relative pressure range. Thus, both equations exhibited their greatest deviation (as graphically observed) from the data in the p/p_{SAT} region above $10^{-2.5}$, as would be expected.

3.3. Sensitivity of equation coefficients to dataset selection

An important time-saving aspect of a modeling equation is the ability to use a relatively small experimental dataset to accurately depict the adsorption behavior. With this in

Table 3

Comparison of Dubinin–Astakhov and Modified Freundlich equations to adsorption data in three relative pressure regions. Equation coefficients have been determined through regression of full adsorption dataset in the region from $p/p_{SAT}=10^{-5}$ to 0.1. Tabulated values represent the coefficients of variation

GAC type: (Adsorbate type)	Coefficients of variation[a]					
	$p/p_{SAT}=10^{-5}$ to 10^{-4}		$p/p_{SAT}=10^{-4}$ to $10^{-2.5}$		$p/p_{SAT}=10^{-2.5}$ to 10^{-1}	
	Modified Freundlich Percent	Dubinin–Astakhov Percent	Modified Freundlich Percent	Dubinin–Astakhov Percent	Modified Freundlich Percent	Dubinin–Astakhov Percent
Using argon						
Coconut BS	3.97	2.13	1.54	1.57	1.84	3.59
Coconut SA	3.22	2.17	1.62	1.41	1.09	2.05
Agro A	4.75	3.11	2.12	2.11	1.63	3.01
F-400 A	3.33	2.04	1.58	1.85	1.96	3.27
F-400 Reac.	3.06	4.04	1.26	2.09	2.19	3.62
Hydro 4000	0.97	2.11	0.74	1.48	1.31	2.20
Using nitrogen						
F-200	2.36	3.65	2.24	3.27	1.89	2.18
F-400 B	3.71	4.62	3.58	4.21	3.28	3.90

[a] Note that a single DA equation and a single MF equation has been applied to the full range of $p/p_{SAT}=10^{-5}$ to 0.1.

mind, we have attempted to examine the dependence of each equation on the distribution and number of experimental data points used for generating each fit. Specifically, we have also modeled each isotherm using both a

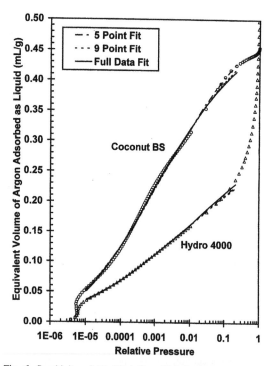

Fig. 6. Sensitivity of Modified Freundlich isotherms to dataset selection for Coconut BS and Hydro 4000.

five- and nine-point dataset whose points were equally distributed on a logarithmic scale. The relative pressures selected for the five-point isotherms were: 10^{-5}, 10^{-4}, 10^{-3}, 10^{-2}, and 10^{-1} and the relative pressures for the nine-point isotherms were: 10^{-5}, $10^{-4.5}$, 10^{-4}, $10^{-3.5}$, 10^{-3}, $10^{-2.5}$, 10^{-2}, $10^{-1.5}$, and 10^{-1}. Equation coefficients for the MF and DA equations that were determined from the five-point, nine-point, and full dataset fits are listed in Table 2 for all the carbons studied herein. For these cases, changing the dataset size and evenly distributing the data produced at most a change of 6% in the v_0 coefficients. The greatest changes in the k coefficients occurred when the optimal λ or n changed for a particular sorbate/carbon system.

The effects of dataset selection on the MF equation has been further illustrated in Fig. 6 for Coconut BS and Hydro 4000. Because these equation fits used evenly distributed datasets, the five- and nine-point fits more closely represented the experimental data than the full set fits, when evaluating relative pressures above $10^{-2.5}$. This pressure range was where the full dataset contained the fewest data points.

The coefficients of variation for the nine-point DA and MF fits (Table 4) show that, in general, both equations fit the experimental data well. Overall, the accuracy of the nine-point fits were roughly equal to their full dataset counterparts for the argon datasets. Similar to the findings for the full dataset in the relative pressure range of 10^{-5} to 10^{-4}, the DA fit the data slightly better than the MF for the more microporous activated carbons. Furthermore, in this pressure range, the MF had slightly lower CV values than did the DA equation for the less microporous activated carbons. Overall, the MF had smaller CV values than did

Table 4
Comparison of nine-point Dubinin–Astakhov and Modified Freundlich equations to full adsorption dataset in three relative pressure regions. Equation coefficients have been determined through regression of adsorption data at $p/p_{SAT} = 10^{-5}$, $10^{-4.5}$, 10^{-4}, $10^{-3.5}$, 10^{-3}, $10^{-2.5}$, 10^{-2}, $10^{-1.5}$, and 10^{-1}. Tabulated values represent the coefficients of variation

| GAC type: (Adsorbate type) | Coefficients of variation[a] | | | | | |
| | $p/p_{SAT} = 10^{-5}$ to 10^{-4} | | $p/p_{SAT} = 10^{-4}$ to $10^{-2.5}$ | | $p/p_{SAT} = 10^{-2.5}$ to 10^{-1} | |
	Modified Freundlich Percent	Dubinin–Astakhov Percent	Modified Freundlich Percent	Dubinin–Astakhov Percent	Modified Freundlich Percent	Dubinin–Astakhov Percent
Using argon						
Coconut BS	4.03	2.34	2.21	2.61	1.34	2.63
Coconut SA	4.00	2.60	1.85	2.12	0.76	1.60
Agro A	5.34	2.74	3.17	3.78	1.24	2.13
F-400 A	3.34	0.91	2.49	3.62	1.48	2.42
F-400 Reac.	5.16	6.84	1.37	2.82	2.20	3.65
Hydro 4000	0.99	1.98	0.56	1.50	0.57	1.46
Using nitrogen						
F-200	1.95	2.72	2.23	2.91	1.64	2.00
F-400 B	2.55	3.54	3.13	3.93	2.66	3.21

[a] Note that a single DA equation and a single MF equation has been applied to the full range of $p/p_{SAT} = 10^{-5}$ to 0.1.

the DA fit for the argon data in the relative pressure range of 10^{-4} to 10^{-1}.

3.4. Similarities of pore volumes to DA and MF data fit parameters: v_{oDA} and v_{oMF}

The v_{oDA} values were almost always lower than their v_{oMF} counterparts (Table 2), suggesting that these two coefficients had different phenomenological meanings. As proposed by Dubinin [5,6], v_{oDA} is defined as the cumulative volume of pores with widths less than 28 Å to 32 Å, and this is the range that Dubinin refers to as micropores. Note that Dubinin's definition of the micropore size range is different than IUPAC's definition (width < 20 Å) [13]. Table 1 presents the cumulative volume in pores less than 30 Å as determined via DFT. Values for the coefficient v_{oDA} matched their corresponding 30 Å cumulative pore volume well for all of the carbons when argon was the adsorbate. In fact, these values were within 5% of each other for all carbons but one. The sole exception was the highly mesoporous Hydro 4000, for which these two values were within 18% of each other.

Similarly, a correlation between the MF coefficient v_{oMF} and its upper pore limit has been evaluated, whereby we found it was best correlated to the cumulative volume in pores with widths less than roughly 60 Å for our argon adsorption data. As shown in Table 1 for argon, the 60 Å cumulative pore volume data and the corresponding v_{oMF} values (in Table 2) were within 10% of each other. These similarities could have revealed evidence for a linkage to an underlying phenomenon, or they could have been partially fortuitous. We temper any interpretation regarding a linkage between these values by acknowledging that

adsorption occurs by means of both condensation in small pores and wall surface coverage in larger pores.

3.5. Consistency of argon, MIBK, and m-xylene adsorption behavior

We have also compared experimental and predicted adsorption data for MIBK, m-xylene, and argon onto Coconut BS GAC. Fig. 7 shows experimental adsorption data for MIBK (at 298 and 378 K) [1], m-xylene (at 338 and 358 K) [31], and argon (at 87.3 K); additionally, the MF full data fit is shown for the argon data. As the temperature was increased toward the MIBK and m-xylene boiling points (390 K and 412 K, respectively), the MIBK and m-xylene adsorption data progressively became closer to the argon adsorption data, which was analyzed at its boiling point. Moreover, as shown in Fig. 8, when MF polytherm fits were used to project MIBK and m-xylene adsorption at their boiling points, the polytherms matched (within 8%) the argon adsorption data over the p/p_{SAT} range of 10^{-4} to 0.1. At their boiling points the three adsorbates would have the same saturation pressures. In this way, we compared adsorption when the relative pressure and adsorption partial pressures matched.

4. Conclusions

For the wide range of activated carbons that we studied, the MF and DA equations both performed well in the relative pressures between 10^{-5} and 0.1. Furthermore, for argon adsorption in the 10^{-5} to 10^{-4} relative pressure range, the DA equation performed slightly better than the

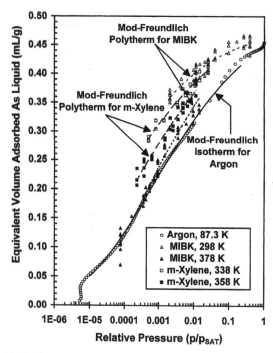

Fig. 7. Comparison of argon, methylisobutylketone, and *m*-xylene adsorption onto Coconut BS at several temperatures.

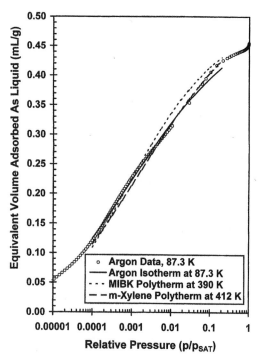

Fig. 8. Comparison of Modified Freundlich equation predictions for argon, methylisobutylketone, and *m*-xylene adsorption onto Coconut BS at their respective boiling points, where saturation pressure equals one atmosphere (101 kPa) for all three adsorbates.

MF equation for the more microporous carbons; conversely, the MF equation performed slightly better than the DA for the less microporous carbons. Above a relative pressure of 10^{-4}, the MF was able to depict argon adsorption better than or equal to the DA equation, for the six activated carbons that were evaluated. As a practical matter, we note that the $>10^{-4}$ relative pressure range also represents the more important range for modeling VOC adsorption onto activated carbon from paint spray booths.

When nitrogen was used as an adsorbate for the two activated carbons that were evaluated, the MF equation performed better than the DA in all three relative pressure ranges considered. Moreover, we found that the best fit λ and n values differed for argon, nitrogen, and MIBK [1] (Table 2), and we surmise the type of adsorbate will be one significant factor in determining the proper value of λ or n. Furthermore, both the λ and n values increased with increasing microporosity.

Parenthetically, we note that for the DA best fit equations herein, the n range of 1.6 to 2.4 bracketed the 2.0 that corresponds to the DR equation. When either the MF or DA equation was fitted to five or nine equally spaced points (on a logarithmic scale) over the 10^{-5} to 10^{-1} relative pressure range, the conformance to the full set of experimental data was almost as good as when the equations were fit to all of the data points (i.e. 34 to 84 points).

The v_{0DA} and v_{0MF} terms were empirically correlated to

the available pore volume for adsorption up to a threshold size. For argon adsorption onto the carbons studied, we found that Dubinin's assessment of v_{0DA} corresponding to the 30 Å cumulative pore volume was usually in close agreement to the 30 Å cumulative pore volume that we determined via the density functional theory. Additionally, the MF parameter v_{0MF} correlated best to the 60 Å cumulative pore volume when argon was used as the adsorbate. However, these correlations may have been partially fortuitous.

The adsorption of argon onto a coconut-based GAC at argon's boiling point was quite similar to the polytherm-predicted adsorption of both MIBK and *m*-xylene at their respective boiling points onto this same GAC. This uniformity of normalized adsorption behavior offers a promising opportunity: namely, to approximate the adsorption of a lightly tested VOC by normalizing it to the adsorption behavior of an extensively tested VOC onto the same activated carbon.

Acknowledgements

This research was funded in part by the Strategic Environment Research and Development Program (US

Marine Corps, DOE, DOD, EPA), by the US Navy through AASERT, and by the City of Cincinnati. Specific thanks are extended to Lewis Watt of ARL, John House and Ron Vargo of the Marine Corps, Chuck Darvin of the EPA, Lynn Shugarman and Jerry Smith of TerrAqua Environmental Systems, and Cliff Shrive of the City of Cincinnati for their input. The interpretations herein represent those of the authors only.

References

[1] Paulsen PD, Cannon FS. Submitted to Carbon 1998.

[2] Rodríguez-Reinoso F. In: Marsh H, Heintz EA, Rodriguez-Reinosos F, editors, Introduction to carbon technologies, University of Alicante, Spain, 1997, p. 35.

[3] Dubinin MM, Zaverina ED, Radushkevich LV. Zh Fiz Khim 1947;21:1351.

[4] Dubinin MM. In: Cadenhead DA, Danielli JF, Rosenberg MD, editors, Progress in surface and membrane science, Vol. 9, Academic Press, New York, 1975, p. 1.

[5] Dubinin MM. Carbon 1985;23:373.

[6] Dubinin MM. Carbon 1989;27:457.

[7] Dubinin MM, Zaverina ED. Zh Fiz Khim 1949;23:1129.

[8] Dubinin MM. Zh Fiz Khim 1965;39:1305.

[9] Stoeckli HF. J Colloid Interface Sci 1977;59:184.

[10] Dubinin MM, Stoeckli HF. J Colloid Interface Sci 1980;75:34.

[11] Jaroniec M, Piotrowska J. Monatsh Chem 1986;117:7.

[12] Aranovich GL, Donohue MD. Carbon 1995;33:1369.

[13] Sing KSW, Everett DH, Haul RAW, Moscou L, Pierotti RA, Rouquerol J, Siemieniewska T. Pure Appl Chem 1985;57:603.

[14] Sing KSW. Carbon 1994;32:1311.

[15] Rodríguez-Reinoso F, Linares-Solano A. In: Thrower PA, editor, Chemistry and physics of carbon, Vol. 21, Dekker, New York, 1989, p. 1.

[16] Marcilla-Gomis A, Garcia-Cortes AN, Martin-Martinez JM. Carbon 1996;34:1531.

[17] Dubinin MM, Astakhov VA. Adv Chem Ser 1970;102:69.

[18] Dubinin MM, Astakhov VA. Izv Akad Nauk SSSR Ser Khim 1971;11:5.

[19] Jaroniec M, Lu X, Madey R, Choma J. Carbon 1990;28:243.

[20] Byrne JF, Marsh H. In: Patrick JW, editor, Porosity in carbons, Halsted Press, New York, 1995, p. 1.

[21] Finger G, Bülow M. Carbon 1979;17:87.

[22] Stoeckli HF. Carbon 1990;28:1.

[23] Bansal RC, Donnet JB, Stoeckli HF. Active carbon, Marcel Dekker, New York, 1988.

[24] Freundlich H. Colloid and capillary chemistry, E.P. Dutton, New York, 1926.

[25] Krupa NE, Cannon FS. AWWAJ 1996;88:94.

[26] Olivier JP. J Porous Mater 1995;2:9.

[27] Webb PA, Olivier JP, Conklin WB. Am Lab 1994;26:38.

[28] McEnaney B, Mays TJ. In: Patrick JW, editor, Porosity in carbons, Halsted Press, New York, 1995.

[29] Lide DR, editor, 73rd ed, CRC handbook of chemistry and physics, CRC Press, Ann Arbor, 1992.

[30] Steel RG, Torrie JH. Principles and procedures of statistics, a biometrical approach, McGraw-Hill, New York, 1980.

[31] Paulsen PD. Mass transfer and adsorption of volatile organic compounds onto granular activated carbon during loading and regeneration. Ph.D. thesis, Penn State University, University Park PA, USA, 1998.

Activated Carbon Compendium
H. Marsh (Editor)

A NEW MODEL FOR THE DESCRIPTION OF ADSORPTION KINETICS IN HETEROGENEOUS ACTIVATED CARBON

D. D. Do,* and K. Wang

Department of Chemical Engineering, The University of Queensland, St Lucia, Qld 4072, Australia

(*Received* 24 *October* 1996; *accepted in revised form* 5 *February* 1998)

Abstract—A new model for the description of adsorption kinetics in heterogeneous activated carbon is presented in this paper. The activated carbon particle is composed of the fluid phase and the adsorbed phase, the latter of which is heterogeneous. This heterogeneity is assumed to be described by a distribution in the energy of interaction between the two phases. This distribution is obtained from the information of the adsorption equilibria of all species. The kinetics model assumes three processes occurring within the porous particle: (1) the pore volume diffusion, (2) the adsorbed phase diffusion and (3) the finite mass interchange between the molecules in the fluid phase and those in the energy distributed adsorbed phase. The distribution of energy of interaction is accounted for in the last two processes. These three processes are found to have rates that are comparable in magnitude, and depending on the adsorbate, the operating conditions and the mode of operation (adsorption or desorption) one or two of these processes dominate the overall uptake. The model is tested with the experimental data collected in our laboratories using the volumetric isotherm apparatus for equilibria and the differential adsorption bed for kinetics. Seven adsorbates were used, and a wide range of parameters as well as operating conditions were also used to validate the mathematical model. It is found that the model proposed describes well the adsorption as well as desorption kinetics. © 1998 Elsevier Science Ltd. All rights reserved.

Key Words—A. Activated carbon, C. adsorption, D. diffusion, D. porosity, D. reaction kinetics, D. equilibria, D. kinetics.

NOMENCLATURE

a	ratio of diffusion energy to adsorption energy (0.5)
b_0	adsorption affinity at zero energy level (kPa^{-1})
b_∞	adsorption infinity at zero energy level and infinite temperature (kPa^{-1})
C_μ	adsorbed phase concentration ($mol\ cm^{-3}$)
$C_{\mu s}$	maximum adsorbed phase concentration ($mol\ cm^{-3}$)
D_p	pore diffusivity ($m^2\ s^{-1}$)
D_μ	surface diffusivity ($m^2\ s^{-1}$)
D_μ^0	surface diffusivity at zero energy level ($m^2\ s^{-1}$)
E_{mean}	average adsorption energy ($J\ mol^{-1}$)
$E(r)$	adsorption energy inside the pores with radius r ($J\ mol^{-1}$)
e	reduced adsorption energy
J_p	pore flux ($mol\ m^{-2}\ s^{-1}$)
J_μ	surface flux ($mol\ m^{-2}\ s^{-1}$)
k_a	adsorption rate constant ($kPa^{-1}\ s^{-1}$)
k_d	desorption rate constant (s^{-1})
$k_{d,\infty}$	desorption rate constant at zero energy level (s^{-1})
k_m	film mass transfer coefficient (m^{-1})
L	mobility constant of adsorbed phase
P	gas phase pressure (kPa)
P^*	hypothetical gas phase pressure (kPa)
P_b	bulk phase pressure (kPa)
Q	distribution parameter $= q^* \sigma_{12}$
q	gamma distribution parameter
r	pore half-width in eqns (43)–(49); particle coordinate elsewhere (m)
r_{min}	minimum pore half-width accessible to adsorbate with zero adsorption potential (m)
T	temperature (K)
t	time (s)
$U_p(z)$	pore potentials at distance z ($J\ mol^{-1}$)
U_s^*	interaction potential minimum for single graphite lattice ($J\ mol^{-1}$)
x	reduced pore half-width (r/σ_{12})
x_{min}	minimum accessible reduced pore half-width ($=0.8885$ in this paper)
z	distance between adsorbate and pore wall (m)

Greek symbols

β	property of solid [$kPa^{-1}(g\ K\ mol^{-1})^{1/2}$]
δ	temperature dependence coefficient of adsorption capacity
σ_{12}	Lennard–Jones collision diameter (m)
ϵ	particle porosity
γ	distribution parameter
μ	chemical potential

1. INTRODUCTION

Adsorption kinetics in activated carbon or other similar microporous solids has posed a very interesting and challenging problem. This is attributed to the complexity of the activated carbon structure. Attempts to model the solid microscopically appear to yield some moderate success, but problems still remain. Until some surface analysis tools are sophisticated enough to give us more information about the structure, modelling of diffusion and adsorption in activated carbon still contains some degree of empiricism. Nevertheless, mathematical models for diffusion and adsorption in activated carbon have achieved good success in simulating kinetics data reasonably well. Parameters such as particle size, bulk concentration, the adsorption affinity etc. and operating conditions of pressure and temperature are

---

*Corresponding author.

Reprinted from *Carbon* **36** (**10**), 1539-1554 (1998)

fully accounted for by the mathematical models. Do and co-workers have developed a number of models over the past five years, and the models were successfully used to simulate numerous kinetics data of hydrocarbons, carbon dioxide and sulphur dioxide in activated carbon [1–4]. Single component systems as well as multicomponent systems were considered by Do and co-workers, and experimental data were carefully collected using the method of the differential adsorber bed. Only with reliable data can we infer some discrimination about the mathematical models.

1.1 Experimental observations

For the successful development of a mathematical model for mass transfer in activated carbon, the model must explain the following experimental observations:

(1) The sorption time scale is proportional to the square of the particle radius when the particle is large. We shall present the criteria of when the particle is considered as large later.

(2) The sorption time scale is proportional to the particle radius raised to some power which is greater than zero and less than 2 when the particle size is intermediate.

(3) When the particle size is very small, the sorption time is independent of the particle size, which means that the sorption rate is limited by the size of the primary unit within the particle. This primary unit does not change with the crushing of the particle.

(4) The sorption time to approach equilibrium is shorter when the bulk concentration increases and when the temperature increases.

(5) The sorption rate decreases as the temperature increases.

(6) The sorption rate into the particle can not be accounted for by the pore diffusion. There must be an additional transport mechanism in parallel with the pore diffusion. This additional flux is attributed to the diffusion of the adsorbed species.

(7) The diffusivity of the adsorbed species increases strongly with the concentration loading on the surface.

(8) The adsorption equilibrium between the fluid and adsorbed phases can not be accounted for by simple equilibrium isotherm equations, such as the Langmuir equation.

(9) The adsorption phase is very heterogeneous as observed by the electron microscopy, or from the analysis of the adsorption equilibrium data or the observation of the desorption behaviour.

(10) The desorption time scale is longer than the adsorption time scale.

(11) The desorption curve exhibits a very long tail, indicating the distribution of energy sites on the surface with high energy sites releasing molecules at a much lower rate.

1.2 The past models

To account for the experimental observations mentioned in the last section, Do and co-workers have developed models such as the Heterogeneous Macropore and Surface Diffusion model, known as the HMSD model in the literature [3,5]. This model was applied with great success to simulate the experimental data of ethane, propane, n-butane, pentane, benzene, toluene, carbon dioxide and sulphur dioxide collected in our laboratories. Single component systems as well as multicomponent systems were tested with the HMSD model. For completeness, we list below the assumptions leading to the formulation of the HMSD model.

(1) The transport mechanisms of adsorbate molecules into the particle are due to the pore and surface diffusions which occur in parallel.

(2) The pore diffusion is characterised by the molecular diffusion and Knudsen diffusion mechanisms.

(3) The surface diffusion is driven by the chemical potential of the adsorbed phase.

(4) The fluid phase and the adsorbed phase are in local equilibrium with each other at any point within the particle and at any time.

(5) The adsorbed phase is heterogeneous in the sense that its heterogeneity is characterised by the distribution of the interaction energy between the adsorbate molecule and the surface.

(6) The energy distribution is taken as either uniform or gamma distribution.

With this HMSD model, we were able to simulate the numerous data collected in our laboratories. This model, however, fails to account for the dependence of the adsorption rate on the particle radius when the particle size is intermediate and it also fails to exhibit the long tail behaviour of the desorption curve. We have resolved the particle size dependence by developing a model called HMSMD [4], which is to account for the Heterogeneity of the particle, the Micropore diffusion, and the Macropore and Surface diffusion, but similar to the HMSD model it fails to explain the long tail desorption curve. Thus, we seem to overlook one or more important processes which become important in the desorption mode. Furthermore, one key assumption in the development of the HMSMD model is that the flux of the adsorbed species in the microparticle coordinate and that along the particle coordinate are defined as follows:

$$J_\mu = -\beta^2 D_\mu(C_\mu) \frac{\partial C_\mu}{\partial r_\mu} \quad (1)$$

$$J_\mu = -D_\mu(C_\mu) \frac{\partial C_\mu}{\partial r} \quad (2)$$

where we argued that the diffusion along the microcoordinate is more restricted than that along the particle; hence the introduction of the β^2 parameter in eqn (1). Although one would expect such an argument is reasonable based on the fact that the activated carbon structure could be non-isotropic.

However, despite the fact that the structure is non-isotropic, the arrangement of various components in activated carbon is quite random. Hence it does not seem logical to assume that the diffusion of the adsorbed phase species, in an average sense, is more restricted along the microparticle coordinate compared to that along the particle coordinate. Furthermore, from the standpoint of the adsorbate molecule, it can not distinguish whether it would diffuse along the coordinate leading into the particle interior or it would diffuse in the direction transverse (microparticle coordinate) to the particle coordinate. Thus, the introduction of the microparticle diffusion mechanism in the HMSMD model seems to resolve the observation of time scale of adsorption following the dependence of R^n, where n is between 0 and 2, but it does introduce an unsound notion of difference rates in diffusion of the adsorbed phase in the two directions.

2. THEORY

Recently, our experimental data of diffusion of the adsorbed species show that the adsorbed diffusivity increases very rapidly with the loading, and a new theory for the diffusion mechanism was proposed by Do [6] to explain such an observation. Basically in this theory, he proposed a model for the structure of activated carbon and then built a diffusion mechanism on such a structured model. The activated structure is basically composed of two distinct regions randomly distributed in space. One region contains the graphitic units, which are basically composed of many graphitic layers, the spaces between these layers form slit-shape channels and such channels have width of molecular dimension such that the dispersive forces between adsorbate molecules and the carbon atoms on the two opposite layers are enhanced, resulting in a strong affinity between the adsorbate molecule and the micropore. The other region is composed of amorphous carbon, and we assume that the void space within this amorphous carbon is large enough to form meso and macropores. In this structure model, the graphitic region is assumed to distribute itself in a maze of amorphous carbon region, that is the graphitic unit is not contiguous. The dimensions of the graphitic unit are about 100 Å in length and 5 to 7 Å in width between layers, and there are about five layers in each graphitic unit.

With this model of structure in place, he then proposed the following processes for the adsorbate molecule to diffuse from the exterior of the particle to the interior. Basically, the adsorbate molecule will diffuse through the mesopore space of the amorphous carbon, and at the same time penetrate into the graphitic layers (a resistance is allowed for). The diffusion along the mesopore and the diffusion of the adsorbed species within the graphitic layers occur in parallel. Once the adsorbed molecule reaches the other end of the graphitic layer, it desorbs (evapo-

rates) and rejoins the other molecules diffusing through the mesopore section which is parallel with the graphitic region. They then diffuse through the amorphous carbon until they encounter another graphitic unit, and the same processes will again repeat until they all reach the particle interior and all the adsorption sites are equilibrated with the surrounding.

2.1 *The rate of adsorption and desorption*

The processes of penetration into and evaporation from the graphitic layer can be viewed macroscopically as the finite mass interchange between the fluid phase and the adsorbed phase. The rates of adsorption and desorption in a graphitic channel of a given width will be assumed to follow the Langmuir kinetics, that is:

$$R_{ads} = k_a P(C_{\mu s} - C_\mu) \tag{3}$$

$$R_{des} = k_d C_\mu \tag{4}$$

where k_a and k_d are rate constants for adsorption and desorption, respectively. These constants are a function of temperature, and also a function of the width of the micropore.

The rates of adsorption and desorption in eqns (3) and (4) describe the rate of collision of the adsorbate molecule to the available adsorption site and the rate of desorption from the site. The site here is viewed as the pore mouth of the micropore where it provides the low potential well for the adsorption to take place.

One could also view the penetration into the micropore as that controlled by the surface resistance of the pore mouth, that is the equilibrium is rapidly established between the gas phase and the adsorbed phase at the pore mouth and that the equilibrated adsorbed molecule has to overcome the surface barrier at the pore mouth to enter the micropore interior, and the rate of evaporation is simply the reverse of the penetration, that is:

$$R_{ads} = k_R(C_\mu^* - C_\mu) \tag{5}$$

$$R_{des} = k_R(C_\mu - C_\mu^*) \tag{6}$$

This form for the surface barrier suggests that the time scale for adsorption is the same as that for desorption, and this is not what we observe experimentally. Equations (3) and (4) on the other hand yield the following net rate of adsorption

$$R_{net} = (k_a C_{\mu s})P - (k_a C + k_d)C_\mu$$

from which we see the time constant for adsorption is:

$$t^0 = \frac{1}{k_a P + k_d}$$

In the adsorption mode with a partial pressure of P_b, the time constant for adsorption is:

$$t_{ads}^0 = \frac{1}{k_a P_b + k_d} \tag{7}$$

On the other hand, for desorption into an environment having a zero adsorbate concentration, the time constant for desorption is:

$$t_{des}^0 = \frac{1}{k_d} \qquad (8)$$

Clearly the finite mass exchange kinetic model of eqns (5) and (6) gives:

$$t_{ads}^0 < t_{des}^0 \qquad (9)$$

which is what we consistently observed experimentally with a long tail in the desorption curve. Thus, the finite mass exchange kinetics eqns (3) and (4) will be accounted for together with the diffusion mechanism of the free species and the adsorbed species.

To allow for the heterogeneity which is characterised by the distribution of the interaction energy between the adsorbate and the adsorbent, we shall assume that the rate constants for adsorption and desorption in eqns (3) and (4) to be a function of this interaction energy, E, between the two phases, that is:

$$R_{ads}(E) = k_a(E)P[C_{\mu s} - C_\mu(E)] \qquad (10)$$

$$R_{des}(E) = k_d(E)C_\mu(E) \qquad (11)$$

where the rates are moles per unit volume of the adsorbed phase per unit time.

This means that the rate of change of the adsorbed species in the absence of the adsorbed phase diffusion for a given micropore having an interaction energy of E with the adsorbate is:

$$\frac{\partial C_\mu(E)}{\partial t} = k_a(E)P[C_{\mu s} - C_\mu(E)] - k_d(E)C_\mu(E)$$

$$(12)$$

The above equation is a linear equation in terms of the adsorbed concentration. For a given fluid phase pressure, we arrange it as follows:

$$\frac{\partial C_\mu(E)}{\partial t} = [k_a(E)P + k_d]\left[C_{\mu s}\frac{b(E)P}{1+b(E)P} - C_\mu(E) \right]$$

$$(13)$$

where b is called the affinity equilibrium constant, which is simply the ratio of the rate constant for adsorption to the rate constant for desorption:

$$b(E) = \frac{k_a(E)}{k_d(E)} \qquad (14)$$

The first term in the second square bracket on the RHS of eqn (13)

$$C_{\mu s}\frac{b(E)P}{1+b(E)P} \qquad (15)$$

is simply the equilibrium adsorbed amount when the rate of adsorption is balanced by the rate of desorp-

tion. We denote this isotherm equation as

$$C_\mu^* = f(P; E) = C_{\mu s}\frac{b(E)P}{1+b(E)P} \qquad (16)$$

eqn (13) then becomes:

$$\frac{\partial C_\mu(E)}{\partial t} = [k_a(E)P + k_d][C_\mu^*(P; E) - C_\mu(E)] \quad (17)$$

The second term on the RHS of the above equation is the deviation from the equilibrium and the first term is the rate proportionality constant. Equation (17) is very similar in form to eqn (5) when we assume surface barrier as the limiting step in the adsorption rate, but in this case we see that the rate proportionality constant

$$[k_a(E)P + k_d] \qquad (18)$$

is function of the fluid phase pressure. The higher is the fluid phase pressure, the faster is the rate towards the equilibrium.

If the energy of interaction is assumed distributed in the sense that $F(E)\,dE$ is the fraction of the adsorption sites having energy between E and $E + dE$, then the amount adsorbed in the adsorbed phase for a given fluid phase pressure is:

$$\langle C_\mu \rangle = \int_{E_{min}}^{E_{max}} C_\mu(E)F(E)\,dE \qquad (19)$$

where E_{min} and E_{max} are the minimum and maximum energies of the energy distribution, dictated by the size of the slit-shaped micropores (the details of which will be discussed in Section 2.5). Thus, keeping the fluid phase pressure constant, solving eqn (17) for the adsorbed concentration of a given pore as a function of time, we get:

$$C_\mu(t, P; E) = C_\mu^*(P; E) + [C_\mu(0; E) - C_\mu^*(P; E)]$$
$$\times e^{-[k_a(E)P + k_d(E)]t} \qquad (20)$$

where $C_\mu(0)$ is the initial concentration of the adsorbed species. The observed amount adsorbed is then given by eqn (19). Let us study this equation a bit further. Take the case where the particle is initially free from any adsorbate, and at time $t = 0+$ the particle is exposed to an environment having a constant fluid phase pressure P. For this case, the amount adsorbed in a pore having an interaction energy of E is:

$$C_\mu(t, P; E) = C_\mu^*(P; E) \times \{1 - e^{-[k_a(E)P + k_d(E)]t}\}$$

$$(21)$$

We further assume that the adsorption rate constant is independent of E, that is the rate of adsorption is controlled by the rate of collision of molecules to the surface, and the rate constant for desorption to follow the Arrhenius relation:

$$k_d = k_{d,\infty}\exp\left(-\frac{E}{RT}\right) \qquad (22)$$

The equilibrium adsorption amount is given in eqn (16) with the equilibrium affinity constant is:

$$b = b_\infty \exp\left(\frac{E}{RT}\right) \qquad (23)$$

where b_∞ is the affinity at infinite temperature.

$$b_\infty = \frac{k_a}{k_{d,\infty}} \qquad (24)$$

The time trajectory of the amount adsorbed given as in eqn (21) is a function of two terms. The first term is the equilibrium term, which increases with an increase in the interaction energy. This means that pores with higher interaction energy will have higher density at equilibrium. The second term in such an equation is the kinetic term, and for a given pressure in the fluid phase, the rate to approach equilibrium is slower for sites having high energy of interaction. What this means is that sites having high equilibrium density will take time to fill and sites having low equilibrium density will approach equilibrium much sooner. We can see this effect more pronouncedly by considering the case of desorption of an initially equilibrated particle into an environment containing no adsorbate species. The amount adsorption as a function of time is:

$$C_\mu(t, P; E) = C_\mu(0; E) \times e^{-|k_d(E)|t} \qquad (25)$$

Since the rate constant for desorption is lower for sites having higher energy of interaction, the rate of desorption will be much lower, and this is what is happening in all desorption data that we have collected in our laboratories for many hydrocarbons on activated carbon. Thus, this means that this process must be included into the overall mass balance equation developed later for the correct description of the phenomena.

The observed amount adsorbed as a function of time is given in eqn (19). Substituting eqn (21) into eqn (19) we get:

$$\langle C_\mu \rangle = \int_{E_{min}}^{E_{max}} C_\mu(t, P; E)F(E)\,dE$$
$$= \int_{E_{min}}^{E_{max}} C_\mu^*(P; E)F(E)\,dE - \int_{E_{min}}^{E_{max}} C_\mu^*(P; E)$$
$$\times e^{-|k_a(E)P + k_d(E)|t}F(E)\,dE \qquad (26)$$

At steady state, the amount adsorbed is given by the first term on the RHS of the above equation. Such an equation is widely used in the literature in the analysis of adsorption isotherms on heterogeneous solids.

2.2 The rate of diffusion of the adsorbed species

So far we have discussed the adsorbed phase kinetic behaviour in the absence of any mobility of the adsorbed species. However, when the adsorbed species is mobile, it will contribute to the overall mass transfer of molecules into the particle interior or out of the particle, depending on whether the process is adsorption or desorption. For any given point, A, within the particle, the chemical potential of the adsorbed phase is denoted as μ, which is assumed to be in equilibrium with a gas phase having a pressure P. All the patches of the adsorbed phase at that point will have the same chemical potential. At another point B, which has a distance Δr from the point A, the chemical potential is μ', which is different from the chemical potential μ of the point A. The gas phase pressure which would have the same chemical potential of the point B is denoted as P'. Due to a difference in the chemical potential between the two points, the diffusion within the adsorbed phase is possible. Let the chemical potential of the point A be greater than that of the point B. The mass transfer from point A to point B along the adsorbed phase contributed by the patch having an interaction energy of E is simply:

$$J_\mu(E) = L(E)C_\mu(E)\frac{\mu - \mu'}{\Delta r} \qquad (27)$$

where $L(E)$ is the mobility constant associated with that patch. Taking the distance between the two points as small as possible, we obtain the following flux equation in differential form:

$$J_\mu(E) = -L(E)C_\mu(E)\frac{\partial \mu}{\partial r} \qquad (28)$$

The adsorbed phase at any local point is assumed to be equal to the chemical potential of an equivalent gas phase, such that:

$$\mu_{ads} = \mu_0 + RT \ln P^* \qquad (29)$$

where P^* is the gas phase pressure which would be in equilibrium with the adsorbed phase concentration $C_\mu(E)$. This hypothetical gas phase pressure is not the same as the gas phase pressure at the point A, because there is a mass transfer resistance between the gas phase and the adsorbed phase as given by eqns (3) and (4). The hypothetical gas phase pressure is given by the following equilibrium isotherm equation:

$$P^* = \frac{C_\mu(E)}{b(E)[C_{\mu s} - C_\mu(E)]} \qquad (30)$$

The above equation simply states that all concentrations of the adsorbed phase $C_\mu(E)$ (for all E) will relate to the hypothetical gas phase pressure in a manner such that the ratio of the two terms, which are a function of interaction energy E, on the RHS of eqn (30) is independent of the interaction energy.

Substitution of eqns (29) and (30) for the chemical potential into eqns (27) and (28) yields the following flux equation for the adsorbed phase written in terms of the adsorbed phase concentration

$$J_\mu(E) = -D_\mu^0(E) \frac{C_{\mu s}}{C_{\mu s} - C_\mu(E)} \frac{\partial C_\mu(E)}{\partial r} \quad (31)$$

where

$$D_\mu^0(E) = RT \times L(E) \quad (32)$$

The units of this flux are the number of moles of adsorbed species diffusing per unit section area of the adsorbed phase per unit time.

It should be reminded again that the adsorbed phase concentration is not locally equilibrated with the fluid phase pressure P. They are only equilibrated with each other when the rates of adsorption and desorption are much faster than the diffusion rate.

What needs to be addressed in eqns (31) and (32) is the expression for the diffusion coefficient corresponding to zero loading conditions, $D_\mu^0(E)$. Since the diffusion in the adsorbed phase is activated, it is usually assumed to follow the Arrhenius law:

$$D_\mu^0(E) = D_{\mu\infty}^0 \exp\left(-\frac{aE}{RT}\right) \quad (33)$$

where $D_{\mu\infty}^0$ is the diffusion coefficient at infinite temperature and zero loading condition. The coefficient "a" in eqn (33) suggests that the activation energy for adsorbed phase diffusion is "a" times the energy of interaction between the adsorbate and the adsorbent. In general, such a coefficient "a" is a function of E, but without the rigorous proof of such dependence, we shall take it as a constant. Literature on surface diffusion has shown that the constant a is between 0.33 to 1.

Knowing the flux equation contributed by the patch of energy E, the overall flux of the adsorbed phase is simply the total contribution of all patches, that is it is the integral of the local flux eqns (31) and (32) over the complete range of the energy distribution

$$\langle J_\mu \rangle = \int_{E_{min}}^{E_{max}} J_\mu(E) F(E)\, dE$$

$$= -\int_{E_{min}}^{E_{max}} \left[D_\mu^0(E) \frac{C_{\mu s}}{C_{\mu s} - C_\mu(E)} \frac{\partial C_\mu(E)}{\partial r} \right] F(E)\, dE \quad (34)$$

2.3 The pore flux

To complete the necessary ingredients for the setting up of the mass balance equation, we need to define the flux for the free species which is occurring in parallel to the adsorbed phase diffusion. The flux of the adsorbate in the fluid phase is given by

$$J_p = -D_p \frac{1}{RT} \frac{\partial P}{\partial r} \quad (35)$$

where D_p is the pore diffusivity, calculated from the Knudsen diffusivity and the molecular binary diffusivity. This pore flux has units of moles of free species

diffusing per unit section area of void space per unit time.

2.4 The mass balance equation

2.4.1 *The overall mass balance equation.* Having obtained all the necessary rate equations (eqns (10), (11), (31), (32) and (35)), the mass balance on a small element of the particle is carried out and we obtain the following equation describing the overall mass balance involving both the fluid and adsorbed phase species:

$$\frac{\epsilon}{RT} \frac{\partial P}{\partial t} + (1-\epsilon) \frac{\partial}{\partial t} \left[\int_{E_{min}}^{E_{max}} C_\mu(E) F(E)\, dE \right]$$

$$= \frac{\epsilon D_p}{RT} \frac{\partial^2 C}{\partial r^2} + (1-\epsilon) \frac{\partial}{\partial r} \left\{ \int_{E_{min}}^{E_{max}} \left[D_\mu^0(E) \frac{C_{\mu s}}{C_{\mu s} - C_\mu(E)} \frac{\partial C_\mu(E)}{\partial r} \right] F(E)\, dE \right\} \quad (36)$$

This equation simply states that the accumulation of mass (LHS) is balanced by the diffusion rates into the particle both by the free as well as the adsorbed species.

To complete the mass balance description, we need to do a mass balance around the adsorbed phase.

2.4.2 *The adsorbed phase mass balance equation.* The mass balance equation of the patch with an interaction energy of E is written as follows:

$$\frac{\partial C_\mu(E)}{\partial t}$$

$$= \frac{\partial}{\partial r} \left[D_\mu^0(E) \frac{C_{\mu s}}{C_{\mu s} - C_\mu(E)} \frac{\partial C_\mu(E)}{\partial r} \right] + R_{ads} - R_{des} \quad (37)$$

where R_{ads} and R_{des} are defined in eqns (10) and (11). Substitution of eqns (10) and (11) into the above equation will yield the mass balance equation for the adsorbed phase:

$$\frac{\partial C_\mu(E)}{\partial t}$$

$$= \frac{\partial}{\partial r} \left[D_\mu^0(E) \frac{C_{\mu s}}{C_{\mu s} - C_\mu(E)} \frac{\partial C_\mu(E)}{\partial r} \right]$$

$$+ k_a(E) P [C_{\mu s} - C_\mu(E)] - k_d(E) C_\mu(E) \quad (38)$$

eqns (36)–(38) define the problem. Solving these equations with some appropriate boundary conditions and initial condition, we will obtain the concentration distributions for the fluid and adsorbed phases.

2.4.3 The boundary conditions of the model equation.
The pertinent boundary conditions and initial condition of the model equations are:

$$r=0, \quad \frac{\partial P}{\partial r}=0 \qquad (39)$$

$$r=R, \quad \epsilon J_p + (1-\epsilon) \int_{E_{min}}^{E_{max}} J_\mu(E) f(E) \, dE = k_m (P - P_b) \qquad (40)$$

$$t=0, \quad P=P_i, \quad C_\mu = C_{\mu i}$$
$$= \int_{E_{min}}^{E_{max}} C_{\mu s} \frac{b(E)P_i}{1+b(E)P_i} f(E) \, dE \qquad (41)$$

where P_b is the bulk pressure and P_i is the initial pressure of adsorbate.

2.5 Equilibria and adsorption energy distribution

The adsorption energy heterogeneity plays an important role in the development of our kinetics model. This section will address the method of determining the energy distribution from adsorption equilibria. This is basically done by fitting the isotherm equation to isotherm data at multiple temperatures of many adsorbates simultaneously.

2.5.1 Energy heterogeneity and porous structure.
The heterogeneous surface of activated carbon can be viewed as patch-wise, that is, it can be treated as the combination of many small patches with different energies, in each patch the adsorption sites with the same energy are grouped together. In adsorption equilibrium, all the patches are in local equilibrium with the gas phase. If this energy distribution can be represented by a function called $F(E)$ and the local isotherm is represented by $C_\mu(E)$, then the amount adsorbed for a given fluid phase pressure is [7]:

$$\langle C_\mu \rangle = \int_{E_{min}}^{E_{max}} C_\mu(E) F(E) \, dE \qquad (42)$$

where E_{min} and E_{max} are the minimum and maximum energies of the energy distribution. Obtaining the appropriate energy distribution has been a great fundamental problem [8], since assuming this energy distribution follows a particular type of distribution function such as uniform or gamma distribution is not fundamentally sound. The approach we use here is the adoption of the intrinsic micropore size distribution, and then through the potential energy of interaction we deduce the proper energy distribution. The carbon structure is assumed to have slit-shaped micropores, in which the adsorption potential is the enhanced dispersive force from both sides of the pore walls. If the pore walls are taken as two parallel graphite layers with infinite length, this potential can

be described by the Lennard–Jones 10–4 potential [9]:

$$U_p(z) = \frac{5}{3} U_s^*$$
$$\left[\frac{2}{5} \left(\frac{\sigma_{12}}{2r-z} \right)^{10} + \frac{2}{5} \left(\frac{\sigma_{12}}{z} \right)^{10} - \left(\frac{\sigma_{12}}{2r-z} \right)^4 - \left(\frac{\sigma_{12}}{z} \right)^4 \right] \qquad (43)$$

where z is the distance between the adsorbate molecule and one of the pore walls, r is the pore half-width ($z < 2r$), σ_{12} is the collision diameter and U_s^* is the well-depth of the Lennard–Jones potential minimum for a single graphite lattice. The adsorption potential is taken as the negative of this potential minimum, which ranges from U_s^* (corresponding to a large pore $r \to \infty$) to $2U_s^*$ (corresponding to $r = \sigma_{12}$). Solving eqn (43) for the minimum will give this interaction energy as a function of pore half-width $E(r)$ and the distribution of this energy is related to the size distribution of slit-shaped micropores (MPSD) by the following relation:

$$F(E) = \frac{f(r)}{dE(r)/dr} \qquad (44)$$

where $f(r)$ is the micropore size distribution.

The micropore size distribution is solid-specific, while the energy distribution is specific to the solid–adsorbate pair. Since the adsorbates used in this paper have similar collision diameters σ_{12}, we shall assume that the micropore size distribution versus the reduced half-width (scaled against the collision diameter) is the same for all adsorbates.

2.5.2 MPSD and energy distribution.
In this paper we assume the micropore size distribution to take the form of the gamma distribution given below:

$$f(r) = \frac{q^{\gamma+1} r^\gamma e^{-qr}}{\Gamma(\gamma+1)} \qquad (45)$$

where q and r are the structural parameters.

If the local isotherm is described by the Langmuir equation, the amount adsorbed as a function of pressure is given by:

$$C_\mu = \int_{x_{min}}^{\infty} C_{\mu s}(T) \frac{Pb[E(x)]}{1+Pb[E(x)]} \frac{Q^{\gamma+1} x^\gamma e^{-Qx}}{\Gamma(\gamma+1)} \, dx \qquad (46)$$

where $x = r/\sigma_{12}$ is the reduced pore half-width, x_{min} is the minimum reduced pore half-width accessible to the adsorbate molecule and $Q = q\sigma_{12}$. The local affinity $b[E(x)]$ has the following functional form in terms of the interaction energy:

$$b = b_0 \exp\left[\frac{E(r)}{RT} \right] \qquad (47)$$

The parameter b_0 is the adsorption affinity at zero energy level, defined in eqn (14). For different species, it can be further represented by:

$$b_0 = \frac{\beta}{\sqrt{MT}} \qquad (48)$$

in which we have assumed that the rate of adsorption is related to the collision rate of molecules towards the surface. This collision rate is inversely proportional to the square root of molecular weight and to the square root of temperature. The parameter β is assumed to be solid-specific and hence independent of adsorbate.

The maximum adsorption capacity of each species $C_{\mu s}$ is assumed to take the following temperature dependence form:

$$C_{\mu s}(T) = C_{\mu 0} \exp[\delta(T - T_0)] \qquad (49)$$

where $C_{\mu 0}$ is the maximum adsorption capacity of

that species at a reference temperature T_0, and the parameter δ is the thermal expansion coefficient.

Equations (45)–(49) are the general isotherm equations applicable to all species in our study. By comparing the model predictions with the experimental data of all species simultaneously, we can derive the unique reduced MPSD which characterises the structure of activated carbon.

2.6 Solution methodology

The model equations (eqns (36)–(38)) are coupled partial differential equations. By normalising the spatial variable using the particle radius as the characteristic length and then applying the orthogonal collocation technique to discretise the normalised spatial variables, we obtain a set of coupled time-derivative ODEs. This set is then solved numerically

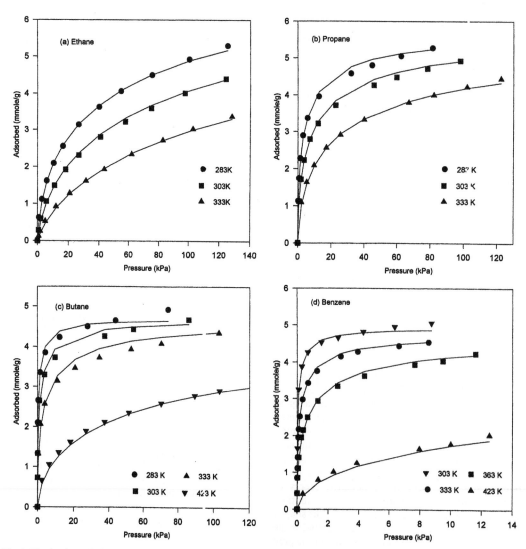

Fig. 1. The isotherm fittings with the unique MPSD parameters for all species: (a) ethane, (b) propane, (c) butane, (d) benzene, (e) toluene, (f) CO_2, (g) SO_2.

by a Runge–Kutta method to obtain the concentration distributions in the particle, from which the mass transfer rates can be calculated.

3. EXPERIMENT

Single component adsorption isotherms on Ajax activated carbon were measured on a high-accuracy volumetric rig for the following adsorbates: (1) ethane, (2) propane, (3) *n*-butane, (4) benzene, (5) toluene, (6) carbon dioxide and (7) sulphur dioxide. For each species, adsorption isotherms at several temperatures were measured. The adsorption/desorption kinetics were studied on a differential adsorber bed rig (DAB). The single component adsorption/desorption kinetics on Ajax carbon were measured under various conditions such as different temperatures, pellet length/geometry and bulk concen-

trations. The detailed rig set-up and experimental procedures can be found in ref. [10].

4. RESULTS AND DISCUSSION

4.1 *Equilibria and energy distribution*

First, the extensive experimental isotherm data of ethane, propane, butane, benzene and toluene are employed simultaneously in eqns (45)–(49) to derive the reduced MPSD for all species. The model fitting (lines) and experimental data (dots) are shown in Fig. 1 and the optimised MPSD parameters for the gamma distribution (Q, γ) and the solid affinity parameter β are listed in Table 1.

It can be seen from Fig. 1 that, with the unique reduced MPSD parameters $(Q = 118.24, \gamma = 143.02)$ the model equations can accurately simulate the adsorption equilibrium of each species under a wide

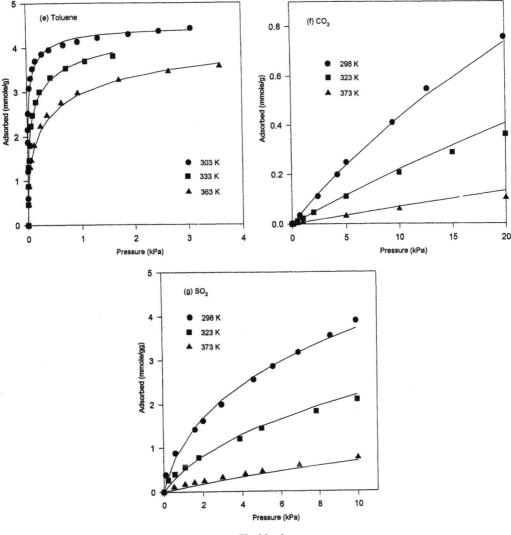

Fig. 1 (e–g).

Table 1. The optimised MPSD and affinity parameters

Q	γ	β (kPa^{-1}/$\sqrt{\text{gK/mol}}$)
118.2	143.2	1.2×10^{-4}

temperature range. The resulted adsorption energy distributions for each species are plotted in Fig. 2, and the average adsorption energy E_{mean} of each species is calculated and tabulated in Table 2. From Table 2, we note that the potential energy of the adsorbate molecule with one single lattice layer U_s^* and the average interaction energy E_{mean} are increasing with the carbon number for a given class of adsorbate family (ethane, propane and butane). Aromatics have higher U_s^* and E_{mean} than those of paraffins, with toluene being the stronger adsorbing species than benzene. The thermal expansion coefficient, δ, of all species is very small, indicating that the saturation capacity is not affected by temperature change.

Although the reduced MPSD is assumed to be the same for all species, the energy distributions of various species are somewhat similar in shape. Their different positions along the adsorption energy coor-

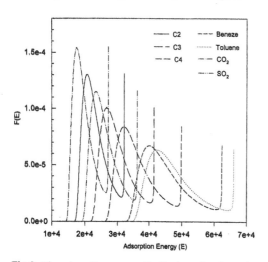

Fig. 2. The adsorption energy distribution of each species converted from the MPSD.

dinate are due to the difference in U_s^*. These energy distributions are then used in the dynamic model equations for the simulations of the kinetics behaviours.

4.2 Kinetics

Having obtained the necessary adsorption energy distributions and isotherm parameters from the equilibrium information, we now study the adsorption kinetics with the extensive single component kinetics data obtained from the DAB experiments. The simulations are performed to examine the model's applicability to each species and the effects of temperature, bulk gas pressure, pellet size etc. on the kinetics behaviours.

4.2.1 *Ethane*. Adsorption kinetics of ethane on Ajax activated carbon were measured on the 4.4 mm full length slab particles. Figure 3 shows the fractional uptake of ethane onto this particle for three mole fractions in the bulk (5, 10 and 20%). The total pressure is 1 atm and the operating temperature is 303 K. The combined diffusivity for the pore volume diffusion is calculated from the combined molecular and Knudsen diffusion processes. This would leave the two kinetic parameters for optimisation which are the surface diffusivity at zero loading and the rate constant for adsorption (k_a), which is assumed to be independent of the interaction energy E. These optimised parameters are tabulated in Table 3. The fitted curves for adsorption and desorption are shown in Fig. 3 as continuous lines, and we see that the fitting is excellent. The mathematical model shows the correct concentration dependence behaviour, that is faster approach to equilibrium for system having higher bulk concentrations. Furthermore, the model shows the correct desorption behaviour, that is it describes the long tail phenomenon well, which the past models have failed.

With the optimised kinetic parameters obtained from the fitting of the adsorption data at 303 K given in Table 3, we use the mathematical model to predict the adsorption kinetics of ethane on the same particle but at two other temperatures, 283 and 333 K. The results are shown in Fig. 3(b and c), respectively. Continuous lines are predicted solutions, while the

Table 2. The optimised isotherm parameters for all species

	Ethane	Propane	Butane	Benzene	Toluene	CO$_2$	SO$_2$
$C_{\mu s}$ (mmol g^{-1})	7.432	5.722	4.717	5.040	4.533	7.394	12.610
U_s^* (J mol^{-1})	1.608×10^4	2.078×10^4	2.493×10^4	3.119×10^4	3.304×10^4	1.359×10^4	1.817×10^4
δ (K^{-1})	9.06×10^{-4}	-4.89×10^{-4}	-1.57×10^{-4}	-6.47×10^{-4}	-4.90×10^{-4}	-4.09×10^{-6}	-3.23×10^{-4}
E_{mean} (J mol^{-1})	2.321×10^4	2.994×10^4	3.592×10^4	4.493×10^4	4.760×10^4	1.959×10^4	2.618×10^4

Table 3. Kinetics parameters for ethane

Species	D_p (10^{-6} m^2 s^{-1})			$D_{\mu\infty}^0$ (10^{-7}) (m^2 s^{-1})	k_a (10^{-5}) (kPa^{-1} s^{-1})	a
	283 K	303 K	333 K			
Ethane	4.03	4.48	5.23	8.48	3.39	0.5

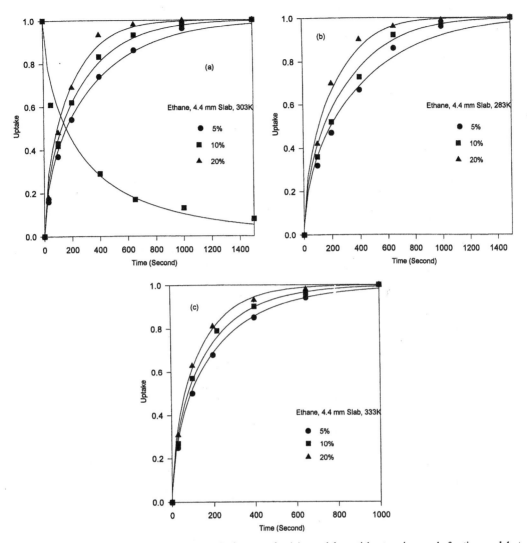

Fig. 3. The model fitting and experimental data of ethane on the 4.4 mm slab particle at various mole fractions and 1 atm total pressure. (a) 303 K, (b) 283 K, (c) 333 K.

symbols are experimental data. We see that the predictions at other temperatures are excellent. The bulk concentration behaviours at these two temperatures are also well predicted by the mathematical model.

4.2.2 *Propane*. The adsorption kinetics of propane were measured on the same particle size and the same operating conditions as those of ethane. The optimised kinetic parameters for propane are obtained by fitting the model equation to the one set of adsorption data at 303 K. The results are tabulated in Table 4, and once the optimised parameters are obtained we use the mathematical model to predict the adsorption kinetics at other conditions. Figure 4(a–c) show such predictions at three different temperatures and three different concentrations at each temperature. As in the case of ethane, we see that the predictions are excellent, notably the prediction of the long tail in the desorption curve. Once

again the validity of the mathematical model is justified.

4.2.3 *Butane*. The adsorption kinetics experiment of *n*-butane was carried out at one temperature (303 K), but on particles with different sizes and geometries to test the model predictability in terms of size and shape of the particle. The particles selected in our experiment are the 2 mm full length slab, 4 mm full length slab and 1/16 in. diameter cylinder. The kinetic data for the 2 mm slab particle with three bulk phase mole fractions (12.9, 20.6 and 31.6%) are used in the fitting. The operating conditions are 1 atm and 303 K. The results of the fitting are shown graphically in Fig. 5(a), and the optimised kinetic parameters are shown in Table 5. As seen in Fig. 5(a), the fit is excellent.

Having obtained the optimised kinetic parameters for *n*-butane, we use them with the mathematical model to predict the other data of the 1/16 in.

Table 4. Kinetics parameters for propane

Species	D_p (10^{-6} m² s⁻¹)			$D_{\mu\infty}^0$ (10^{-7}) (m² s⁻¹)	k_a (10^{-5}) (kPa⁻¹ s⁻¹)	a
	283 K	303 K	333 K			
Propane	3.20	3.47	5.23	8.31	2.23	0.5

cylindrical particle and 4 mm slab particle. The predictions are shown in Fig. 5(b) and 5(c), where we again note the excellent predictability of the mathematical model.

4.2.4 *Benzene and toluene*. To further examine the model applicability and predictability, we apply it to another class of adsorbate, the aromatics. We chose benzene and toluene in our study, and studied the adsorption kinetics for the 2 mm slab and 1/16 in. cylinder with bulk mole fractions of 0.415 and 0.912%. The operating conditions are 1 atm and 303 and 333 K. The fitting between the model and

the data is shown in Figs 6 and 7 for benzene and toluene, respectively. The fit is excellent for both aromatics, and the optimised kinetic parameters are tabulated in Table 6.

4.2.5 CO_2. The adsorption kinetics of CO_2 were studied on the 6 mm full length slab particles at three different temperatures: 298, 323 and 343 K and a total pressure of 1 atm [2]. Adsorption as well as desorption modes were carried out. In the adsorption mode, a mole fraction of 20% was maintained in the bulk, while in the desorption mode the particle was equilibrated with 20% carbon dioxide and then

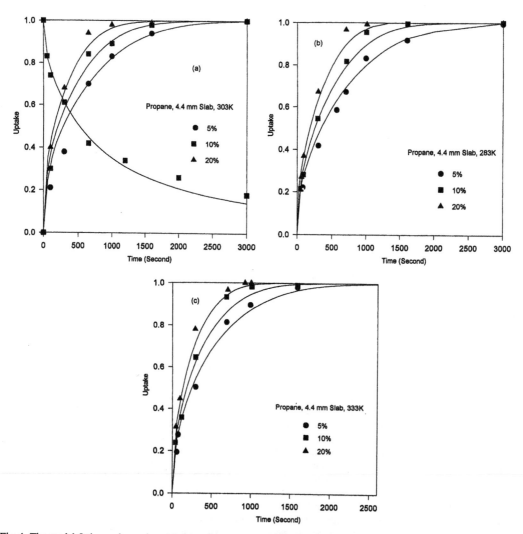

Fig. 4. The model fitting and experimental data of propane on the 4.4 mm slab particle at various mole fractions and 1 atm total pressure. (a) 303 K, (b) 283 K, (c) 333 K.

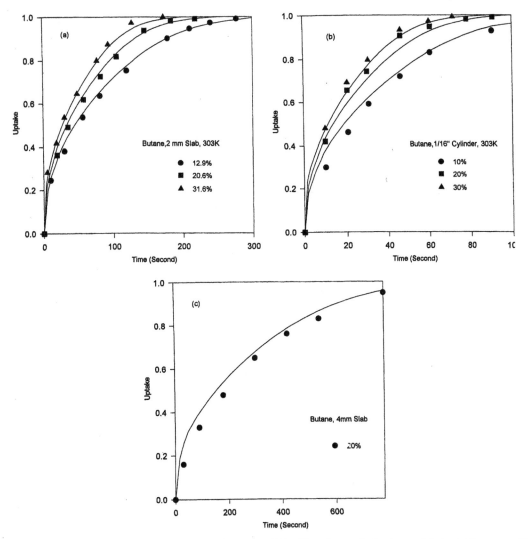

Fig. 5. The model fitting and experimental data of butane at 303 K with various mole fractions and 1 atm total pressure. (a) 1 mm Slab, (b) 1/16 in. cylinder, (c) 2 mm slab.

Table 5. Kinetics parameters for butane

Species	D_p $(10^{-6} \, m^2 \, s^{-1})$			$D^0_{\mu\infty}$ (10^{-6}) $(m^2 \, s^{-1})$	k_a (10^{-5}) $(kPa^{-1} \, s^{-1})$	a
	283 K	303 K	333 K			
Butane	2.75	3.04	3.57	1.10	1.92	0.5

was desorbed in a stream of inert gas. To obtain the kinetic parameters, we used the adsorption data of three temperatures in the fitting. The fit is shown in Fig. 8(a), and the optimised parameters are tabulated in Table 7. The fit is excellent. These parameters are then used with the mathematical model to predict the desorption behaviour. The results are shown in Fig. 8(b). We see that the predictions are excellent, and we note again the excellent ability of the model to describe the long tail behaviours.

4.2.6 SO_2. We finally test the model applicability to another class of adsorbate, sulphur dioxide [1]. The particle used in this test is the 4 mm full length slab particle. The adsorption and desorption runs were conducted at three temperatures: 298, 323 and 373 K and a total pressure of 1 atm. The bulk mole fractions used were 0.5, 2 and 5%. To obtain the kinetic parameters, we use the desorption data of three runs in which the particles were equilibrated with three different initial bulk mole fractions of 0.5,

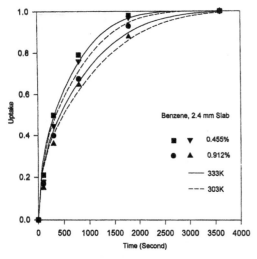

Fig. 6. The model fitting and experimental data of benzene on the 2.4 mm slab at different temperatures and mole fractions. The total pressure is 1 atm.

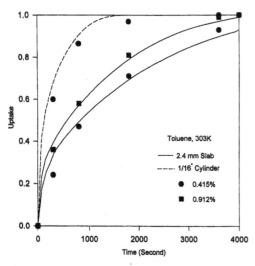

Fig. 7. The model fitting and experiment data of toluene at 303 K on different-sized particles. The total pressure is 1 atm.

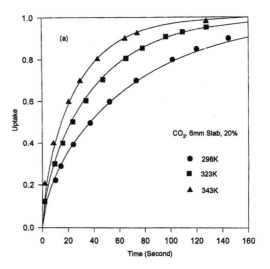

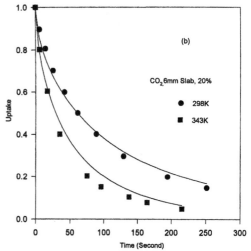

Fig. 8. The model fitting and experimental data of CO_2 on the 6 mm slab particle at different temperatures. The total pressure is 1 atm. (a) Adsorption, (b) desorption.

2 and 5%. The fit is shown in Fig. 9(a) and the optimised parameters are tabulated in Table 8. The long tail in desorption is well described by the model.

The optimised parameters in Table 8 are then used in the model to predict the desorption at two other temperatures, 283 and 373 K. The predictions are shown in Fig. 9(b), where we see the excellent predic-

tion and again the long tail phenomenon is well described by the model. The predictability of the model is tested with a set of adsorption data at three temperatures (Fig. 9(c)). Again, we note the very good predictability, although the experimental data exhibit a slightly faster kinetics than what was predicted by the model.

4.2.7 *Discussion of the extracted optimised parameters.* We have successfully tested the pro-

Table 6. Kinetics parameters for benzene and toluene

Species	D_p (10^{-7} m^2 s^{-1})			$D_{\mu\infty}^0$ (10^{-7}) (m^2 s^{-1})	k_a (10^{-6}) (kPa^{-1} s^{-1})	a
	283 K	303 K	333 K			
Benzene	10.6	10.8	11.0	8.42	7.78	0.5
Toluene	8.67	8.85	9.95	5.51	6.05	0.5

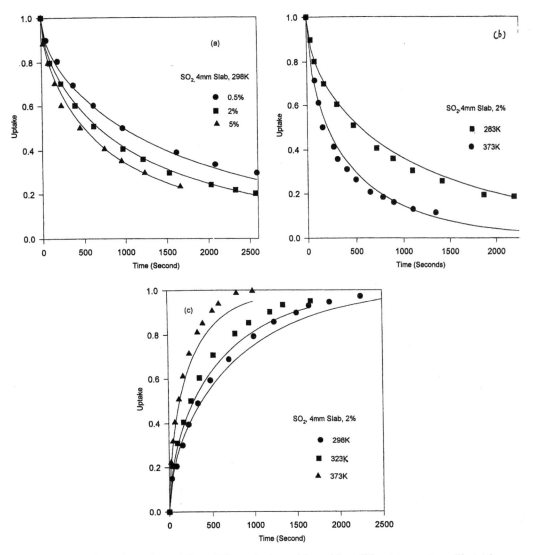

Fig. 9. The model fitting and experimental data of SO_2 on the 4 mm slab particle at different temperatures. The total pressure is 1 atm. (a) Desorption, (b) desorption, (c) adsorption.

posed mathematical model with the extensive kinetic data of adsorption of seven adsorbates onto activated carbon. The two kinetics parameters extracted from the fitting are the surface diffusivity at zero loading and at infinite temperature $D_{\mu\infty}^0$ and the rate constant for adsorption k_a.

Table 7. Kinetics parameters for CO_2

	D_p ($10^{-6}\,m^2\,s^{-1}$)			$D_{\mu\infty}^0$ (10^{-6}) ($m^2\,s^{-1}$)	k_a (10^{-5}) ($kPa^{-1}\,s^{-1}$)	a
Species	298 K	323 K	343 K			
CO_2	4.70	5.20	5.60	1.18	3.08	0.5

Table 8. Kinetics parameters for SO_2

	D_p ($10^{-6}\,m^2\,s^{-1}$)			$D_{\mu\infty}^0$ (10^{-7}) ($m^2\,s^{-1}$)	k_a (10^{-5}) ($kPa^{-1}\,s^{-1}$)	a
Species	298 K	323 K	373 K			
SO_2	3.80	4.20	5.00	7.33	8.88	0.5

246

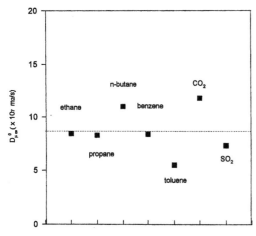

Fig. 10. The plot of surface diffusivity at zero loading and zero energy level D_μ^0 of each species.

The parameter $D_{\mu\infty}^0$ represents the mobility of adsorbed molecules on the surface at infinite temperature, and one can accept the view that this parameter is independent of adsorbate. The values of $D_{\mu\infty}^0$ for the seven adsorbates tested are plotted in Fig. 10, and we note that within experimental errors these values are of the same order of magnitude with an average of $8.7 \times 10^{-7} \, \text{m}^2 \, \text{s}^{-1}$. This supports our hypothesis that $D_{\mu\infty}^0$ is intrinsic and only specific to the solid concerned. It may contain a weak dependence on the molecular weight, analogous to the rate of collision of gaseous molecules towards a surface, where the rate constant of collision is inversely proportional to the molecular weight. Further fundamental work is needed to test this hypothesis.

The rate constant for adsorption, k_a, shows more scatter than the surface diffusivity at zero loading and at infinite temperature $D_{\mu\infty}^0$, but it seems to have an order of magnitude of $2-3 \times 10^{-5} \, \text{kPa}^{-1} \, \text{s}^{-1}$.

5. CONCLUSION

A new adsorption kinetics model was proposed for activated carbon. The model incorporates pore diffusion, surface diffusion and the finite mass interchange rate between the gas and adsorbed phase. The driving force for surface diffusion is taken as the chemical potential gradient and the mass interchange rate is described by Langmuir kinetics. The size distribution of slit-shaped micropores is considered as the main source of surface heterogeneity. This distribution is obtained by fitting isotherm data of many adsorbates simultaneously through the use of the Lennard–Jones potential theory. The induced energy distribution due to the micropore size distribution is used in the kinetic model to predict the kinetic data of ethane, propane, n-butane, benzene, toluene, carbon dioxide and sulphur dioxide over a wide range of operating conditions and parameters. The predictions have been found to be excellent in general, and in particular the model is able to predict the long tail behaviour observed in the desorption mode. Such long tail behaviour has been unable to be accounted for by any previous models, and this has testified for the potential applicability of the model proposed in this paper.

Acknowledgements—This project is supported by the Australia Research Council (ARC).

REFERENCES

1. Gray, P. and Do, D. D., Adsorption and desorption of dynamics of sulphur dioxide on a large activated carbon particle, *Chem. Eng. Commun.*, 1990, **96**, 141–154.
2. Gray, P. and Do, D. D., Dynamics of carbon dioxide sorption on activated carbon particle, *AIChE J.*, 1991, **37**, 1027–1034.
3. Do, D. D. and Hu, X., An energy distributed model for adsorption kinetics in large heterogeneous microporous particle, *Chem. Eng. Sci.*, 1993, **48**, 2119.
4. Hu, X. and Do, D. D., Role of energy distribution in multicomponent sorption kinetics in bidispersed solid, *AIChE J.*, 1993, **39**, 1628–1635.
5. Hu, X. and Do, D. D., Effect of surface heterogeneity on the sorption kinetics of gases in activated carbon, pore size distribution vs energy distribution, *Langmuir*, 1994, **10**, 3296–3302.
6. Do, D. D., A model for surface diffusion of ethane and propane in activated carbon, *Chem. Eng. Sci.*, 1996, **51**, 4145–4158.
7. Myers, A. L., Adsorption of pure gases and their mixtures on heterogeneous surface, in *Fundamentals of Adsorption*, American Institute of Chemical Engineers, New York, 1984.
8. Russell, B. P. and LeVan, M. D., Pore size distribution of BPL activated carbon determined by different methods, *Carbon*, 1994, **32**, 845–885.
9. Everett, D. H. and Powl, J. C., Adsorption in slit-like and cylindrical micropores in the Henry's law region, *J. Chem. Soc., Faraday Trans.*, 1976, **72**, 619–636.
10. Hu, X., King, B. and Do, D. D., Ternary adsorption kinetics of gases in activated carbon, *Gas Sep. and Purif.*, 1994, **175**, 3.

247

DYNAMICS OF NATURAL GAS ADSORPTION STORAGE SYSTEMS EMPLOYING ACTIVATED CARBON*

J. P. Barbosa Mota,[a] A. E. Rodrigues,[b] E. Saatdjian[c] and D. Tondeur[d]

[a]Instituto de Biologia Experimental e Tecnológica, Apartado 12, 2780 Oeiras, Portugal
[b]Laboratory of Separation and Reaction Engineering, School of Engineering, University of Porto, 4099 Porto Codex, Portugal
[c]Laboratoire d'Energétique et de Mécanique Théorique et Appliquée, ENSEM, 2 avenue de la Forêt de Haye, B.P. 160, 54504 Vandoeuvre Cedex, France
[d]Laboratoire des Sciences du Génie Chimique – CNRS, ENSIC, 1 rue Grandville, B.P.451, 54001 Nancy Cedex, France

(*Received* 20 *June* 1996; *accepted in revised form* 31 *March* 1997)

Abstract—Various aspects of the dynamics of natural gas adsorption storage systems employing activated carbon are studied theoretically. The fast charge of the storage system is the first subject addressed. Emphasis is given to thermal effects and hydrodynamics of flow through the carbon bed. In order to study the influence of diffusional resistances on charge dynamics, an intraparticle transport equation governed by a diffusion law is added to the computational model. Lastly, the slow discharge process and proposed solutions for reducing the adverse effect of the heat of adsorption on storage capacity, including *in situ* thermal energy storage, are discussed. © 1997 Elsevier Science Ltd

Key Words—A. Activated carbon, C. adsorption, D. gas storage.

1. INTRODUCTION

Natural gas (NG) is a potentially attractive fuel for vehicle use. It is cheaper than gasoline or diesel, and NG vehicles have a less adverse effect on the environment than liquid-fuelled vehicles, emitting less carbon dioxide, a major greenhouse gas, as well as several other air pollutants [1]. The technical feasibility of NG vehicles is well-established. Several countries already have natural gas fleets and ongoing research and development programmes [2]. One point stressed by the task force "The car of Tomorrow" of the European Commission's Green Paper on Innovation is that "···in the medium term, vehicles running on compressed natural gas will have a major role" [3].

Natural gas is about 95% methane, a gas that cannot be liquefied at ambient temperature ($T_c = -82.6°C$). Therefore, its compact on-board storage has required the use of expensive and heavy, high-pressure compression technology (16.5–20.7 MPa) [4]. All efforts to improve this technology have focused on lowering the storage pressure in order to decrease capital and operating costs of compression stations, and allow the use of lighter on-board gas storage reservoirs. A promising low-pressure (<4 MPa) system for storing NG is adsorption storage, it is a good compromise between compression costs and on-board storage capacity [1,5].

Most research on adsorbed natural gas (ANG) has focused on the development and evaluation of economical adsorbents with storage capacities comparable to that of compressed NG [6–14]. To date, the most promising adsorbents for NG storage are highly microporous carbons with high packing density.

A common ANG storage performance indicator is volumes of stored NG, measured at standard conditions, per storage volume (v/v). The highest experimental storage capacities obtained to date (at 3.5 MPa and 25°C) are for the high performance, but expensive, activated carbon AX-21®: 101 v/v for granulated particles and 144 v/v for a mixture of carbon with polymeric binder, pressed into the desired geometrical shape under mechanical pressure and then dried [14]. However, good capacities (82 and 103 v/v) were also obtained by the same authors with a much cheaper, commercially available carbon (CNS®). The theoretical maximum storage capacity predicted by molecular simulations is 209 v/v for monolithic carbon and 146 v/v for pelletized carbon [15]. These values may be compared to 240 v/v for compressed NG at 20.7 MPa.

Several problems that affect the success of ANG technology have been addressed in the literature. One of them is the capacity loss due to the residual amount of NG left at depletion, which can be as high as 30% of the amount stored at charge conditions [1,15–17]. This is a consequence of the shape of the adsorption isotherm and of the infeasibility of lowering the storage pressure below atmospheric.

Another problem is the management of the thermal effects, related to the heat of adsorption, which adversely affect ANG storage [4,14,16,18]. If the heat of adsorption released during charge is not removed from the storage system, less methane is

---

*Presented at the American Carbon Society Workshop "Carbon Materials for the Environment", Charleston, SC, June 1996.

Reprinted from *Carbon* 35 (9), 1259-1270 (1997)

adsorbed as the substrate heats up. If the heat of adsorption is not resupplied during discharge, the bed temperature drops, increasing the residual amount of NG that remains in storage at depletion. The penalty in both phases can surpass 25%.

Yet another problem that has received attention in the literature is the gradual contamination of the adsorbent with heavy hydrocarbons and water vapor present in natural gas [1,14,16,17]. If these are allowed to enter the on-board storage system, they can adsorb preferentially to high equilibrium residual levels and decrease storage capacity. However, the presence of impurities is not always prejudicial; net storage capacity might even be increased with the introduction into the reservoir of a judiciously chosen additive [17]. This is true if the impact of the additive on the amount of NG adsorbed at depletion pressure is higher than at charge pressure.

The authors' effort in this field of research has focused on the development of theoretical tools providing an accurate description of the dynamics of ANG storage systems with realistic geometries and dimensions. These tools are essential for the development and manufacture of on-board storage reservoirs.

The fast charge of the storage system is the first subject addressed in the present work. Emphasis is given to thermal effects and hydrodynamics of flow through the carbon bed. In order to study the influence of diffusional resistances on charge dynamics, an intraparticle transport equation governed by a diffusion law is added to the computational model. Lastly, the slow discharge process and some proposed solutions for reducing the adverse effect of the heat of adsorption on storage capacity, including *in situ* thermal energy storage, are discussed.

2. DESCRIPTION OF THE STORAGE SYSTEM

The ANG storage system under study is depicted in Fig. 1. It is a 50 liter cylindrical reservoir (85 cm × 28 cm) filled with the G216 carbon from North American Carbon Inc. of Columbus, Ohio. The storage system is charged through a 10 mm opening located in the center of the cylinder's front face. Despite the fact that the quantitative results presented here are specific for this system, the general pattern should be the same for storage systems employing other carbons with similar storage capacities.

The G216 carbon, characterized by data listed in Table 1, has a methane storage capacity of about 80 v/v at 25°C and 3.5 MPa. Experimental methane adsorption isotherms for this carbon were reported by Remick and Tiller [18], they are well approximated by the following Langmuir type relation:

$$q = (q_{m}bP)/(1+bP)$$

$$q_{m} = 55920T^{-2.3}, \ b = 1.0863 \times 10^{-7} \exp (806/T)$$

$$(1)$$

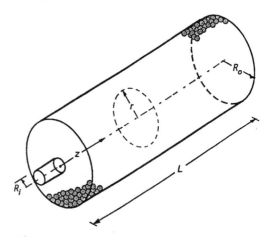

Fig. 1. Schematic diagram of storage cylinder and coordinate system.

Table 1. Characterization data for G216 carbon

Residue after ignition	1.2%
pH	9.6
Specific surface area (BET)	$1370 \ m^2 \ g^{-1}$
Packing density	$0.50 \ g \ cm^{-3}$
Particle size	$R_p = 0.5 \ mm$
Cumulative pore volume (< 200 Å)	$0.026 \ ml \ g^{-1}$

where q is the adsorbed phase concentration ($kg \ kg^{-1}$), P is pressure (bar) and T is temperature (K).

The effective bed thermal conductivity in the presence of fluid flow is given by

$$\lambda_e = \lambda_e^0 + 2\gamma C_{pg}R_p c\|\boldsymbol{v}\|, \ \gamma = 0.3 \qquad (2)$$

where $\|\boldsymbol{v}\| \equiv (v_z^2 + v_r^2)^{1/2}$ is the gas superficial velocity vector modulus. The other symbols are defined in the nomenclature. For a nonconsolidated porous medium, the static contribution, λ_e^0, is a function of gas conductivity, particle/gas conductivities ratio and bed porosity:

$$\lambda_e^0 = \lambda_g f(\bar{\lambda}_s/\lambda_g, \epsilon) \qquad (3)$$

Symbol $\bar{\lambda}_s$ denotes the effective thermal conductivity of a carbon particle, which depends on intraparticle porosity, ϵ_p. The λ_e^0 value considered in this work, $1.2 \ W \ m^{-1} \ K^{-1}$, is based on the correlation proposed by Zehner and Schlünder [21] for nonconsolidated isotropic packed beds. The relevant values of gas, carbon and reservoir properties employed in the numerical simulations are listed in Table 2.

3. FAST CHARGE

In this section, intraparticle and film resistances to heat and mass transfer are neglected in order to focus attention on the hydrodynamics of flow through the carbon bed. These simplifications will be removed later when discussing particle kinetics.

Table 2. Data used in numerical simulations

$C_{pg} = 2450$ J kg^{-1} K^{-1}	$T_{ext} = 285$ K
$C_{ps} = 650$ J kg^{-1} K^{-1}	$T_i = 285$ K
$\Delta H = -1.1 \times 10^6$ J kg^{-1}	$T_\infty = 285$ K
$L = 0.85$ m	$\epsilon = 0.35$
$M_g = 16.04 \times 10^{-3}$ kg mol^{-1}	$\epsilon_p = 0.6$
$P_h = 3.5$ MPa	$\lambda_e^0 = 1.2$ W m^{-1} K^{-1}
$R_i = 0.005$ m	$\mu = 1.25 \times 10^{-5}$ kg m^{-1} s^{-1}
$R_0 = 0.14$ m	$\rho_b = 410$ kg m^{-3}

3.1 Model equations

The model equations and boundary conditions are listed below without detailed discussion. Main assumptions are considering natural gas to be pure methane, whose thermodynamic behavior is modeled as ideal, and taking an average value of 4 kcal mol^{-1} for the heat of adsorption. The reader is referred to Barbosa Mota [19] and Barbosa Mota et al. [20] for further details.

Continuity equation:

$$\frac{\partial}{\partial t}(\epsilon_t c + \rho_b q) + \nabla \cdot (c\mathbf{v}) = 0 \tag{4}$$

Ideal gas law:

$$c = PM_g/(RT) \tag{5}$$

Momentum balance (Ergun's equation [22] extended to two dimensions):

$$\mathbf{v} = -\frac{2\nabla P}{\alpha + \sqrt{\alpha^2 + 4\beta c \|\nabla P\|}} \tag{}$$

$$\|\nabla P\| = [(\partial P/\partial z)^2 + (\partial P/\partial r)^2]^{1/2} \tag{6}$$

$$\alpha = 150(1-\epsilon)^2\mu/(4\epsilon^3 R_p^2)$$

$$\beta = 1.75(1-\epsilon)/(2\epsilon^3 R_p)$$

Energy equation:

$$C_{pg}\frac{\partial}{\partial t}((\epsilon_t c + \rho_b q)T) + \rho_b \Delta H \frac{\partial q}{\partial t} - \epsilon_t \frac{\partial P}{\partial t} + \rho_b C_{ps}\frac{\partial T}{\partial t}$$

$$+ C_{pg}\nabla \cdot (Tc\mathbf{v}) - \nabla \cdot (\lambda_e \nabla T) = 0 \tag{7}$$

Boundary conditions:

$$\left.\begin{array}{l} \mathbf{v} \cdot \mathbf{n} = 0 \\ \lambda_e \nabla T \cdot \mathbf{n} = h_w(T_\infty - T) \end{array}\right\} \text{ for } \begin{cases} z = 0, R_i \leq r \leq R_0 \\ 0 \leq z \leq L, r = R_0; \\ z = L, 0 \leq r \leq R_0 \end{cases}$$

$$\tag{8}$$

$$\frac{\partial P}{\partial r} = 0, \frac{\partial T}{\partial r} = 0 \text{ for } 0 \leq z \leq L, r = 0 \tag{9}$$

$$P = P_h, C_{pg}(T_{ext} - T)c\mathbf{v} = -\lambda_e \partial T/\partial z \tag{10}$$

for $z = 0$, $0 \leq r \leq R_i$ (opening).

Symbol $\mathbf{n}$ denotes the outwards unit vector normal to each point of the wall and T_{ext} is the temperature

of the external source gas. Before charging, the storage system is empty, i.e. free of gas, and is at temperature T_i. The wall heat transfer coefficient, h_w, is set to zero (adiabatic operation), which is a good approximation for fast charges.

3.2 Results and discussion

The pressure field during charge is shown in Fig. 2 for two different times. The pressure drop is located in a small region surrounding the opening. There, the gas velocity decreases rapidly due to the sudden expansion and frictional drag. The decrease in velocity reduces the pressure drop allowing the rest of the cylinder to fill uniformly.

The bed temperature decreases as the gas expands near the opening. However, this is a local effect caused by the opening and by the fact that gas is being charged at high pressure. In fact, the bed temperature increases in the rest of the reservoir due to pressurization of the storage system, and the overall effect is an increase in average bed temperature. This is depicted in Fig. 3, which shows the temperature field during charge of a bed of inert particles with characteristics similar to the G216 carbon.

The temperature field during charge when the storage system is filled with G216 carbon is shown in Fig. 4 for two different times. In this case, the heat of adsorption released compensates for the local temperature drop near the opening and is responsible for a significant increase in average bed temperature.

The charges simulated occur under adiabatic conditions since h_w was set to zero. In this case, the predicted average bed temperature rise for G216 carbon is 79°C. This value is in agreement with experimental values of 78 and 82°C observed in two independent adiabatic charge experiments. The former was carried out by Sejnoha et al. [14] on a 71 liter reservoir filled with CNS carbon; the latter was performed by Remick and Tiller [18] on a 1 liter cylinder filled with G216 carbon. If there is no adsorption, the predicted average bed temperature rise is 7.7°C.

A simple formula for estimating the filling time when the carbon bed is the only resistance to gas flow (fast charge) is

$$\frac{W_f}{\pi R_i^2 \sqrt{P_h M_g(P_h - P_1)/(RT_{ext}\beta\zeta)}} \tag{11}$$

where W_f stands for the adiabatic storage capacity, β is defined in eqn (6) and ζ is a measure of the mean penetration distance from the opening of the reservoir where the major pressure drop is located (see Fig. 2). The formula is based on two conclusions drawn from the fast charge simulations: i) the filling time depends on reservoir volume and diameter of the opening but not on the system's geometry, and ii) the gas velocity is sufficiently high to make inertial forces prevail over viscous effects.

The ζ values were obtained by matching the filling times given by eqn (11) with those predicted by the numerical simulations for a wide range of bed porosi-

250

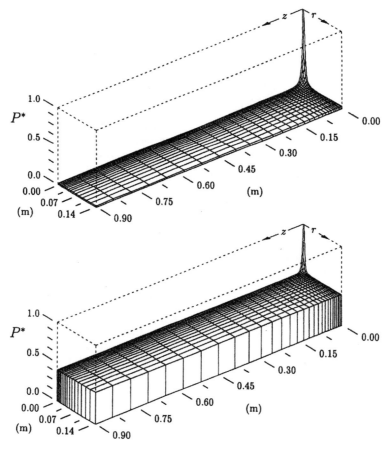

Fig. 2. Dimensionless pressure field, $P^* = P/P_h$, during charge of the G216 carbon bed. Top figure: $t = 10$ s; bottom figure: $t = 30$ s.

ties and particle diameters. For the opening radius considered here ($R_i = 5$ mm), ζ takes the value 4.8 mm if there is gas adsorption and 7.2 mm otherwise. The reader is referred to Barbosa Mota *et al.* [20] for further details.

4. INTRAPARTICLE DIFFUSIONAL RESISTANCES

In order to study the effect of diffusional resistances on charge dynamics, an intraparticle transport equation, governed by a diffusion law, is added to the model described earlier.

High storage densities can only be achieved if practically all stored methane is in the adsorbed phase. The void space where methane is at gas density can be minimized by producing the carbon as a monolith or as shaped pieces that nest together. Since methane adsorption at ambient temperature is mainly a micropore phenomenon, most of the intraparticle void volume should be micropore. This, in turn, may increase significantly the intraparticle diffusional resistances to mass transport and, consequently, lead to filling times that are too long.

The diffusional model might also be useful for

studying the performance of storage systems employing zeolites, since these adsorbents have diffusional time constants, $D/R_p^2 \approx 10^{-3}$ s^{-1}, which are smaller than those for activated carbons, $D/R_p^2 \approx 10^{-1}$ s^{-1} [24]. Despite the fact that carbons are better adsorbents for natural gas on a weight basis, the difference between carbon and zeolite performance is much smaller on a volumetric basis since zeolites have higher density than carbons [1].

4.1 *Model equations*

It is assumed that the carbon particles can be approximated by equivalent spheres of radius R_p. This makes the intraparticle profiles dependent only on the radial coordinate, r_p, of the particles. In this case, the differential mass balance on a spherical shell element of a carbon particle gives

$$\frac{\partial}{\partial t}(c_p^* + q^*) = \frac{D/R_p^2}{x^2}\frac{\partial}{\partial x}\left(x^2\frac{\partial q^*}{\partial x}\right) \quad (12)$$

$$0 < x \equiv r_p/R_p < 1$$

where $c_p^* \equiv \epsilon_p c_p$ and $q^* \equiv \rho_b q/(1-\epsilon)$. The effective diffusion coefficient, D, is based on the adsorbed

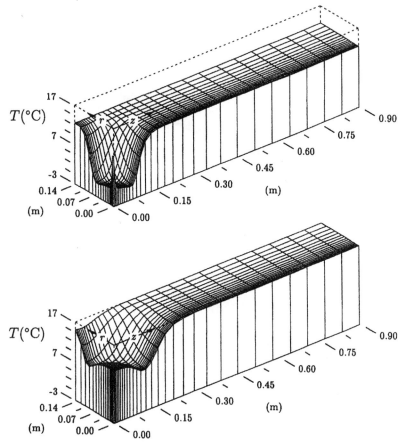

Fig. 3. Temperature field during charge of a bed of inert particles with characteristics similar to those of the G216 carbon particles. Top figure: $t = 10$ s; bottom figure: $t = 25$ s.

phase concentration gradient and is considered constant. Equation (12) is subjected to boundary conditions

$$\begin{cases} \partial q^*/\partial x = 0 & \text{for } x = 0 \\ q^* = \rho_b q(P,T)/(1-\epsilon) & \text{for } x = 1 \end{cases} \quad (13)$$

Heat transfer is assumed to be sufficiently rapid, relative to sorption rate, so that intraparticle temperature gradients are negligible, i.e. $\partial T_p/\partial x \approx 0$ inside every particle. Previous simulations employing a separate intraparticle energy balance with the heat transfer resistance concentrated in the external film showed that, for the system studied, the heat transfer resistance introduced by the external film can be neglected. Thus, $T_p = T$ for each particle, where T is the temperature of the gas surrounding it.

The continuity and energy equations for the carbon bed are modified to take the intraparticle profiles into account:

$$\frac{\partial}{\partial t}[\epsilon c + (1-\epsilon)(\bar{c}_{p*} + \bar{q}^*)] + \nabla \cdot (cv) = 0 \quad (14)$$

$$\frac{\partial}{\partial t}[(\epsilon c + (1-\epsilon)(\bar{c}_{p*} + \bar{q}^*))C_{pg}T] + \Delta H \frac{\partial \bar{q}^*}{\partial t} - \epsilon \frac{\partial P}{\partial t}$$

$$-(1-\epsilon)\epsilon_p \frac{\partial \bar{P}_p}{\partial t} + \rho_b C_{ps} \frac{\partial T}{\partial t}$$

$$+ \nabla \cdot (C_{pg}Tcv - \lambda_e T) = 0 \quad (15)$$

In these equations $\bar{c}_p^*(t,z,r)$ and $\bar{q}^*(t,z,r)$ are averaged values of c_p^* and q^* for time t over a particle located at point (z,r) in the reservoir. These average values are related to the intraparticle profiles by

$$(\bar{c}_p^*, \bar{q}^*) = \frac{1}{V_p}\int\int\int_{V_p}(c_p^*, q^*)dV_p = 3\int_0^1(c_p^*, q^*)x^2 dx \quad (16)$$

4.2 Results and discussion

The two-dimensional pressure and temperature fields, during charge, are similar to the figures presented earlier regarding the equilibrium model. However, temperature spatial variations are less pronounced when intraparticle diffusional resistances are

252

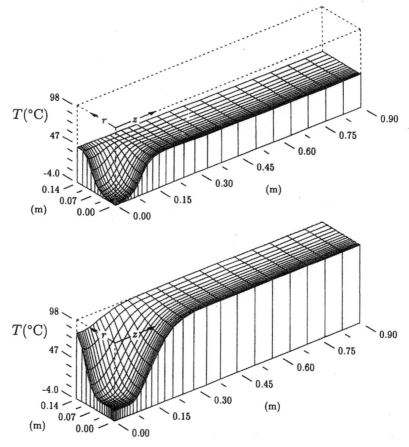

Fig. 4. Temperature field during charge of the G216 carbon bed. Top figure: $t = 10$ s; bottom figure: $t = 50$ s.

not negligible, because the carbon heats up at a slower rate and the storage system has more time to smooth the temperature front by conduction.

Figure 5 shows the pressure history in the center of the reservoir, during charge, for different values of D/R_p^2. In the presence of important diffusional

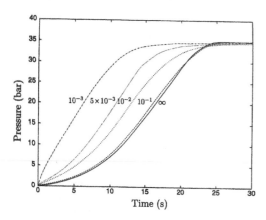

Fig. 5. Bulk pressure history in the center of the reservoir, during charge, for different values of D/R_{p^2} (s^{-1}).

resistances, the gas penetrates the particles very slowly and accumulates initially in the bed void volume. This results in a rapid pressure increase up to practically the nominal charge value, whose dynamics are governed by the hydrodynamics of flow through the carbon bed and the void volume between particles. The subsequent filling proceeds with a rate controlled entirely by the diffusion process. The pressure remains at its charge value during this period, since all methane that adsorbs is compensated quickly by the external pressure source. As a result, the pressure history is not representative of the charge process and cannot be used to quantify it. In practice, fuel supply at natural gas refuelling stations will be controlled by mass flow measurements so that the amount of gas delivered to the consumer is known.

The pressure history curve is always bounded by two limiting cases: the lower one is the equilibrium model ($D/R_p^2 \to \infty$), the upper one corresponds to the charge of a bed of nonporous particles with the same diameter as the carbon particles ($D/R_p^2 = 0$).

Figures 6 and 7 show, respectively, histories of average temperature and total amount of methane stored per unit reservoir volume, during charge, for

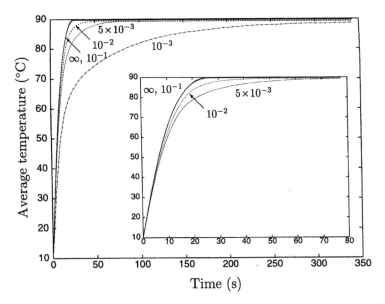

Fig. 6. Average bed temperature history, during charge, for different values of D/R_{P^2} (s^{-1}).

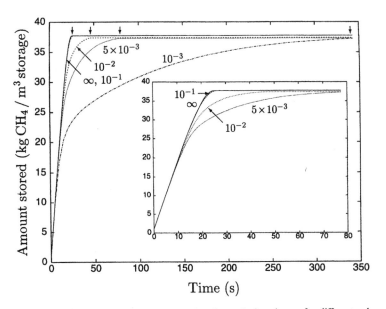

Fig. 7. History of amount of methane stored per unit reservoir volume, during charge, for different values of D/R_{P^2} (s^{-1}).

different values of D/R_p^2. These average values are calculated using the following formula:

$$\bar{\bar{\phi}} = \frac{2}{R_0^2 L} \int_0^L \int_0^{R_0} \phi(r,z) r \, dr \, dz \qquad (17)$$

The dummy variable ϕ stands for T in the average temperature calculation. If the quantity of interest is the amount of stored methane, ϕ is replaced by $\epsilon c + (1-\epsilon)(\bar{c}_{p^*} + \bar{q}^*)$.

Both figures show a similar trend, this is expected since the adsorbed phase concentration is strongly affected by temperature. It is clear that for D/R_p^2 values

smaller than $10^{-2}\,\text{s}^{-1}$, the time needed to completely fill the reservoir is considerably longer than for a system with the same characteristics but without diffusional restrictions. For example, according to Fig. 7 the filling time when $D/R_p^2 = 10^{-3}\,\text{s}^{-1}$ is nearly 14 times greater than the time required when the storage system has no diffusional limitations.

5. SLOW DISCHARGE

The following work addresses the situation of practical relevance where the NG admission to the motor is controlled by power requirements of the running vehicle. In this case, discharge duration is increased considerably by the slow rate at which it is carried out. The heat of adsorption consumed during discharge lowers the bed temperature and, consequently, increases the amount of NG that remains in storage at depletion. However, the process is not adiabatic and the heat transferred from the surroundings partially compensates for the consumed heat of adsorption.

5.1 Model equations

It may be shown [19] that if the cylinder is sufficiently long, so that its front and rear faces have negligible influence on the thermal field, then the problem is reduced to the study of a cross section of the cylinder as shown in Fig. 8. Moreover, during a slow discharge the temperature field is dependent only on the radial coordinate of the reservoir and the spatial pressure variation is negligible [19].

The differential equation governing the radial temperature profile during a slow discharge is

$$(\rho_b C_{ps} + C_{pg}(\epsilon_t c + \rho_b q)) \frac{\partial T}{\partial t} + \rho_b \Delta H \frac{\partial q}{\partial t}$$
$$- \epsilon_t \frac{d\bar{P}}{dt} - \frac{\lambda_e}{r} \frac{\partial}{\partial r}\left(r \frac{\partial T}{\partial r}\right) = 0 \qquad (18)$$

subjected to boundary conditions

$$\partial T/\partial r = 0 \text{ for } r = 0 \qquad (19)$$

and

$$e_w C_w \frac{\partial T}{\partial t} + \lambda_e \frac{\partial T}{\partial r} + h_w(T - T_\infty) = 0 \text{ for } r = R_0 \qquad (20)$$

The latter condition is an energy balance to the stainless steel cylinder wall ($C_w = 3.92 \times 10^6$ J m^{-3} K^{-1}, $e_w = 1$ cm).

The system is driven by the pressure history imposed on the reservoir during discharge. The bar on top of P emphasizes the fact that during discharge,

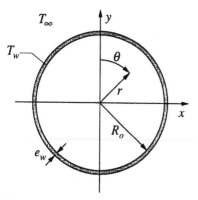

Fig. 8. Schematic diagram of storage cylinder's cross section; the dark region identifies the lateral wall.

pressure does not vary along the r coordinate, i.e. $\bar{P} = \bar{P}(t)$. The pressure histories considered in this work lead to constant discharge mass flow rates.

The external wall heat transfer coefficient, h_w, is obtained from the correlation [24,25]

$$\frac{h_w R_0}{\lambda} = a(GrPr)^m \qquad (21)$$

where $Gr = R_0^3 \rho^2 g\alpha(T_\infty - T|_{r=R_0})/\mu^2$ is the Grashof number and $Pr = C_p\mu/\lambda$ is the Prandtl number; the thermodynamic properties refer to air. For horizontal cylinders with $R_0 < 20$ cm the values of a and m are, respectively, 1.09 and 1/5.

5.2 Results and discussion

Figure 9 shows radial temperature profiles for a discharge with a duration of two hours. The initial and final pressures are $P_h = 3.5$ MPa and $P_1 = 0.101$ MPa, respectively. At the end of discharge the wall temperature has decreased 30°C, whereas in the center of the bed the temperature drop is 55°C. This gives an average bed temperature decrease of 40°C. The amount of methane delivered is 23% higher than the amount delivered by the adiabatic discharge, whose average bed temperature decrease is 64°C.

The temperature drop during discharge can be reduced by increasing the heat transfer rate from the surroundings. Following, is a brief description of three ways to accomplish this:

Increase the bed thermal conductivity by mixing the carbon particles, or replacing part of the binder in the case of a carbon monolith, with a highly thermal conductive material. This has the disadvantage, at least in the case of granular beds, of reducing the isothermal storage capacity since part of the carbon bed is replaced by an inert material. Moreover, this technique is highly dependent on the number of new contact points created between particles.

Maintain the reservoir wall at the highest possible temperature in order to maximize temperature gradients in the bed and increase the heat transfer by conduction. This could be accomplished by making the exhaust gases that leave the motor at high temperature flow through a jacket around the reservoir.

Increase the wall area per unit reservoir volume, since this leads to an increase in heat transfer rate from the surroundings. In the case of cylindrical reservoirs, this strategy favors long cylinders with short radii.

A series of simulations were carried out to determine the effect of discharge duration on wall temperature, average bed temperature and capacity ratio, CR, defined by

$$CR = \frac{(\text{methane delivered})_{\text{real operation}}}{(\text{methane delivered})_{\text{isothermal operation}}}$$
$$= \frac{(\epsilon_t c + \rho_b q)(P_h, T_i) - (\epsilon_t c + \rho_b q)(P_1, \bar{T}_e)}{(\epsilon_t c + \rho_b q)(P_h, T_i) - (\epsilon_t c + \rho_b q)(P_1, T_i)}$$
$$(22)$$

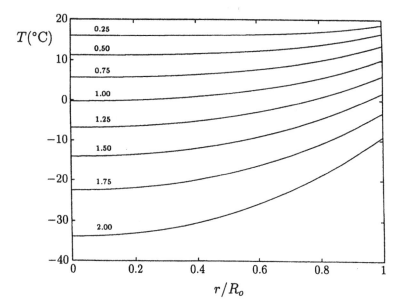

Fig. 9. Radial temperature profiles for a constant mass flow rate discharge with a duration of 2 hours. The time values are expressed in hours on top of each curve.

where $\bar{T}_e$ is the average bed temperature at the end of discharge. Parameter CR gives the ratio of methane delivered by discharging between pressures P_h and P_l at constant mass flow rate and the amount obtained from discharging isothermally between the same two pressures. Results are depicted in Fig. 10, which shows that the thermal effects remain important for the discharge duration range considered in this study. When the discharge duration is four hours, which is a considerably long one, the wall temperature decreases 29°C, the average bed temperature drop is 37°C, and the capacity ratio is 0.85, which means that the amount of methane delivered is 15% lower than the amount delivered by discharging the gas isothermally.

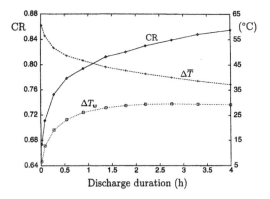

Fig. 10. Wall temperature decrease, ΔT_w, average bed temperature decrease, $\Delta \bar{T}$, and capacity ratio, CR, as a function of discharge duration.

6. *IN SITU* THERMAL ENERGY STORAGE

The Institute of Gas Technology and the Gas Research Institute developed an interesting method to reduce the adverse effect of the heat of adsorption on storage capacity. Basically, the method consists of introducing in the carbon bed an encapsulated phase change material that has a relatively high heat of fusion at ambient temperature. This material would be capable of absorbing the heat of adsorption released during charge and resupplying it during discharge.

There is an interesting optimization problem related to this method, it is the determination of the optimal amount of phase change material that maximizes the amount of gas delivered during discharge. For the amount of phase change material to be optimum the heat of fusion must have been totally consumed at the end of discharge, otherwise the amount of phase change material is in excess. Moreover, if the method is to work properly, the phase change material must be in the liquid state at the beginning of the discharge. This forces its fusion temperature to be lower than the initial bed temperature.

In the more general case, the temperature of the superheated phase change material decreases up to the fusion value, T_f, then the heat of fusion is released during solidification at constant temperature. If the heat of fusion is completely consumed before the end of discharge, the bed temperature decreases below T_f. This is depicted in Fig. 11. In the drawing W_k ($k = 1,2,3$) is the amount of gas delivered per unit storage volume on each stage of the discharge.

256

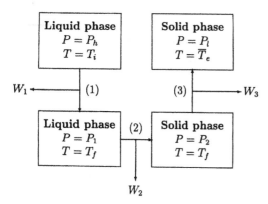

Fig. 11. Evolution of phase change material's physical state during discharge.

In order to simplify the discussion, the storage system is assumed to be initially at the phase change material's fusion temperature ($T_i = T_f$). This means that only stages 2 and 3 in Fig. 11 are considered.

Overall mass and energy balances to the storage system for stage 2 give

$$\alpha \Delta_2(\epsilon_t c + \rho_b q) = W_2 \qquad (23)$$

$$\alpha \Delta_2(C_{pg} T(\epsilon_t c + \rho_b q) + \rho_b q \Delta H - \epsilon_t P)$$
$$+ (1 - \alpha)\rho_f \Delta H_f = C_{pg} T_f W_2 \qquad (24)$$

Subscript f relates to the phase change material, ΔH_f is the heat of fusion and α is the fraction of storage volume occupied by the carbon bed, the remaining $1 - \alpha$ being occupied by phase change material. The operator $\Delta_2(\cdot)$ is defined as

$$\Delta_2(\phi) = \phi(P_h, T_f) - \phi(P_2, T_f) \qquad (25)$$

Eliminating W_2 from eqns (23) and (24) gives an equation for calculating P_2:

$$\alpha \Delta_2(\rho_b q \Delta H - \epsilon_t P) + (1 - \alpha)\rho_f \Delta H_f = 0 \qquad (26)$$

Similarly, the overall mass and energy balances for stage 3 are

$$\alpha \Delta_3(\epsilon_t c + \rho_b q) = W_3 \qquad (27)$$

$$\alpha \Delta_3 \left(C_{pg} T(\epsilon_t c + \rho_b q) + \rho_b q \Delta H + \rho_b C_{ps} T - \epsilon_t P \right.$$
$$\left. + \frac{1 - \alpha}{\alpha} \rho_f C_{pf} T \right) = \frac{1}{2} C_{pg}(T_f + \bar{T}_e) W_3 \qquad (28)$$

with $\Delta_3(\cdot)$ given by

$$\Delta_3(\phi) = \phi(P_1, T_f) - \phi(P_1, \bar{T}_e) \qquad (29)$$

The equation for calculating $\bar{T}_e$ is obtained by eliminating W_3 from eqns (27) and (28):

$$\Delta_3 \left(C_{pg} T(\epsilon_t c + \rho_b q) + \rho_b q \Delta H + \rho_b C_{ps} T - \epsilon_t P \right.$$
$$\left. + \frac{1 - \alpha}{\alpha} \rho_f C_{pf} T \right) = \frac{1}{2} C_{pg}(T_f + \bar{T}_e)\Delta_3(\epsilon_t c + \rho_b q)$$
$$\qquad (30)$$

The optimum α value is the one that maximizes the amount of methane delivered during discharge, i.e.

$$\alpha(\epsilon_t c + \rho_b q)(P_h, T_i) - \alpha(\epsilon_t c + \rho_b q)(P_1, \bar{T}_e) \qquad (31)$$

The phase change material considered in the simulations is the same employed by Jasionowski et al. [4], it is a salt hydrate ($Na_2HPO_4 \cdot 12H_2O$) for which $\rho_f = 1000 \text{ kg m}^{-3}$, $T_f = 35°C$, $\Delta H_f = 2.8 \times 10^5 \text{ J kg}^{-1} \text{ K}^{-1}$ and $C_{pf} = 1693 \text{ J kg}^{-1} \text{ K}^{-1}$.

Figure 12 shows the effect of parameter α on $\bar{T}_e$ and CR, the latter being defined by

$$CR = \alpha \frac{(\epsilon_t c + \rho_b q)(P_h, T_i) - (\epsilon_t c + \rho_b q)(P_1, \bar{T}_e)}{(\epsilon_t c + \rho_b q)(P_h, T_i) - (\epsilon_t c + \rho_b q)(P_1, T_i)}$$
$$\qquad (32)$$

The slope of the $\bar{T}_e$ curve has a discontinuity that identifies the minimum amount of salt that maintains $\bar{T}_e$ equal to T_f. In the present case, this value is greater than the optimum amount of salt, which means that $\bar{T}_e$ is smaller than T_f by a few degrees. The loss in net storage due to this temperature decrease is smaller than the gain due to the additional carbon inserted.

Curiously, the optimum value, $\alpha_{opt} = 0.86$, corresponding to 14% of storage volume occupied by the

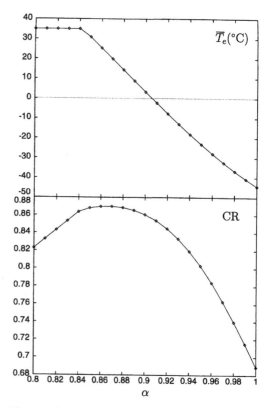

Fig. 12. Average bed temperature at the end of discharge, $\bar{T}_e$, and capacity ratio, CR, as a function of parameter α for the experimental setup of Jasionowski et al. [4].

salt, is close to the value employed experimentally by Jasionowski *et al.* [4] (17.5%). The gain relative to the adiabatic discharge without salt is about 25%.

7. CONCLUDING REMARKS

Great improvements have been made in the development of carbon adsorbents for NG storage. Performance levels required for commercial viability of on-board ANG storage have already been achieved with methane under isothermal conditions. However, ANG storage is not isothermal as demonstrated by the pronounced effect of the heat of adsorption on storage capacity. The problem seems to be solved for charge at a refuelling station. For example, the gas can be cooled before it enters the reservoir and any non-adsorbed gas recirculated back into the refrigeration unit using a compressor [26]. An economic solution for eliminating the temperature drop during discharge will increase the delivered storage capacity under real conditions above the performance level necessary for economic feasibility.

The effect of diffusional resistances on charge dynamics is a subject that has received little attention in the literature. It was discussed here for a granular activated carbon bed. If, in order to reduce void space, the carbon adsorbent is packed as a monolith of high density, an additional resistance to gas flow is introduced and it increases even further the filling time. A measure of this resistance, equivalent to D/R_p^2, is given by $K(P_1+P_h)/(2\mu L^2)$, where K is the permeability of the carbon monolith. Is there a limit where further optimization of the micropore structure of carbon particles and of the density of the monolith to increase the adsorptive capacity lead to filling times that are too long?

8. NOMENCLATURE

c Concentration of gas ($kg\ m^{-3}$)

c_p Concentration of intraparticle gas ($kg\ m^{-3}$)

C_{pg} Heat capacity of gas at constant pressure ($J\ kg^{-1}\ K^{-1}$)

C_{ps} Heat capacity of carbon particle ($J\ kg^{-1}\ K^{-1}$)

C_w Volumetric heat capacity of wall ($J\ m^{-3}\ K^{-1}$)

D Effective diffusivity ($m^2\ s^{-1}$)

e_w Wall thickness (m)

h_w Wall heat transfer coefficient ($W\ m^{-2}\ K^{-1}$)

$-\Delta H$ Heat of adsorption ($J\ kg^{-1}$)

ΔH_f Heat of fusion ($J\ kg^{-1}$)

K Permeability of carbon monolith (m^2)

L Reservoir length (m)

M_g Molecular mass of gas ($kg\ mol^{-1}$)

P Gas pressure (Pa)

P_h Charge pressure (Pa)

P_1 Depletion pressure (Pa)

P_p Intraparticle gas pressure (Pa)

q Adsorbed phase concentration ($kg\ kg^{-1}$)

r Radial coordinate in reservoir (m)

r_p Radial coordinate in particle (m)

R Ideal gas constant ($8.31441\ J\ mol^{-1}\ K^{-1}$)

R_i Radius of opening (m)

R_0 Cylinder radius (m)

R_p Particle radius (m)

t Time (s)

T Temperature of gas (K)

$\bar{T}_e$ Average bed temperature at the end of discharge (K)

T_{ext} Temperature of external source gas (K)

T_f Fusion temperature of phase change material (K)

T_i Initial temperature in reservoir (K)

T_p Intraparticle gas temperature (K)

v (v_z,v_r), superficial velocity vector ($m\ s^{-1}$)

V_p Particle volume (m^3)

x Dimensionless radial coordinate in particle

z Axial coordinate in reservoir (m)

Greek symbols

ϵ Bed porosity

ϵ_p Particle porosity

ϵ_t $\epsilon+(1-\epsilon)\epsilon_p$

λ_e Effective thermal conductivity of the carbon bed ($W\ m^{-1}\ K^{-1}$)

λ_g Thermal conductivity of gas ($W\ m^{-1}\ K^{-1}$)

$\bar{\lambda}_s$ Thermal conductivity of carbon particle ($W\ m^{-1}\ K^{-1}$)

μ Viscosity of gas ($kg\ m^{-1}\ s^{-1}$)

ρ_b Bulk density of carbon bed ($kg\ m^{-3}$)

Acknowledgements—Financial support received from JNICT (PBIC/P/QUI/2415/95), Program PRAXIS XXI (BPD/6066/95), and FLAD (#192/96) is gratefully acknowledged.

REFERENCES

1. Talu, O., in *Proceedings of International Conference on Fundamentals of Adsorption*, ed. M. Suzuki, Kyoto, Japan, 1992, pp. 655–662.
2. Penilla, R. N., Paez, J. M. O. and Plata, G. A., The use of natural gas. Paper presented at 14th World Petroleum Congress, Stavanger, Norway, 1994.
3. European Commission, *Bulletin of the European Union*, No. 5/95, Annex 1, 1995.
4. Jasionowski, W. J., Tiller, A. J., Fata, J. A., Arnold, J. M., Gauthier, S. W. and Shikari, Y. A., Charge/discharge characteristics of high-capacity methane adsorption storage systems. Paper presented at International Gas Research Conference, Tokyo, Japan, November 1989.
5. Remick, R. J. and Tiller, A. J., in *Proceedings of Nonpetroleum Vehicles Fuels Symposium*, Chicago, IL, 1985, pp. 105–119.
6. Barton, S. S., Dacey, J. R. and Quinn, D. F., in *Proceedings of International Conference on Fundamentals of Adsorption*, ed. A. L. Myers and G. Belfort, United Engineering Trustees, 1984, p. 65.
7. Barton, S. S., Holland, J. A. and Quinn, D. F., in *Proceedings of International Conference on Fundamentals of Adsorption*, ed. I. Liapis, United Engineering Trustees, 1986, p. 99.

258

8. Chahine, R. and Bose, T. K., Solidification of carbon powder adsorbents for NGV storage. Paper presented at 20th Conference on Carbon, Santa Barbara, CA, June 1992.
9. Czepirski, L., *Indian J. of Technology*, 1991, **29**, 266.
10. Quinn, D. F., Presentation at GURF *Meeting*, London, July 1990.
11. Quinn, D. F. and MacDonald, J. A., *Carbon*, 1992, **7**, 1097.
12. Quinn, D. F., MacDonald, J. A. and Sosin, K., Presentation at *ACS National Meeting*, San Diego, CA, March 1994.
13. Remick, R. J., Elkins, R. H., Camara, E. H. and Bulicz, T., *Advanced Onboard Storage Concepts for Natural Gas-Fueled Automotive Vehicles*. Technical report DOE/NASA/0327-1, 1984.
14. Sejnoha, M., Chahine, R., Yaïci, W. and Bose, T. K., Adsorption storage of natural gas on activated carbon. Paper presented at AIChE Annual Meeting, San Francisco, CA, November 1994.
15. Matranga, K. R., Myers, A. L. and Glandt, E. D., *Chem. Eng. Sci.*, 1992, **47**, 1569.
16. BeVier, W. E., Mullhaupt, J. T., Notaro, F., Lewis, I. C. and Coleman, R. E., Adsorbent-enhanced methane storage for alternate fuel powered vehicles. Paper presented at SAE Future Transportation Technology Conference and Exposition, Vancouver, British Columbia, August 1989.
17. Talu, O., in *Proceedings of First Separations Division Topical Conference on Separations Technologies: New Developments and Opportunities*, AIChE, 1993, pp. 409–414.
18. Remick, R. J. and Tiller, A. J., Heat generation in natural gas adsorption systems. Paper presented at Gaseous Fuels for Transportation International Conference, Vancouver, Canada, August 1986.
19. Mota, J. P. B., Modélisation des Transferts Couplés en Milieux Poreux, Ph.D. thesis, Institut National Polytechnique de Lorraine, Nancy, France, 1995.
20. Mota, J. P. B., Saatdjian, E., Tondeur, D. and Rodrigues, A. E., *Adsorption*, 1995, **1**, 17.
21. Zehner, P. and Schlünder, E. U., *Chemie. Ingr.-Tech.*, 1970, **47**, 933.
22. Ergun, S., *Chem. Eng. Prog.*, 1952, **42**, 89.
23. Yang, R. T., *Gas Separation by Adsorption Processes*, Butterworths Publishers, Boston, MA, 1987.
24. Cess, R. D., in *handbook of Heat Transfer*, McGraw–Hill, New York, 1973.
25. Perry, R. H., *Perry's Chemical Engineers' handbook*, 5th edn. McGraw–Hill, New York, 1984.
26. Jasionowsky, W. J., Kountz, K. J., Blazek, C. F., Tiller, A. J., Gauthier, S. W. and Takagishi, S. K., Thermal energy storage system for adsorbent low-pressure natural gas storage. Paper presented at International Gas Research Conference, 1992.

RADIAL DISTRIBUTION FUNCTION ANALYSIS OF THE STRUCTURE OF ACTIVATED CARBONS

A. Burian,[a],* A. Ratuszna,[a] J. C. Dore[b] and S. W. Howells[c]

[a]Instytut Fizyki, Uniwersytet Śląski, ul. Uniwersytecka 4, 40-007 Katowice, Poland
[b]School of Physical Sciences, University of Kent, Canterbury, CT2 7NR, U.K.
[c]Neutron Division, Rutherford Appleton Laboratory, Didcot, OX11 0QX, U.K.

(*Received* 22 September 1997; *accepted in revised form* 4 March 1998)

Abstract—The structure of a series of activated carbon materials prepared from a polymer of phenol formaldehyde resin with different degree of "burn-off" from 0 to 32% and variable porosity has been studied by wide-angle neutron scattering. The intensity measurements, extended to Q values of $50\,\text{Å}^{-1}$ yielded the radial distribution functions with the real space resolution of about $0.12\,\text{Å}$. The obtained results show that the structural features of the investigated materials are not affected by the activation process. All well-defined peaks of the radial distribution functions can be reproduced by the model consisting of the defective graphite-like layers with weak inter-layer correlations. © 1998 Elsevier Science Ltd. All rights reserved.

Key Words—A. Activated carbon, C. neutron scattering, D. defects, D. microstructure.

1. INTRODUCTION

Activated carbons are a class of carbonaceous materials with a very large internal surface area which allows many applications. These materials are usually prepared by carbonization and activation of organic precursors such as phenolic resins, polyfurfuryl alcohol, cellulose, anthracene, saccharose, and natural products such as olive or peach stones. The atomic scale structure of different carbonaceous materials has been studied by electron, X-ray or neutron scattering. It has been established that different forms of disordered carbons, despite different precursors, preparation techniques and thermal treatments, exhibit some common features. Their structure can be described in terms of disordered graphite-like layers with very weak inter-layer correlations or random stacking. Such a structure is termed turbostratic, according to the Warren terminology (see review paper [1] and refs. [2–9]). The size of ordered regions within a single layer and number of layers in a sequence are limited and depend on the thermal treatment of carbonized materials [7–9]. It has been found that degree of disorder increases with interatomic distances [4,7–9]. In the present paper the structural studies on a series of activated carbons by analysis of the radial distribution functions derived from the wide-angle neutron scattering measurements are described.

2. EXPERIMENTAL

2.1 *Sample preparation*

The activated carbons studied in the present work were obtained from a polymer of phenol formaldehyde resin by carbonization in nitrogen flow ($80\,\text{ml\,min}^{-1}$) at 1273 K (carbon A) and then activated with carbon dioxide under a pressure of 0.1 MPa at 1073 K for varying time periods. The degree of activation, increasing with the amount of time, measured as "burn-off", i.e. the percentage weight loss, are 14% (carbon A14) and 32% (carbon A32). The carbonization process, carried out at the above indicated temperature, removes the greatest amount of the non-carbon components from the samples (within ~ 1 at% accuracy—no additional correction procedure was necessary to remove contribution of residual hydrogen to measured intensity) whereas activation causes creation of pores by removing some of the carbon. The studies of the surface area and the pore volumes were performed using the nitrogen absorption (BET) and mercury porosimetry methods. The BET surfaces are 580, 840 and $1000\,\text{m}^2\,\text{g}^{-1}$ for the carbons A, A14 and A32, respectively. The pore volume comes mainly from micro- and mesopores, the volume coming from macropores is smaller. More detailed information about the pore volumes is given in our previous paper [10].

2.2 *Wide-angle neutron scattering experiment*

The earlier X-ray [4,6] or neutron scattering studies [5] yielded the radial distribution functions of limited resolution, resulting from relatively low value of a scattering vector available in these experiments. The data presented in this paper were collected using the Liquids and Amorphous Materials Diffractometer (LAD) at the ISIS pulsed neutron facility of the Rutherford Appleton Laboratory (U.K.). The intensities were measured in the range of the scattering vector Q ($Q = 4\pi \sin(\theta)/\lambda$, 2θ is the scattering angle and λ is the wavelength of the

*Corresponding author. Tel: 48 32 58 84 31; Fax: +48 32 58 84 31; e-mail: burian@us.edu.pl

Reprinted from *Carbon* **36** (11), 1613-1621 (1998)

incident neutrons) from about 0.25 to 50 Å^{-1}, providing the data with high real resolution ($\simeq 0.12$ Å). In the present experiments neutrons were scattered from the cylindrical sample to fixed angle detectors on each side of the diffractometer. The scattered intensities were measured as a function of time-of-flight (TOF), which is directly related to the scattering vector Q. Additional measurements were made for a background, the empty sample container and a standard vanadium bar in order to put the experimental intensities on an absolute scale. The samples were heated at about 473 K before the intensity measurements in order to remove water, which might be the source of the very strong incoherent scattering because of the high incoherent cross-section of the hydrogen atom.

3. THEORETICAL OUTLINE

The intensity of scattered neutrons is related to the radial distribution function (RDF) as follows:

$$4\pi r^2 \rho(r) b^2 = 4\pi r^2 \rho_0 b^2 + rG(r) \qquad (1)$$

where the reduced radial distribution function $G(r)$ is the sine Fourier transform of the reduced intensity $i(Q)$ multiplied by Q and terminated at Q_{max} (the maximum value of the scattering vector).

$$G(r) = \frac{2}{\pi} \int_0^{Q_{max}} Q i(Q) \sin(Qr)\, dQ \qquad (2)$$

The reduced intensity $i(Q)$ is expressed as the difference $I(Q) - b^2$ of the corrected and normalized intensity function $I(Q)$ and square of the coherent scattering length of carbon b^2. The experimental RDF provides information about the probability of finding an atom in a spherical shell at a distance r from an arbitrary origin. Successive peaks correspond to nearest-, second- and next-neighbour atomic distribution. The structural parameters: the interatomic distances r_i, their standard deviations σ_i and the coordination numbers N_i can be obtained from the experimental RDF by the curve-fitting method. Assuming three-dimensional Gaussian distribution of the interatomic distances with the standard deviation σ

$$F(x, y, z) = \frac{1}{(2\pi\sigma^2)^{3/2}} \exp\left(-\frac{x^2 + y^2 + z^2}{2\sigma^2}\right) \qquad (3)$$

the shape of the one-dimensional pair distribution function can be derived by averaging F over all orientations and shifting to the position defined by the interatomic distance. These procedures can be readily done by convoluting F with the shell function, defined by Dirac's delta [11–13]

$$\frac{1}{4\pi r_i^2} \delta(r - r_i) \qquad (4)$$

where r_i is the radius of the ith coordination sphere and $r = (x^2 + y^2 + z^2)^{1/2}$. Using this approach the theoretical form of the RDF is finally expressed as follows:

$$4\pi r^2 \rho(r) b^2 = r \sum_i \int_0^\infty \frac{1}{\sqrt{2\pi}\sigma_i} \frac{N_i}{r_i}$$
$$\left\{ \exp\left[-\frac{(r'-r_i)^2}{2\sigma_i^2}\right] - \exp\left[-\frac{(r'+r_i)^2}{2\sigma_i^2}\right] \right\}$$
$$\cdot [P(r-r') - P(r+r')]\, dr',$$

$$P(r) = b^2 \frac{Q_{max}}{\pi} \frac{\sin(Q_{max}r)}{Q_{max}r} \qquad (5)$$

The sum in eqn (5) is taken over all coordination spheres.

Based on eqn (5) one may model the atomic distributions in terms of a series of peaks convoluted with a peak-shape function $P(r)$ to include the effects of data termination at Q_{max}. In order to compensate for the edge effects in a finite-sized model the correction term $\epsilon(r)$, which corresponds to the probability of finding an atom at a distance r from another atom lying inside a considered model, is used. According to ref. [14], for a rectangular microcrystallite this is given by

$$\epsilon(r) = 1 - \frac{1}{abc}\left(\frac{r^3}{4\pi} - \frac{Br^2}{3\pi} + \frac{Cr}{2}\right) \qquad (6)$$

for $r < a, b, c$ and 0 for $r^2 \geq a^2 + b^2 + c^2$, where a, b, c are the dimensions of the microcrystallite, $B = 2(a + b + c)$ and $C = (ab + bc + ac)$.

In the case of the turbostratic structure the RDF can be calculated taking into account interatomic correlations within a single layer (RDF_{sl}) and adding the term resulting from random stacking of layers [5]. Now the RDF can be written as

$$4\pi r_2 \rho(r) b^2 = (RDF_{sl}) + 4\pi r^2 \rho_0 b^2 + 4\pi r n_a b^2 \cdot$$
$$\left\{ \frac{\left[1 - \exp\left(-\frac{d_c}{L_c}\right)^{[r/d_c]}\right]}{\exp\left(\frac{d_c}{L_c}\right) - 1} - \frac{L_c}{d_c}\left[1 - \exp\left(\frac{r}{L_c}\right)\right] \right\} \qquad (7)$$

where n_a is the number of atoms per unit area on a plane ($\rho_0 = n_a/d_c$), d_c is the plane spacing, L_c is the mean microcrystallite size along the direction perpendicular to planes and $[r/d_c]$ denotes the greatest integer in r/d_c.

4. RESULTS AND DISCUSSION

The scattered intensities for the three investigated samples were corrected for background, container and multiple scattering, dead time, attenuation and then normalized using standard procedures [15]. The resulting normalized intensity functions are shown in Fig. 1. The enlarged small-angle parts ($0 \leq Q \leq 1$ Å^{-1}) of the plots presented in Fig. 1 are replotted in Fig. 2. From comparison of these curves it is apparent that the intensities for all samples are

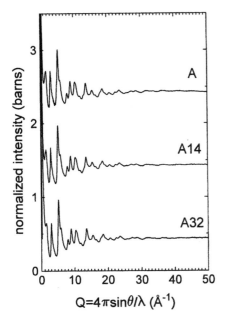

Fig. 1. The normalized intensities for the carbons A, A14 and A32.

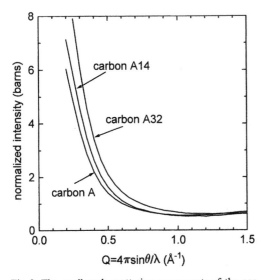

Fig. 2. The small-angle scattering components of the normalized intensities for the carbons A, A14 and A32 (zoom of Fig. 1 in the Q range 0–1.5 Å^{-1}).

very similar above a Q-value of 2 Å^{-1}. The presented diffraction patterns resembles those typical for amorphous materials rather than sharp Bragg peaks. The differences are visible in the small-angle region, in which the contribution to the intensity is due to the small-angle scattering coming from the pores created in the samples during the preparation and thermal treatment processes. The variations are related to the degree of "burn-off", i.e. the greater the percentage burn-off the higher the small-angle intensity. It is noteworthy that the small-angle contribution extends

beyond the range typical for heterogeneous systems. Such a feature can be explained by the presence of the small-size pores in the investigated carbons. The very weak variations in the intensity functions above 2 Å^{-1} suggest that the local structure is not affected by the burn-off process. The same conclusion was also reached by Gardner et al. [16] for carbons prepared from olive stones. In order to calculate the RDFs from the experimental intensities the small-angle contributions were subtracted by Lorentzian extrapolation of the peak at about 1.6 Å^{-1} to $Q=0$. The reduced intensities multiplied by Q were then Fourier transformed to the $G(r)$ functions. The resulting curves are shown in Figs 3 and 4.

4.1 Microcrystalline modelling

The first attempts to simulate the experimental RDFs were made with a crystallite model based on the graphite structure. The model RDFs were calculated using eqn (5). If the small-angle scattering in the experimental data for porous materials is ignored by extrapolating to $Q=0$ the approximate average number density is that of the solid part of the investigated sample [17,18]. However this value is often not well known. The experimental RDFs were computed taking the value of $\rho_0=0.112$, obtained by analogy with graphite using the position of the first $G(r)$ peak ($\simeq 1.411$ Å) and the layer spacing of 3.45 Å. The interatomic distances and the coordination numbers were taken as for the infinite graphite structure ($a=2.456$ Å, $c=6.696$ Å). The standard deviation of the interatomic distances $\sigma=0.06$ Å was adjusted to reproduce the amplitude of the first RDF peak. In Fig. 5(a) the calculated and experimental curves are compared for the A32 sample. From this comparison

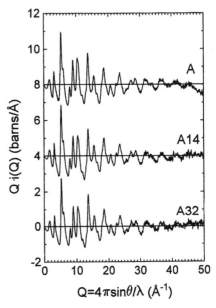

Fig. 3. The reduced intensity functions multiplied by Q for the carbons A, A14 and A32.

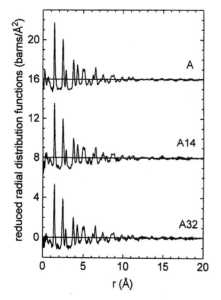

Fig. 4. The reduced radial distribution functions for the carbons A, A14 and A32.

it can be concluded that the model based on the infinite and perfectly ordered graphite structure does not account for the experimental data. The calculated function beyond the first peak region exhibits sharp structure, typical for crystalline materials. The peaks in the range of 2–12 Å have much higher amplitudes than those of the experimental RDF. Additionally in the calculated curve the peaks at $r = 3.35$, 4.15, 5.41, ~ 6 and ~ 7 Å are observed. These peaks are due to the inter-layer correlations in the graphite structure and their absence in the experimental curve provides evidence that –A–B–A–B– stacking is not preserved in the investigated samples. In order to attenuate the inter-layer correlations the value of the standard deviation $\sigma = 0.12$ Å was chosen for atoms lying in different layers. The result of such a simulation is shown in Fig. 5(b). The peak at 3.35 Å is almost completely damped out but the model RDF has too much structure in the range of 3–12 Å.

In the next step of modelling it was assumed that the model has a limited size. Taking the form of the correction term $\epsilon(r)$, given by eqn (6), the RDF was simulated for a rectangular microcrystallite of $15 \text{ Å} \times 15 \text{ Å} \times 12 \text{ Å}$ in size. The values of σ, 0.05 and 0.10 Å, were chosen for the intra- and inter-layer correlations, respectively. The resulting model curve is shown in Fig. 5(c) as the dotted line. From inspection of Fig. 5(c) one can see that simple damping of the calculated RDF by the exponential (eqn (5)) and $\epsilon(r)$ terms is not sufficient to match the experimental RDF. It is also necessary to point out that the first peak of the experimental function is shifted slightly towards shorter distances from the position characteristic of the carbon–carbon bond in graphite (1.418 Å) and the second smaller peak at about

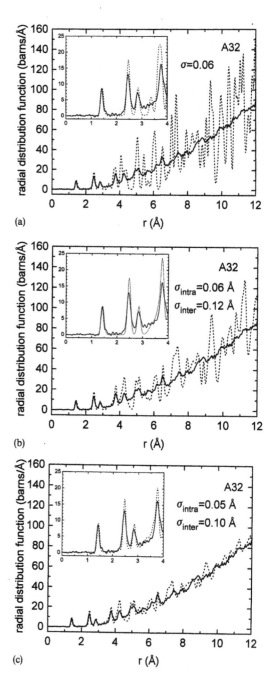

Fig. 5. Comparison of the calculated RDFs for graphite (dotted line) with the experimental data (solid line) for the carbon A32. (a) The simple model with one σ parameter; (b) the model with two σ parameters for intra- and inter-layer correlations, respectively; (c) the limited size model ($15 \text{ Å} \times 15 \text{ Å} \times 12 \text{ Å}$) with two σ parameters for intra- and inter-layer correlations, respectively.

1.55 Å is clearly distinguishable. A comparison of the curves presented in Fig. 5 shows that the model consisting of uncorrelated small microcrystallites of graphite does not properly reproduce the experimental RDF, in spite of the relatively high value of σ

parameters for the inter-layer correlations and limitation of the model size. In other words no simple way exists to convert the graphite RDF into the experimental one. The sharp first few peaks confirm that the local correlations are well defined but additional disorder should be superimposed on the model. The present results suggest that the structure of the investigated carbons is related to the turbostratic one rather than to that of graphite. Although these findings are expected from results of previous work, obtained for different carbons [1–9], we include microcrystalline modelling to make the next sections more clear and further conclusions more conscious.

4.2 Turbostratic modelling

Using eqns (5) and (6) the experimental RDFs can be analyzed. In the present studies the Hook–Jeeves least-squares type procedure [19] was adapted to derive the structural parameters: the interatomic distances, their standard deviations and the coordination numbers. This procedure minimizes the sum of squares of residuals between the calculated and experimental RDFs normalized by squares of the experimental points (the R factor).

$$R = \left\{ \frac{\sum_i \left[RDF_i^{(exp)} - RDF_i^{(cal)} \right]^2}{\sum_i \left[RDF_i^{(exp)} \right]^2} \right\}^{1/2} \quad (8)$$

The turbostratic correction term (see eqn (7)) proposed in ref. [5], shown in Fig. 6 as the dotted line, was averaged assuming the Gaussian distribution of the inter-layer spacing with the mean value 3.45 Å and the standard deviation 0.25 Å. This averaged curve is shown also in Fig. 6 as the solid line. Such a modification leads to better agreement between the calculated and experimental RDFs in the vicinity of the steps of the Mildner and Carpenter function (no

step-like features are observed in the experimental curves).

First we analyze the near-neighbour peaks of the RDFs, i.e. the r range from 1 to 2 Å. The fitting procedure converged to the structural parameters which are listed in Table 1. The experimental and best-fit curves are shown in Fig. 7(a–c) together with enlarged parts of the first peak regions. It can be

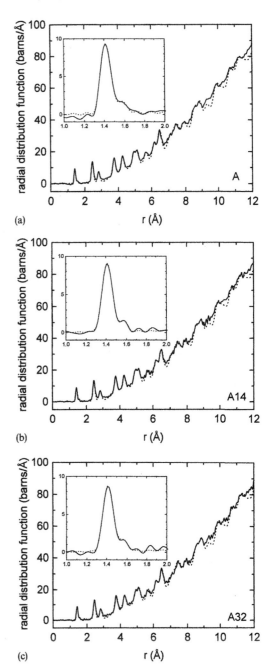

(a)

(b)

(c)

Fig. 7. Comparison of the calculated RDFs for the graphite-like model (dotted line) with the experimental data (solid line): (a) carbon A, (b) carbon A14 and (c) carbon A32, together with the enlarged parts of the first peak regions.

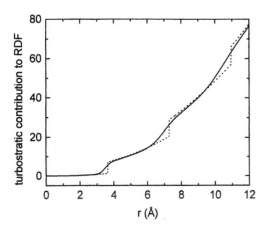

Fig. 6. The turbostratic contributions to RDF according to ref. [5] (dotted line) and with Gaussian distribution of inter-layer spacing (inter-layer spacing = 3.45 Å and its standard deviation = 0.25 Å).

Table 1. RDF peak positions r, coordination numbers N and standard deviations of the interatomic distances σ for the carbons A, A14 and A32

Carbon A			Carbon A14			Carbon A32			Graphite	
r	N	σ	r	N	σ	r	N	σ	r	N
1.394	2.42	0.046	1.396	2.59	0.050	1.397	2.41	0.048	1.4183	3
1.526	0.63	0.070	1.555	0.30	0.036	1.538	0.41	0.050	–	–
2.443	5.12	0.068	2.443	5.17	0.070	2.440	5.10	0.070	2.456	6
2.826	2.91	0.077	2.822	3.00	0.080	2.820	2.90	0.078	2.836	3
3.740	4.05	0.075	3.739	4.00	0.074	3.740	4.10	0.075	3.752	6
4.242	3.98	0.080	4.240	3.60	0.080	4.240	3.60	0.080	4.256	6
4.887	4.25	0.093	4.890	4.50	0.088	4.887	4.17	0.093	4.912	6
5.105	5.25	0.100	5.110	5.00	0.095	5.105	5.35	0.101	5.113	6
5.650	3.90	0.110	5.650	4.20	0.105	5.650	4.50	0.105	5.672	3
6.152	6.25	0.110	6.154	6.30	0.110	6.153	6.26	0.110	6.181	6
6.485	9.75	0.114	6.485	9.40	0.114	6.485	9.75	0.114	6.498	12
7.080	3.80	0.120	7.080	3.80	0.120	7.080	3.80	0.120	7.090	3
7.350	4.00	0.125	7.350	4.20	0.125	7.350	4.20	0.125	7.368	6
7.500	4.00	0.125	7.500	4.20	0.125	7.500	4.20	0.125	7.503	6
7.890	4.70	0.130	7.890	4.90	0.130	7.890	4.70	0.130	7.895	6
8.486	4.50	0.130	8.487	4.50	0.130	8.487	4.50	0.130	8.508	6
8.593	4.50	0.133	8.594	4.50	0.133	8.594	4.50	0.133	8.625	6
8.860	9.00	0.135	8.860	9.00	0.135	8.860	9.00	0.135	8.855	12
9.300	5.00	0.135	9.300	5.00	0.135	9.300	5.00	0.135	9.298	6
9.800	4.00	0.140	9.800	4.20	0.140	9.800	4.20	0.140	9.824	6
9.910	7.00	0.140	9.910	7.20	0.140	9.910	7.20	0.140	9.926	6
10.200	5.00	0.145	10.200	5.00	0.145	10.200	5.00	0.145	10.225	6
10.700	10.00	0.150	10.700	10.00	0.150	10.700	10.00	0.150	10.706	12
11.070	3.50	0.160	11.070	3.80	0.160	11.070	3.80	0.160	11.075	6
11.250	7.00	0.160	11.250	7.80	0.160	11.250	7.80	0.160	11.255	12
11.340	2.00	0.165	11.340	2.00	0.165	11.340	2.00	0.165	11.344	3
11.600	4.00	0.165	11.600	4.00	0.165	11.600	4.00	0.165	11.607	6
12.100	6.00	0.170	12.100	6.00	0.170	12.100	6.00	0.170	12.115	6

The uncertainty estimates are: 0.01 for r, 0.25 for N and 0.02 for σ. The interatomic distances and the coordination numbers for graphite refer to one atomic layer.

concluded that in the vicinity of the first peaks the two-shell model accounts very well for the experimental data. The main peak positions (1.394–1.397 Å) are shorter than the carbon–carbon distance in graphite (1.418 Å) and exactly the same as for the aromatic compounds (1.395 Å) [20]. The second interatomic distances (1.526–1.555 Å) are very close to the C–C single bond length in the paraffinic compounds (1.541 Å) and the near-neighbour distance in diamond (1.544 Å). This splitting of the first RDF peak was not revealed in the previous paper [16] because of lower resolution ($\sim$0.45 Å) resulting from truncation of the data at about 15 Å^{-1}. However, the average values of the near-neighbour C–C distances (weighted by the coordination numbers) calculated from the present results (1.41–1.42 Å) agree very well with those reported in ref. [16].

Now the question arises whether this smaller peak is due to the tetrahedral coordination of some of the carbon atoms or is the length of a single C–C bond for a threefold coordinated carbon atom. Since the total near-neighbour coordination numbers are 3.05, 2.89 and 2.82 for the carbons A, A14 and A32, respectively, a high proportion of tetrahedrally coordinated atoms is not expected to be the case for the investigated samples. Similar conclusion has been drawn by Mildner and Carpenter [5,17] and by Gardner et al. [16] from the neutron scattering

studies on the disordered carbons prepared from furfuryl alcohol and olive stones, respectively.

Recently the concentration ratio of the sp^3 hybridized bonds in amorphous carbon films has been analyzed by the photoemission spectroscopy method. It has been found that in the films obtained by pulsed laser evaporation of a graphite target, in which about 40% of the sp^3 hybridized bonds was detected after deposition, annealing at 1300 K (and above this temperature) leads to decreasing of the percentage of the sp^3 hybrids to zero [21]. It means that the sp^3 bonds are not stable at temperatures above 1300 K and it is reasonable to assume that in the A, A14 and A32 carbons the amount of tetrahedrally bonded carbon atoms is practically negligible, if present.

The occurrence of the small peaks in the RDFs at about 1.54 Å can be explained referring to the concept of the quinoid structure, proposed by Pauling [22]. The quinoid hypothesis is based on an assumption that each carbon atom forms two single (longer) and one double (shorter) bonds. In hexagonal graphite the double bond is assumed to resonate equally among the three positions leading to ideal hexagonal symmetry. The quinoid structure is not a completely resonating structure and together with graphite can be regarded as extreme cases. Ergun and Schehl [4] and Mildner and Carpenter [5,17] have considered four structures of a graphite layer as possible variants of the disordered carbon model. The authors have shown that the experimental RDF, obtained from the X-ray scattering experiment, can be reproduced by the disordered graphite or quinoid models. However it is essential to note that such a conclusion has been drawn based on the data of much poorer real space resolution achieved in the X-ray (~ 0.28 Å) or previous neutron (~ 0.25 Å) scattering experiment, insufficient to separate the single and double bond lengths. The present resolution (~ 0.12 Å) is lower then the difference of the longer and shorter bond and we can distinguish between them. The amplitudes of the longer distance peaks, which are clearly lower then the shorter ones indicate that the structure of the investigated carbons is not quinoid but rather a superposition of the graphite-like and quinoid structures in which the quinoid elements are randomly distributed. The presence of two bond lengths leads to distortion of the hexagonal rings and is the source of disorder within a single layer. This point will be discussed later. On the other hand the maximum value of the scattering vector, achieved using the pulsed neutron source, $Q_{max} = 50$ Å^{-1} is insufficient to enable a spatial resolution which would differentiate between the double (or partial double) bond length (~ 1.34–1.35 Å) and the aromatic bond length 1.395 Å. We have attempted to fit the experimental RDF in the region of the first peak with the three shell model. For the A32 carbon the fitting procedure converged to the values of the interatomic distances 1.355, 1.399 and 1.545 Å and the coordination numbers 0.15, 2.31 and 0.38, respectively. However this result cannot be regarded as a deciding argument for the quinoidal model because of insufficient resolution of the experimental data.

In a higher r region the experimental RDFs can be satisfactorily fitted with the model (dotted lines in Fig. 7), which assumes geometrical correlations between atoms lying in one graphite-like layer and the turbostratic contribution as indicated above. The misfit, visible in a higher r range ($r > 8$ Å) may be related to small inaccuracies in subtraction of the small-angle contribution from the intensity before the Fourier transform (see refs. [5,17,18]). The best fit has been obtained for the model composed of four layers. This conclusion is in agreement with previous findings. Analysis of the experimental diffraction data by Carrion and Dore [23] has shown that only a few (~ 4) layers are compatible with the observed diffraction patterns for most non-graphitized activated carbons investigated so far. From inspection of Table 1 one can see that the obtained interatomic distances are in a good agreement with those of a single graphite layer. All peak positions of the experimental RDFs can be reproduced assuming only correlations within the layer. Almost all coordination numbers are lower in comparison with the single layer graphite structure but such a behaviour is expected for a finite-sized model. For the peaks at 5.65, 6.152 and 7.08 Å the coordination numbers are greater. We

cannot find any satisfactory explanation of this feature—it seems that it may be the result of some remaining geometrical correlations between layers or problems with the residual small-angle contribution to the experimental intensity.

It is apparent from an analysis of the values of the σ parameters that a spread of the subsequent coordination spheres within a single layer increases with the interatomic distance. The plots shown in Fig. 8 exhibit the linear dependence of σ on $\sqrt{r}$. Increase in σ with the interatomic distances indicates that the structure of a single layer is disordered. This kind of disorder cannot be explained by random displacing of atoms from their equilibrium positions as high amplitude thermal vibration or random static disorder. The distance independent σ parameters proved to be ineffective as damping factors in the modelling procedure described in Section 4.1 (see Fig. 5(a)). Such a characteristic form of disorder is predicted by the paracrystalline theory, developed by Hosemann and his group [24–26].

The paracrystalline model is based on the assumption that the distances from any atom to adjacent atoms fluctuate without statistical correlations. As a result, in a reasonable approximation, these fluctuations propagate proportionally to the square root of

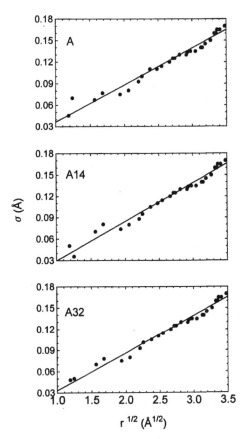

Fig. 8. σ versus $r^{1/2}$ plots for the carbons A, A14 and A32.

the interatomic distance, according to the combination law of independent probability distributions of the Gaussian type. Now the presence of two interatomic distances between nearest neighbours and increasing disorder within a single layer can be correlated. The vectors of adjacent atoms may vary in magnitude and direction due to displacements of the atoms from their positions in the ideal two-dimensional hexagonal network. Such displacements are realized if a certain number of the atoms of the hexagonal network, chosen statistically, are shifted to the positions defined by the quinoid-type structure [4]. This kind of disorder causes an increase in σ with $\sqrt{r}$ and results in a loss of long-range ordering, as predicted by the paracrystalline theory.

The paracrystalline theory implies the so-called "α^*-relation", which has been empirically established [23–25]. This relation is expressed as

$$\sqrt{N}g = \alpha^* \qquad (9)$$

where $0.15 \leq \alpha^* \leq 0.20$, N indicates the average number of the netplanes in the paracrystal and $g = [(\langle d^2 \rangle - \langle d \rangle^2)/\langle d \rangle^2]^{1/2}$ is the relative statistical deviation of the interplanar spacing d in the paracrystal (d is the interplanar distance for the netplanes with the highest planar density of atoms. $\langle \rangle$ indicates the statistical average value of the quantity in the brackets. The relation given by eqn (9) has the physical meaning that real paracrystals have limited sizes controlled by a degree of disorder indicated by g and that the surface netplanes have a statistical roughness of 0.15–0.20 of the netplane distance. In other words, large paracrystals are not strongly distorted.

Applying this approach to the two-dimensional graphite-like arrangement, in which the two interatomic distances are randomly distributed in proportions resulting from the coordination numbers, the distortion parameter g can be calculated in analogy to the two-dimensional coin model presented in refs. [24–26]. For the investigated carbons one obtains $g = 0.034 - 0.038$ and $N = 16–19$. In the planar graphite structure the line rows [11] satisfy the criterion of the higher atomic density and finally the size of the planar paracrystal is estimated to be 19–23 Å. This corresponds to the fact that the oscillations in the $G(r)$ functions beyond the range of 12–15 Å are practically nonobservable and $G(r)$ approaches 0. This value is a rough estimation of the size of the ordered regions within the single layer. Additionally the quinoid deformation of the hexagonal structure has an effect on stacking of the layers. Pauling concluded that the quinoid structure permits the better packing of the superimposed layers, a consequent decrease in the inter-layer spacing and increased stabilization of this structure through van der Waals interaction between adjacent layers [22]. Local decreasing in the interplanar spacing, related to random distribution of the quinoid elements in the hexagonal network suggests local curvature of the layers. It has been also suggested that the distorted quinoid rings are likely to be found in the vicinity of defects because their distortion is less under strain that complete resonating rings. This leads to the additional stabilization because of smaller bond-angle and bond-compression strain [17,22]. Taking into account possible curvature of the layers it seems to be interesting to consider whether the presence of pentagonal carbon rings, as in fullerenes, in activated carbons can be deduced from the present data. The RDF obtained by Li et al. for C_{60} [27] from the neutron scattering data of the same resolution exhibits peaks with the maxima at 1.44, 2.44 and 2.85 Å. The position of these peaks can be reproduced assuming the double and single bond lengths of 1.39 and 1.46 Å, respectively. These values agree very well with predictions of ab initio molecular-dynamic simulations yielding the average lengths of 1.40 and 1.45 Å [28]. The average first and third interatomic distances 1.41–1.42 Å and 2.82–2.83 Å obtained from our experiment are slightly shorter, the second one is exactly the same. But all remaining peaks of the RDF reported in ref. [27] are significantly shifted towards shorter values of the interatomic distances. This shift is related to curvature of the fullerene particle and the presence of five-membered rings. Therefore the present neutron scattering data do not provide evidence that the investigated carbons are constructed from fullerene-like elements. However, the presence of such elements in a small proportion cannot be completely ruled out within the precision of the method. The positions of the first, second and third RDF peaks of C_{60} and A, A14 and A32 agree well and it is likely that the remaining peaks are merged in the turbostratic contribution and hence are unrecognizable.

5. SUMMARY AND CONCLUSIONS

The neutron wide-angle scattering data collected using the pulsed neutron source in the range of the scattering vector 0–50 Å enable the real space resolution of 0.12 Å, much higher than that achieved in any other diffraction studies on graphite-like carbons. The intensity curves show the small-angle scattering contributions from all investigated samples, which extends up to the region of the first diffraction peaks. This behaviour suggests that the porous structure, which is the origin of the strong small-angle scattering, influences the local atomic arrangement of the investigated carbons in the region of medium-range ordering. The simulation studies show that the structure of the carbons A, A14 and A32 cannot be described in terms of the simple microcrystalline model, based on the graphite structure. A satisfactory fit to the experimental data was achieved for the model with the graphite-like arrangement within the single layer and the paracrystalline-type distortion of the two-dimensional hexagonal lattice. We explain the paracrystalline nature of the lattice deformations by the presence of quinoid bonding arrangements, randomly distributed in the two-dimensional graphite-like network. The simulations indicate that on average only a few (~ 4) of these layers are associated. The inter-layer correlations are very weak and can be practically ruled out. This very open structure leads to a very large surface area of all the investigated carbons. However, additional studies on other forms of activated carbons are necessary to test the applicability of this model.

Acknowledgements—We thank Dr D. Cazorla-Amoros of the Department of Inorganic Chemistry, University of Alicante for providing the samples and Mr M. Śliwiński of the University of Silesia for participation in the experiment and the data processing. Beam-time for this experiment on the ISIS facility was supported by the neutron programme of the EPSRC.

REFERENCES

1. Robertson, J., *Adv. Phys.*, 1986, **35**, 317.
2. Warren, B. E., *Phys. Rev.*, 1941, **9**, 693.
3. Warren, B. E. and Bodenstain, P., *Acta Cryst.*, 1965, **18**, 282.
4. Ergun, S. and Schehl, R., *Carbon*, 1973, **11**, 127.
5. Mildner, D. F. R. and Carpenter, J. M., in *Proceedings of 5th International Conference on Amorphous and Liquid Semiconductors*, Vol. 1, ed. J. Stuke and W. Brening. Garmisch-Partenkirchen, 1973, pp. 463–477.
6. Ergun, S., *Carbon*, 1976, **14**, 139.
7. Rousseaux, F. and Tchoubar, D., *Carbon*, 1977, **15**, 55.
8. Rousseaux, F. and Tchoubar, D., *Carbon*, 1977, **15**, 63.
9. Kodera, S., Minami, N. and Ino, T., *Jpn. J. Appl. Phys.*, 1986, **25**, 328.
10. Burian, A., Śliwiński, M., Ratuszna, A., Dore, J. C. and Cazorla-Amoros, D., submitted to *Fuel*.
11. Hosemann, R. and Bagchi, S. N., in *Direct Analysis of Diffraction by Matter*. North-Holland, Amsterdam, 1962, p. 69.
12. Bagchi, S. N., *Adv. Phys.*, 1970, **19**, 119.
13. Bagchi, S. N., *Acta Cryst. A*, 1972, **28**, 560.
14. Bell, J., *Nature*, 1968, **218**, 985.
15. Howells, W. S., Soper, A. K. and Hannon, A. C., Report RAL-89-046. Rutherford Appleton Laboratory, May 1989.
16. Gardner, M. A., Dore, J. C., North, A. N., Cazorla-Amoros, D., Salinas-Martinez de Lecea, C. and Bellisent-Funel, M. C., *Carbon*, 1996, **34**, 857.
17. Mildner, D. F. R. and Carpenter, J. M., *J. Non-Cryst. Solids*, 1982, **47**, 391.
18. Mildner, D. F. R. and Carpenter, J. M., *J. Non-Cryst. Solids*, 1984, **69**, 27.
19. Kunzi, H. P., Tzschach, H. G. and Zehnder, C. A., in *Numerical Methods of Mathematical Optimization with ALGOL and FORTRAN Programs*. Academic Press, New York, 1971.
20. *International Tables for X-Ray Crystallography*, Vol. III. The Kynoch Press, Birmingham, 1969, p. 276.
21. Diaz, J., Paolicelli, G., Ferrer, S. and Comin, F., *Phys. Rev. B*, 1996, **54**, 8064.
22. Pauling, L., *Proc. Natl. Acad. Sci. U.S.A.*, 1966, **56**, 1646.
23. Carion, M. and Dore, J. C., to be submitted to *J. Appl. Cryst.*
24. Hindeleh, A. M. and Hosemann, R., *J. Phys. C: Solid State Phys.*, 1988, **21**, 4155.
25. Hindeleh, A. M. and Hosemann, R., *J. Mater. Sci.*, 1991, **26**, 5127.
26. Hindeleh, A. M. and Hosemann, R., *J. Macromol. Sci. – Physics B*, 1995, **34**, 327.
27. Li, F., Ramage, D., Lannin, J. S. and Conceicao, J., *Phys. Rev. B*, 1991, **44**, 13167.
28. Zhang, Q. M., Yi, J.-Y. and Bernholc, J., *Phys. Rev. Lett.*, 1991, **66**, 2633.

Determination of fractal dimensions of solid carbons from gas and liquid phase adsorption isotherms

Nasrin R. Khalili[a,*], Minzi Pan[a], Giselle Sandí[b]

[a]*Department of Chemical and Environmental Engineering, Illinois Institute of Technology, Chicago, IL 60616, USA*
[b]*Chemistry Division, Argonne National Laboratory, Argonne, IL 60439, USA*

Received 24 September 1998; accepted 7 July 1999

Abstract

The total surface area, micropore volume, and fractal dimensions of five different carbons (Sorbonorite 4, GAC 1240, and three amorphous carbons) were evaluated from analysis of gas (N_2) and liquid (phenanthrene) adsorption isotherm data. The modified BET and fractal Frenkel–Halsey–Hill (FHH) models were used to estimate surface fractal dimensions. Micropore volumes were estimated from Dubinin–Radushkevich (DR) plots and were compared to those calculated from standard N_2 adsorption isotherm data using de Boer's *t*-method. The estimated surface fractal dimensions using the modified BET and FHH models ($D_s = 3 + 3h$, and P/P_0 from 0.0 to 0.4) were (2.7, 2.6, 2.1, 2.4, and 2.1) and (2.5, 2.6, 1.9, 2.4, and 1.9), respectively. The FHH fractal analysis suggested that van der Waals forces are the dominant interaction forces between nitrogen and carbon surfaces. Depending on the method of analysis, the fractal dimensions of the carbons with suggested micropore structure, Sorbonorite 4 and GAC 1240, were 2.5–2.9 and 2.6–2.9, respectively. Analysis of the adsorption-desorption data suggested that amorphous carbons with fractal dimensions of 2.1 (from the modified BET model) have smooth surfaces, with respect to their micropore structure. Further analysis of the adsorption data showed that the slopes of the linear segment of the plots of adsorption potential versus relative amount adsorbed are dependent on the pore size range and surface structure (fractal dimension) of the carbons. © 2000 Elsevier Science Ltd. All rights reserved.

Keywords: A. Activated carbon; Amorphous carbon; C. Adsorption; D. Surface properties; Microporosity

1. Introduction

Activated carbon is by far the most frequently used adsorbent and has virtually displaced most other materials in solvent recovery, gas refining, air purification, exhaust desulfurization, deodorization, and gas separation and recovery. The application of activated carbon for water treatment includes, but is not limited to: decolorization of solutions (removal of color, odor, taste, and other undesirable organic impurities from water), treatment of domestic and industrial wastewater, and collection and recovery of solutes (recovery of gold and silver). Most recently activated carbon has found application as catalyst and catalyst supports [1–3].

Activated carbon (AC) is known to be a superior adsorbent because of its excellent surface properties in-

cluding surface area and pore structure. Recent studies showed that the unique adsorption capability of activated carbon is related to its extended surface area, microporous structure, high adsorption capacity, and high degree of surface reactivity [1–4]. Since pore structure and surface irregularity are important properties of activated carbons, advanced activation techniques have been used to produce carbons that have specific surface area, micropore volume and pore size distribution. These carbons have extensive application in pollution control as adsorbents and as catalyst support. The accessible surface area of a given carbon may differ dramatically for small and large molecules. For example, to remove large molecules, a carbon should have a high surface area in mesopores because the micropores are inaccessible [5].

The adsorption capacity and catalytic capability of activated and amorphous carbons can be identified from their surface properties such as surface area and pore structure. The most common and recognized methods of surface analysis are the Brunauer–Emmett–Teller (BET)

*Corresponding author. Tel.: +1-312-567-3534; fax: +1-312-567-8874.
E-mail address: envekhalili@minna.acc.iit.edu (N.R. Khalili).

model (to estimate surface area), de Boer's 't-method' and Dubinin–Radushkevich (DR) plots (for micropore analysis). All these methods use standard gas phase N_2 adsorption isotherms to estimate the surface area and the extent of microporosity [4]. The DR plots are constructed according to the DR equation (Eq. (1)) and are used to estimate the micropore volume from the low and medium pressure parts of the adsorption isotherm. This method mainly relies on the Polanyi theory of adsorption [4]. The essential parameter of this theory is the quantity $A' = RT \ln (P_0/P)$, which is equated to the adsorption potential of the sample.

According to this theory, the DR plot of $\log W$ (the amount adsorbed expressed as liquid volume) against $\log^2 (P_0/P)$ should be a straight line with an intercept equal to the total micropore volume W_0 that is:

$$\log W = \log W_0 - D \log^2 \left(\frac{P_0}{P}\right). \tag{1}$$

In addition to gas adsorption, many other techniques have been developed to characterize the AC surface. With the development of surface imaging techniques, such as scanning tunneling microscopy (STM) [6], high resolution electron microscopy [7], scanning electron microscopy (SEM) [8], and atomic force microscopy (AFM) [9], carbon surface studies have experienced great advances. Image analysis techniques, such as fast Fourier transform (FFT) and inverse FFT (IFFT), are also used to derive AC pore size distribution information from optical data [10].

While the existence of a strong correlation between the surface area and the micropore volume of AC has been shown to be significant, quantitative data representing the surface irregularity (i.e. fractal dimension) of carbons and its relation to the adsorption capability of a given carbon is limited. In 1983, Pfeifer et al. proposed a promising approach to describe the complex porous surface of ACs from their fractal geometry [11]. Using fractal theory, Avnir and Farin [12] and Segars and Piscitelle [13] showed that the fractal concept can be adapted to explain the relationship between AC surface structure and its adsorption capacity. These studies yielded many new adsorption isotherms through which the surface fractal dimension (that is the level of surface irregularity) can be derived conveniently.

In this study the gas and liquid phase isotherm adsorption data of five different carbons were used to identify their total surface area, micropore volume, surface fractal dimensions and adsorption capacities. These carbons were synthesized at different conditions. The modified BET, fractal Frenkel–Halsey–Hill (FHH), t and DR models [4] were used to analyze standard nitrogen adsorption isotherms. The fractal dimensions were evaluated using Visual Basic programs that were written as an Excel spreadsheet. The best least squares fit of the adsorption isotherm data to the modified BET model was used to estimate the surface fractal dimensions.

1.1. Application of the modified N_2-BET model

Fractal geometry, first coined by Benoit Mandelbrot in 1982, is a mathematical tool that deals with complex systems that have no characteristic length scale [9,11,14–16]. The observed potential advantage of fractal theory in adsorbent surface analysis and adsorption behavior prediction was first proposed in 1983 [11,14]. In 1996, Segars and Piscitelle provided an innovative angle to derive the fractal dimension and other surface properties such as the specific surface area of an adsorbent from a single adsorption isotherm [13]. Their model was based on the original adsorption model, Eq. (2), proposed by Brunauer, Emmett and Teller in 1938 [4]:

$$\frac{V}{V_m} = \frac{cX}{1-X} \sum_{n=1}^{\infty} \beta_n \left(\frac{1-(n+1)X^n + nX^{n+1}}{1+(c-1)X - cX^{n+1}}\right). \tag{2}$$

In this model, V is the volume of adsorbate per mass of adsorbent at equilibrium, V_m is the volume of monolayer adsorbate per mass of adsorbent, c is a dimensionless constant related to the difference in free energy between adsorbate on the first and successive layers, n is the number of layers adsorbed, β_n is the fraction of the adsorbent surface that is covered by n and only n layers of adsorbate, and X represents relative pressure, P/P_0, where P is the equilibrium pressure of adsorbate and P_0 is the saturated vapor pressure.

This model assumes that the energy of adsorption is the same for all surface sites and does not depend on the degree of coverage. With assumptions of $\beta_n = 1$ and $n = \infty$, Eq. (2) is simplified to the well-known BET model, Eq. (3):

$$\frac{X}{V(1-X)} = \frac{1}{V_m c} + \frac{c-1}{V_m c} X. \tag{3}$$

Using the BET model and fractal concept for surfaces, Segars and Piscitelle proposed an expression for β_n as follows:

$$\beta_n = \left(\frac{r}{L}\right)^{D_S - D_{n-1}} - \left(\frac{r}{L}\right)^{D_S - D_n} \tag{4}$$

$$D_{n+1} = \left[1 - \ln\left(\frac{r}{L}\right)\right]$$
$$- \sqrt{\left[1 - \ln\left(\frac{r}{L}\right)\right]^2 + 2\left[D_n \times \ln\left(\frac{r}{L}\right) + (D_S - 2) \times \ln 2 - \ln(3 - D_n)\right]} \tag{5}$$

where r is the size of the adsorbed molecule, L is the linear size of the fractal system, and D_S is the fractal dimension of the adsorbent surface. It is important to note that Eq. (5) describes the smoothing of a fractal surface on adsorption of molecules of size r.

Segars and Piscitelle [13] plotted a series of isotherms calculated from the model by changing one of the parameters while keeping the others unchanged. They found that the model parameters were sensitive to different concentration regions of the isotherm. The steepness of the rise at the lowest concentrations gave a good measure of the interaction potential. The intermediate concentration was sensitive to the change of fractal dimension and the monolayer volume, and the tail part changed with the number of layers adsorbed. Based on these results, they claimed that with a complete isotherm (ranging from zero concentration to saturation), all model parameters could be determined from a single isotherm plot.

1.2. Application of the fractal Frenkel–Halsey–Hill (FHH) model

The BET model incorporates the assumption that the energy of adsorption is the same for all surface sites and does not depend on the degree of coverage. Another approach, which takes into account the heterogeneity of adsorption sites, is the Polanyi adsorption potential theory [17,18]. In this model, the net potential energy at a point above the surface, ε, is equal to the free energy released in bringing a mole of adsorbate from solution to the point of adsorption. The condition for condensation of adsorbate at any point is:

$$\varepsilon \geq RT \ln \frac{1}{X} \tag{6}$$

where R is the universal gas constant, T is the absolute temperature, and X is the relative pressure, P/P_0. The term on the left, ε, is the free energy required to condense a mole of adsorbate from solution to the solid or liquid phase. The Frenkel–Halsey–Hill adsorption isotherm applies the Polanyi adsorption potential theory and is expressed as:

$$\ln N' = \text{constant} + h \ln A' \tag{7}$$

where N' is the amount adsorbed, and A' is the adsorption potential defined as:

$$A' = -\Delta G = RT \ln \left(\frac{1}{X} \right). \tag{8}$$

For a smooth surface, the parameter h is assumed to be equal to $-1/3$ [19], while for a fractal surface, h is a function of the surface fractal dimension, D_s. If the van der Waals attractive forces are dominant between adsorbent and adsorbate, then h is equal to $(D_s - 3)/3$ [15]. For higher surface coverage where the adsorbent–adsorbate interface is controlled by the gas/liquid surface tension, h would be equal to $(D_s - 3)$ [20,21]. According to the FHH model, on a $\ln N'$ vs. $\ln A'$ plot, the slope of the straight-line part should be equal to h. In this paper both the $(D_s - 3)/3$ and $(D_s - 3)$ values were used to evaluate fractal dimensions.

Jaroniec et al. [22] proposed two models that were derived based on the FHH model and an assumption termed the condensation approximation (CA). These models relate the pore size distribution and the amount adsorbed to the fractal dimension and adsorption potential, A', as follows:

$$\Theta(A') = \frac{A'^{D-3} - A'^{D-3}_{\max}}{A'^{D-3}_{\min} - A'^{D-3}_{\max}} \tag{9}$$

$$N'(A') = N'_{\max} \Theta(A'). \tag{10}$$

In these equations $\Theta(A')$ is the degree of pore filling, D is the fractal dimension, A' is the adsorption potential, $N'(A')$ is the amount adsorbed and $N'_{\max}$ is the maximum amount adsorbed. The $A'_{\min}$ refers to the capillary condensation in the larger pore sizes, and $A'_{\max}$ corresponds to the pores of the minimum size. According to Eq. (9), calculated adsorption isotherms will deviate from linearity at higher values of fractal dimension. Basically, plots of $\ln \Theta$ vs. $\ln A'$ were found to be more linear at higher $A'_{\max}/A'_{\min}$ ratios (lower D values). In this study, we have derived the adsorption isotherms for all five carbons using Eq. (9) to show the effect of D value on the linearity of the adsorption isotherms. We also have used Eq. (10) to test conditions under which a linear relationship exists for the plot of $\Theta(A')$ vs. $N'(A')/N'_{\max}$. Results of this analysis are provided in Section 3.

The techniques introduced for fractal analysis are only some of the major techniques used in surface studies. Different techniques provide surface information in different scales. For example, the imaging techniques can provide surface information only within their resolution ranges, while the nitrogen adsorption isotherms can provide only micro- and mesopore information. Choosing a proper technique to study a given surface is the key for successful surface analysis.

2. Experimental

2.1. Materials

To characterize the surface of the activated carbon using isotherm data, we considered the general mechanism of adsorption. It is well recognized that the total volume of gases adsorbed at any temperature T, is directly related to the relative pressure (P/P_0), the volume adsorbed by the micropores, and the thickness of the layer adsorbed on the non-microporous surface [1,16].

In this study, two types of granular activated carbons (GACs) and three types of synthetic carbons (amorphous carbons) were analyzed. The GACs were commercial carbons provided by American NORIT Company. They were extruded peat-based Sorbonorite 4 and coal-based GAC 1240. The synthetic carbons were provided by the Argonne National Laboratory. Sorbonorite 4 is a steam activated extruded carbon prepared from peat and has an apparent density of 390 kg/m^3. Due to its superior mechanical hardness and favorable adsorption and desorption properties, this is an excellent carbon type for solvent recovery. GAC 1240 is a granular activated carbon produced by steam activation of selected grades of coal. As a result of a unique patented activation process and stringent quality control, these materials offer superior adsorption properties and are recommended for the removal of impurities from water and industrial process applications. The synthesized carbons were derived from a natural Montmorillonite clay, Bentolite L, supplied by Southern Clay Products, Inc. The approach was to use pillared clays (PILCs) as inorganic templates. These modified clays have aluminum oxide supports between the layers that help prevent the collapse of the layers upon heat treatment. Four organic compounds were used to produce the carbons: pyrene, styrene, ethylene, and propylene. Carbon from pyrene as the precursor (referred to as amorphous carbon 1) is produced by a mechanism described by Sandí et al. [23] in which the alumina pillars in the clay should act as acid sites to promote condensation similar to the Schöll reaction [24]. Trioxane/pyrene (amorphous carbon 2) was produced by the incorporation of liquid monomer in the PILC, followed by a low temperature polymerization reaction. The polymerization of styrene produces a linear polymer while the trioxane and pyrene reaction produces a condensation polymer similar to phenoplasts. Carbons from ethylene (amorphous carbon 3) and propylene were synthesized by a mechanism similar to that described by Copley [25] in which the gaseous hydrocarbon is deposited in the PILC layers and subsequently pyrolyzed. After elimination of the inorganic matrix via demineralization, the resulting layered carbons show holes due to the pillaring Al$_{13}$ cluster unit.

2.1.1. Measurement of the gas-phase adsorption–desorption isotherms

The commercial activated carbons and amorphous carbons were crushed and powdered to shorten the time required for reaching equilibrium in the isotherm study. Nitrogen isotherms at 77 K were obtained in an Autosorb 6 instrument from Quantachrome. Approximately 0.10 g of material was weighed in a Pyrex tube and evacuated at 80 mTorr overnight at room temperature. After backfilling with He, the sample tube was briefly exposed to air prior to analysis. The static physisorption experiments were conducted using the Autosorb 6 instrument to determine the amount of liquid nitrogen (LN$_2$) adsorbing to or desorbing from the material as a function of pressure ($P/P_0 = 0.025$–0.999, increments of 0.025). Data were obtained by admitting or removing a known quantity of adsorbate gas into or out of a sample cell containing the solid adsorbent maintained at a constant temperature (77 K) below the critical temperature of the adsorbate. As adsorption or desorption occurs, the pressure in the sample cell changes until equilibrium is established. The quantity of gas adsorbed or desorbed at the equilibrium pressure is equal to the difference between the amount of gas admitted or removed and the amount required to fill the space around the adsorbent (void space).

2.1.2. Measurement of the liquid-phase adsorption isotherm

Liquid phase isotherms were evaluated for the aqueous phase systems containing phenanthrene-9-^{14}C (13.3 mCi/mmol of solution) and a selected mass of carbons and disordered carbons. Radioactive labeled phenanthrene was used to identify equilibrium concentrations of phenanthrene in a manner that was easy, quick and accurate. The total concentration of phenanthrene in the stock solution was 1.0 mg/l, near its saturation concentration (1.18 mg/l at 25°C [26]). To measure adsorption isotherms, selected masses of each carbon were put into 20-ml glass scintillation vials. Vials were labeled and capped loosely, and autoclaved for 30 min at 121°C and 103 kPa of pressure to prevent the growth of bacteria during analysis. After autoclaving, 5 ml of phenanthrene stock solution and 10 ml of deionized water were added into each vial. The vials were shaken in a horizontal shaker for 48 h to reach equilibrium conditions. During shaking, the vials were wrapped with aluminum foil to prevent the effect of light. After the shaking period, vials were centrifuged at least twice to avoid transport of any carbon particle when the solution was taken out for radioactivity measurement. Three millilitres of liquid from each vial were removed and put into a 7-ml scintillation vial containing 3 ml of Beckman Scintillation Ready Gel. The vials were shaken to maintain a completely mixed system. To measure the equilibrium concentration of phenanthrene-9-^{14}C in solution, the radioactivities of each sample were measured using a Beckman LS6000SC scintillation counter. Using calibration curves and measured DPM values for each sample, the initial and equilibrium concentrations of phenanthrene-9-^{14}C were determined.

The fractal dimensions, micropore volumes, and adsorption capacities of the carbons were determined using gas and liquid phase adsorption isotherm data and modified N$_2$-BET and FHH models and DR plots. Due to the obvious differences between the origin and applied treatment procedures, it was anticipated that the surface area, micropore volume, and degree of surface irregularity (fractal dimensions), as well as the adsorption behavior of the amorphous carbons would be different than those of the activated carbons.

3. Results and discussion

3.1. Analysis of gas-phase adsorption isotherms

Nitrogen adsorption and desorption isotherms for carbons were evaluated using the standard N$_2$-BET test. Fig.

1a–e shows the isotherm data for adsorption in nitrogen at 77 K. The adsorption–desorption isotherm loops of Sorbonorite 4 and GAC 1240 were shown to be H_4 type isotherms [4]. In these loops, most of the adsorption branches were horizontal and parallel to the desorption branches. These isotherms indicated the possibility of existing micropores and slit pores, as the dominant forms of the pore structure for the activated carbons. The isotherms of amorphous carbons 1 and 3 showed H_3 loops, which represent the existence of the slit shape pores, with the possibility of the formation of micropores. The isotherm loop of amorphous carbon 2 showed a steep change in the middle of the desorption branch. The adsorption and desorption curves were not parallel for this carbon. By showing a H_2 loop, it was inferred that amorphous carbon 2 was composed of micro-particles (a micro-particle system such as silica gel usually shows a H_2 loop [16]).

The reported total surface area (from the BET model), micropore area and micropore volume (from de Boer's t-method) for the carbons studied are presented in Table 1.

As the analysis of the isotherms suggested, the microporosity, surface area, and micropore volumes were much higher for the Sorbonorite 4 and GAC 1240 compared to those of amorphous carbons 1, 2, and 3. The (DR) plots were also constructed to estimate the micropore volume from the low and medium pressure parts of the adsorption isotherms. Fig. 2 shows the DR plots for the activated carbons and the three amorphous carbons studied. As shown in Table 1, although strong similarities were observed between the micropore volumes calculated using DR plots and de Boer's t-method, the DR plots predicted much higher values for the amorphous carbon micropore volumes. The observed differences could be related to the absence of a substantial microporosity, the lack of achieving complete pore coverage, or simply to the existence of the amorphous structure that resulted in the measured micropore volumes to be representing only the variability of the measurements.

The following section presents the results of the fractal analysis. To identify fractal dimensions, adsorption iso-

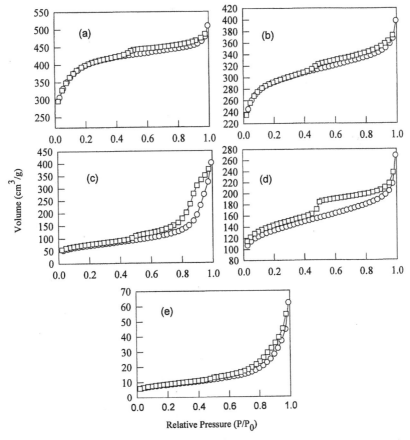

Fig. 1. N_2 adsorption and desorption curves for (a) Sorbonorite 4, (b) GAC 1240, (c) amorphous carbon 1, (d) amorphous carbon 2, and (e) amorphous carbon 3 (○ adsorption; □ desorption).

Table 1
Measured surface area and calculated micropore volume from N_2 isotherms using de Boer's t-plot and DR plots

Carbons	BET surface area (m^2/g)	t-Plot micropore area (m^2/g)	t-Plot micropore volume (cm^3/g)	DR micropore volume (cm^3/g)	Modified N_2-BET fractal dimension
Sorbonorite 4	1207	982	0.79	0.79	2.7
GAC 1240	863	702	0.61	0.63	2.6
Amorphous carbon 1	235	75	0.04	0.16	2.1
Amorphous carbon 2	417	269	0.14	0.25	2.4
Amorphous carbon 3	28	4	0.002	0.02	2.1

therms were analyzed by the modified BET and the Frenkel–Halsey–Hill (FHH) models. From different theoretical approaches, both models could derive fractal dimensions from single adsorption isotherms.

3.1.1. Estimation of the fractal dimensions from the modified BET model

Table 2 shows the surface fractal dimensions of five carbons estimated from modified and traditional BET

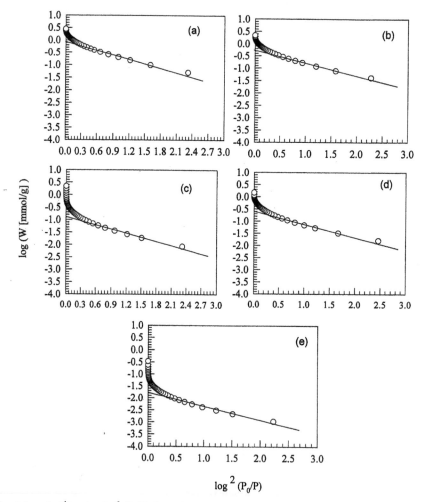

Fig. 2. Plots of log (n/mmol g^{-1}) against $\log^2$ (P_0/P) for (a) Sorbonorite 4, (b) GAC 1240, (c) amorphous carbon 1, (d) amorphous carbon 2, and (e) amorphous carbon 3.

Table 2
Calculated parameters of the modified and traditional BET models

Carbons	Modified BET model					Traditional BET model		
	D	R/L	c	N	V_m	V_m	c	R^2
Sorbonorite 4	2.7	0.1	100	11	405.0	277.3	41.2	0.994
GAC 1240	2.6	0.1	250	45	282.5	198.3	32.9	0.996
Amorphous carbon 1	2.1	1E−5	90	90	48.0	54.1	127.3	0.999
Amorphous carbon 2	2.4	0.01	400	95	115.0	96.0	50.0	0.997
Amorphous carbon 3	2.1	1E−6	90	90	4.6	4.6	849.9	0.999

models (Eqs. (3) and (5)). The modeled fractal dimensions of Sorbonorite 4 and GAC 1240 were 2.7 and 2.6, respectively. These high values suggested that these surfaces are irregular and have strong micropore structures. Amorphous carbons 1 and 3 had a fractal dimension of 2.1, which indicated that these two carbons have smooth surfaces and could have small pores in the micro-scale range. The amorphous carbon 2 had a fractal dimension of 2.4, suggesting that this carbon could have a partially irregular surface due to its micro-particle structure. In general, the results of fractal analysis were consistent with the observations of the adsorption–desorption loops.

Fig. 3a–e shows fair agreement between adsorption isotherms calculated according to the modified BET model

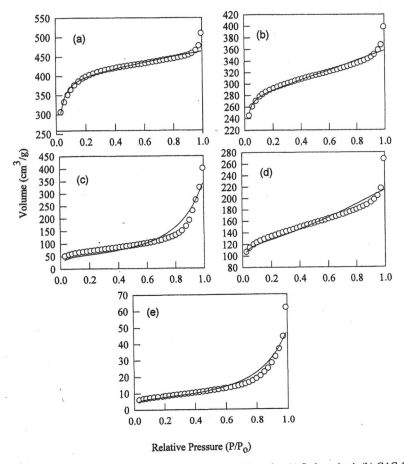

Fig. 3. Measured and calculated (using modified BET model) adsorption isotherms for: (a) Sorbonorite 4, (b) GAC 1240, (c) amorphous carbon 1, (d) amorphous carbon 2, and (e) amorphous carbon 3.

and estimated parameters, and those constructed based on the experimental data in the $P/P_0 < 0.6$ range. As shown in Fig. 3, calculated isotherms from the modified BET model deviated from real isotherms at the saturation parts. Deviation could be caused by the changes of molecular arrangement and inter-forces, which the BET theory cannot account for (BET theory ignores capillary condensation in meso- and macropores). These deviations can result in predicting inaccurate values for N, the maximum number of adsorption layers.

The predicted values of V_m from the modified BET model were higher than those calculated with the traditional model by about 46%, 42%, and 20% for Sorbonorite 4, GAC 1240, and amorphous carbon 2, respectively. These data suggested that surface microporosity could restrain the significance of deviation between the predicted and measured V_m. In fact, the $\Delta V_{m\ (Sorbonorite\ 4)}$ was 2.3 times higher than the $\Delta V_{m\ (amorphous\ carbon\ 2)}$. For other carbons with smooth surfaces, the predicted values of V_m (from the modified model) were similar to those calculated using the traditional BET model. The measured lower V_m values, particularly for microporous carbons, can be a result of ignoring capillary condensation in meso- and macropores and modeling theoretical assumptions. For example, the model assumes that the surface area of each upper layer is the same as that of the adsorbent surface, e.g. $\beta_n = 1$. This assumption works well for smooth surfaces, but can cause a major deviation for irregular surfaces. For irregular surfaces, it is safe to assume that the upper layers have a smoothing effect and $\beta_n < 1$. If the adsorbent surface fractal dimension is high enough, then the surface area of the second layer can be less than 50% of that of the first layer (modeling results showed that an inverse exponential type relationship exists between β_n and n). If the fractal factors are introduced in the BET model, the V_m values for the fractal surfaces will show higher values than those calculated from the traditional model.

3.1.2. Estimation of fractal dimensions using the fractal FHH model

According to the fractal FHH model (Eq. (7)), on the plot of $\ln N'$ vs. $\ln A'$, the slope of the straight-line portion should equal h. Fig. 4a–e shows the FHH plots of five different carbons. As is shown, most of the FHH plots approach straight lines. The expressions of $h = (D_S - 3)/3$ (which is derived based on the assumption that van der Waals forces are dominant between the adsorbate and the adsorbent), and $h = D_S - 3$ (which assumes a higher surface coverage with gas/liquid surface tension as the controlling factor at the adsorption interface) [20,21] were used to calculate D values. Table 3 summarizes the slopes of the regression lines, h, and the D values calculated for the relative pressures (P/P_0) of 0.0–1.0 and 0.0–0.4 (where micropore filling is assumed to be complete [4]). For the purpose of comparison, fractal dimensions calculated from the modified BET model are also provided.

These results showed that: (a) considering the approximations used during the modeling, the range of the Ds calculated from the FHH model are similar to those estimated from the modified BET model, (b) it seems that the expression $D = 3 + 3h$ provides a more consistent result with those offered by the modified BET model, so it can be assumed that only one kind of interaction (van der Waals forces) is dominant during adsorption, (c) the calculated Ds were not significantly different for both ranges of pressures, (d) Sorbonorite 4 and GAC 1240 had the highest fractal dimensions, and (e) the h values for amorphous carbons 1 and 3 were about $-1/3$, indicating the presence of a smooth surface [1,22].

Comparing the results of adsorption–desorption isotherm loop analysis with those of fractal analysis (Table 3) showed that: (a) carbons with a proposed micropore structure have fractal dimensions of about 2.5–2.9, (b) amorphous carbons which did not indicate any strong microporous structure presented D values of about 2 (smooth surface), and (c) amorphous carbon 2 had a fractal dimension of 2.4, representing a semi-microporous surface. This was consistent with the results of the adsorption-desorption loop study that presented a H_2 adsorption loop for this carbon.

3.1.3. Fractal analysis of the isotherms based on the extended FHH model

According to the fractal theory of adsorption on porous solids, one can relate the fractal dimension of the surface to the degree of pore filling, adsorption potential A', pore size and pore volume distribution [22]. To identify the relationship between the above parameters, we used two models defined by Jaroniec et al. (Eqs. (9) and (10)) to construct adsorption isotherms for different values of D at relative pressures (P/P_0) of 0.0–1.0. Fig. 5 shows the constructed adsorption isotherms for fractal dimensions of 2, 2.7, and 2.9. As is shown, none of these curves is linear and deviation from linearity increases as the D value increases, regardless of the range of adsorption potential. It was concluded that, regardless of the pore structure or size, if the value of D increases, deviation from linearity is expected for adsorption isotherms calculated based on the relative adsorption and adsorption potentials. The structure of the carbons did not influence the assumptions held for Eq. (9).

Plots of $\Theta(A')$ vs. $N(A')/N_{max}$ identified two different regions (Fig. 6a): (a) at relative pressures ranging from 0.0 to 0.4 where mostly micropore filling occurs, no linear relationship was identified between $\Theta(A')$ and $N(A')/N_{max}$ for small values of D (i.e. $D = 2$). The slopes of the lines increased as D increased (Fig. 6b). Linear relationships with slopes close to one were identified for all carbons studied (except amorphous carbon 3) at $D = 2.7$, (b) at $N(A')/N_{max}$ corresponding to the P/P_0 of 0.4–1.0 (at this range meso- and macropore filling are expected to be

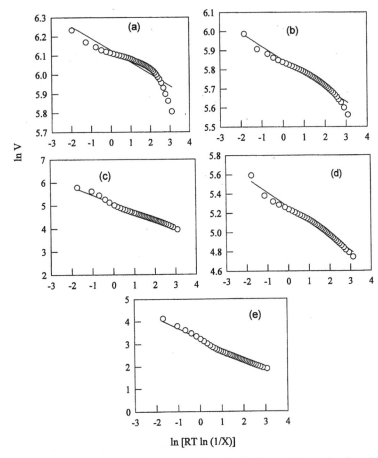

Fig. 4. FHH plots for: (a) Sorbonorite 4, (b) GAC 1240, (c) amorphous carbon 1, (d) amorphous carbon 2, and (e) amorphous carbon 3.

predominant [4]), all of the plots had parallel lines regardless of the D values. At this region, Sorbonorite 4, GAC 1240, amorphous carbon 1, amorphous carbon 2, and amorphous carbon 3 had slopes of 5, 4, 1, 2, and 1, respectively. As shown in Fig. 6a, plots of Sorbonorite 4, GAC 1240 and carbon 2 (which had fractal dimensions

greater than 2), were similar in shape. Similarity was also observed between plots of amorphous carbons 1 and 3 with fractal dimensions of 2.1.

These observations suggested that the proposed relationship between $\Theta(A')$ and $N(A')/N_{max}$ is more complex than was expected. Clearly, this model is influenced by the

Table 3
Fractal dimensions derived from FHH and modified BET model

Carbons	Fractal FHH model						Modified BET model
	h		$D = 3 + h$		$D = 3 + 3h$		D
	P/P_0 0.0–1.0	P/P_0 0.0–0.4	P/P_0 0.0–1.0	P/P_0 0.0–0.4	P/P_0 0.0–1.0	P/P_0 0.0–0.4	P/P_0 0.0–1.0
Sorbonorite 4	−0.06	−0.17	2.94	2.83	2.82	2.49	2.70
GAC 1240	−0.07	−0.12	2.93	2.88	2.79	2.64	2.60
Amorphous carbon 1	−0.36	−0.36	2.64	2.64	1.92	1.92	2.10
Amorphous carbon 2	−0.15	−0.20	2.85	2.80	2.55	2.40	2.40
Amorphous carbon 3	−0.46	−0.36	2.54	2.64	1.62	1.92	2.10

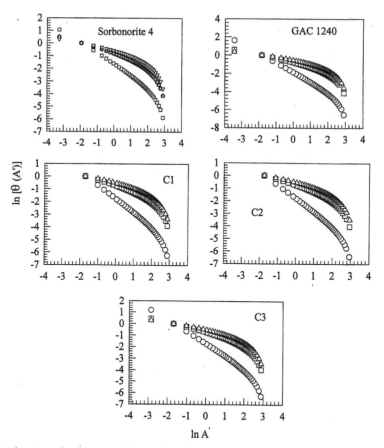

Fig. 5. Calculated adsorption curves for the assumed fractal dimensions ($D = 2$, 2.7 and 2.9). The C_1, C_2 and C_3 represent amorphous carbons 1, 2, and 3, respectively. (○, $D = 2$; □, $D = 2.7$; △, $D = 2.9$).

patterns of pore size range. Further investigation is needed in order to correct Eq. (10) for the effect of pore size range.

3.2. Liquid-phase adsorption isotherms and surface fractal dimensions

Modeling of liquid phase adsorption of phenanthrene at 298 K provided fractal dimensions, monolayer adsorption capacities, and other adsorption parameters, from the modified BET model, as discussed below.

The liquid phase adsorption isotherms were analyzed using the Freundlich isotherm model:

$$q = kC^{1/n'} \qquad (11)$$

where q is the amount of adsorption on the unit amount of adsorbent (mg/mg-adsorbent), C is the adsorbate concentration in the solution at equilibrium (mg/ml), and k and n' are constants. The regression lines of the log–log plots of q vs. C were used to calculate the Freundlich

isotherm parameters k and n'. These values are listed in Table 4. As shown, activated carbons' adsorption capacities were much higher than those calculated for amorphous carbons ($k_{\text{Sorbonorite 4}}$ and $k_{\text{amorphous carbon 3}}$ were 332 and 3.196, respectively). The measured and modeled isotherms for the five carbons tested are presented in Fig. 7.

As was expected, fractal modeling using phenanthrene adsorption isotherm data estimated a lower fractal dimension for carbons with fractal surfaces. Compared with the nitrogen adsorption isotherm, adsorption from aqueous solutions is more complex. Water molecules can surround the adsorbate and adsorbent and result in vibrations of adsorbate molecules on the adsorbent surface. Therefore, the adsorption layer may not be attached to the adsorbent surface. Knowing that the presence of water molecules and the size of the adsorbate used can interfere with the modeling results, we have attempted to investigate the extent of the change for the predicted model parameters when liquid phase isotherm data are used. To meet our objective the modified BET model was used to study phenanthrene adsorption layers. Modeling was carried out

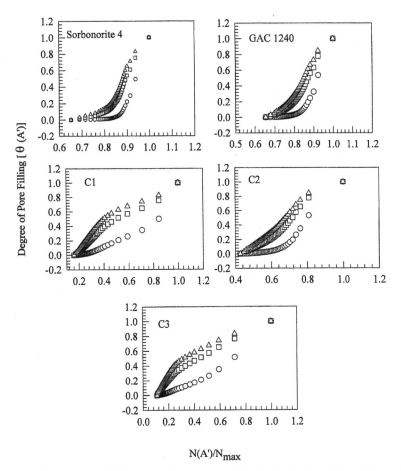

Fig. 6. (a) Variations of the $\Theta(A')$ versus the $N(A')/N_{max}$ for different values of D. ($\bigcirc$, $D=2$; $\square$, $D=2.7$; $\triangle$, $D=2.9$). (b) Variations of the $\Theta(A')$, versus the $N(A')/N_{max}$ for different values of D at $P/P_0 = 0.0$–0.4. ($\bigcirc$, $D=2$; $\square$, $D=2.7$; $\triangle$, $D=2.9$).

by analyzing phenanthrene adsorption isotherm data using Excel macros. The important variables were the mass of adsorption per unit mass of adsorbent, q, and the relative concentration, (C/C_0). It was assumed that:

$$\frac{q}{q_m} = \frac{V}{V_m}. \tag{12}$$

Therefore, the model for liquid phase adsorption was expressed as:

$$\frac{q}{q_m}$$

$$= \frac{c(C/C_0)}{1-(C/C_0)} \sum_{n=1}^{\infty} \beta_n \left(\frac{1 - (n+1)(C/C_0)^n + n(C/C_0)^{n+1}}{1 + (c-1)(C/C_0) - c(C/C_0)^{n+1}} \right) \tag{13}$$

where C_0 is the solubility of the adsorbate, and q_m is the monolayer adsorption mass capacity. The results of the

best least squares fit are presented in Table 5. The derived fractal dimensions are identified as $D_{s\text{-ph}}$ to separate them from the fractal dimension derived from the corresponding nitrogen adsorption isotherm. Within model parameters, the fractal dimension and the monolayer adsorption capacity are the most important parameters to be discussed in the context of a surface and adsorption layer study. Results of fractal analysis showed that the fractal dimension of the adsorption layer on Sorbonorite 4 was 2.3, which was lower than the carbon's surface fractal dimension (2.7) derived from the nitrogen adsorption isotherm. The fractal dimensions of the adsorption layer on the other carbons, including GAC 1240, were between 2.1 and 2.2.

The monolayer adsorption mass capacity, q_m, was calculated in two ways. In the first approach, the adsorption mass on the unit mass of carbon, q, was replaced by the adsorption volume, V, in Eq. (13), and then q_m was calculated. Another approach to predict the monolayer adsorption mass capacity was to use some of the parame-

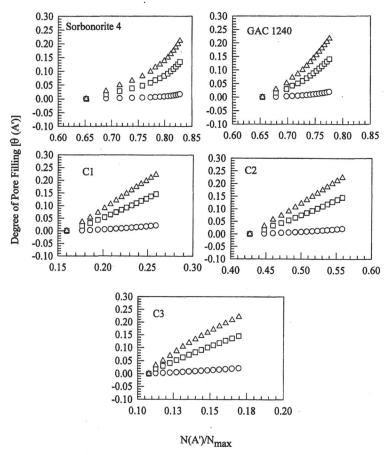

Fig. 6. *(continued)*

ters estimated in the gas phase modeling. According to the fractal power law, for a fractal adsorbent surface, the specific surface area, A, and the adsorbate molecule's cross section, σ, are related by:

$$A \propto \sigma^{(2-D_s)/2}. \tag{14}$$

The specific surface area estimated from nitrogen ad-

sorption isotherms (using the modified BET model, $A(N_2)$), was used to calculate the specific surface area available to the phenanthrene molecules as follows:

$$A = A(N_2)\left(\frac{\sigma}{\sigma_0}\right)^{(2-D_s)/2} \tag{15}$$

where σ_0, the cross section of nitrogen, is 16.2 $\mathring{A}^2$, and σ,

Table 4
Freundlich adsorption isotherm parameters

	Sorbonorite 4	GAC 1240	Amorphous carbon 1	Amorphous carbon 2	Amorphous carbon 3
k	332	278	0.090	0.079	3.196
$1/n'$	0.774	0.791	0.366	0.349	0.593
R^2	0.919	0.957	0.919	0.899	0.962

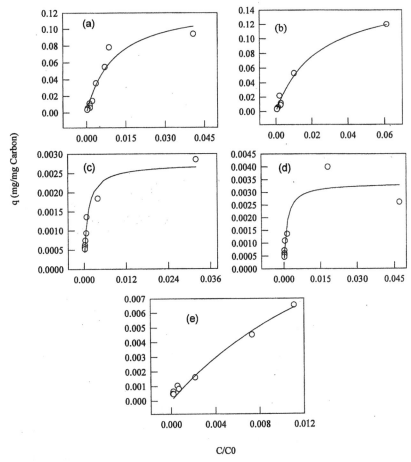

C/C0

Fig. 7. Measured and calculated liquid-phase phenanthrene adsorption isotherms using modified BET model for: (a) Sorbonorite 4, (b) GAC 1240, (c) amorphous carbon 1, (d) amorphous carbon 2, and (e) amorphous carbon 3. Liquid phase adsorption study was conducted at room temperature.

in this application, the cross section of phenanthrene, is 68.8 Å^2 [14]. Knowing the adsorbent surface area available for the phenanthrene molecules, the amount of monolayer coverage was calculated from the following equation:

$$q'_m = \frac{AM}{N_a \sigma} \qquad (16)$$

where M is the molecular weight of phenanthrene, and N_a is Avogadro's number. The calculation of q'_m was based on

Table 5
Modified BET model parameters for phenanthrene isotherms obtained from a room temperature adsorption study

Carbons	$D_{s\text{-ph}}$	r/L	C	N	Q_m	S	Q'_m
Sorbonorite 4	2.3	0.001	98	1	0.57	8.15E−3	0.46
GAC 1240	2.1	0.010	49	6	0.36	5.18E−3	0.34
Amorphous carbon 1	2.1	0.001	958	1	0.03	1.97E−4	0.08
Amorphous carbon 2	2.1	0.100	818	1	0.02	4.49E−4	0.15
Amorphous carbon 3	2.2	0.01	50	16	0.02	3.80E−4	0.01

the assumptions that the fractal dimension is the same in the range of scales from nitrogen to phenanthrene molecules, and that the adsorption layer is attached to the carbon surface. However, the real carbon pore structure is not ideally self-similar at different scales; therefore, the fractal dimension can be different between the micropores and mesopores. Also, under the effect of the surrounding water molecules, the adsorption layer may not be attached to the carbon surface. These factors can cause the calculated monolayer capacity, q'_m, to be different from the experimental value, q_m. The calculated q_m and q'_m of the carbons are listed in Table 5. For amorphous carbon 2, the value of q_m was significantly smaller than that of q'_m, while for the other carbons, their q_m and q'_m values were similar.

From the results shown above, the carbons investigated were placed in three groups: $D_{s\text{-ph}}$ was close to D_s while q_m was close to q'_m (amorphous carbons 1 and 3), $D_{s\text{-ph}}$ was lower than D_s while q_m was close to q'_m (activated carbons), and $D_{s\text{-ph}}$ was lower than D_s while q_m was significantly lower than q'_m (amorphous carbon 2). Carbons in the first group, with $D_{s\text{-ph}}$ close to D_s, and q_m near to q'_m, are all amorphous carbons with low D_s values. As shown through the modified N_2-BET isotherm study, their surfaces are smooth without a significant amount of micropores. The phenanthrene adsorption layers on such carbons should also be smooth. As was expected for these carbons, the specific surface areas of the adsorption layers were close to those of the carbon surfaces.

Carbons in the second group, with low $D_{s\text{-ph}}$ values, and q_m near to q'_m, are all activated carbons with high D_s values (2.6 and 2.7). The results of modified N_2-BET isotherm data and fractal analysis suggested that these carbons have irregular surfaces and their micro- and mesopore structures are well developed (Table 1). The fractal dimensions of the micro- and meso-scale are almost the same, so the calculated surface areas available to phenanthrene molecules are close to the measured values. During the process of adsorption in the liquid phase, water molecules can fill the small pores entirely and make them inaccessible to the phenanthrene molecules, they could influence the estimated values of $D_{s\text{-ph}}$. At the phenanthrene adsorbent interface, the surrounding small water molecules can also smooth the adsorbent surfaces and result in the prediction of a lower surface fractal dimension. Our data suggested that, although adsorbent surfaces in the presence of water molecules could be smoothed for the big phenanthrene molecules, the values of specific surface area of the adsorption layers did not change significantly, and the q_m value was near to q'_m.

The results of fractal analysis for the third group of carbons suggested that these carbons have different fractal dimensions at different scales. Since q_m was shown to be significantly lower than q'_m it was suggested that the mesopore structure changes dramatically from the micropore structure in these carbons.

4. Conclusions

In this study, the adsorptive properties (total surface area, micropore volume, surface fractal dimension, and adsorption capacity) of five different carbons, including Sorbonorite 4, GAC 1240, and three amorphous carbons, were evaluated using gas and liquid phase isotherm adsorption data. The major goal of this study was to evaluate and compare the surface fractal dimensions of these carbons using modified BET and FHH models. The surface structure of carbons can vary from graphite-like to extremely porous structure depending on the method of preparation and the degree of activation. Therefore it was expected that the fractal dimensions of these carbons would present a range between 2 and 3, depending on the surface characteristics of the carbons. This assumption was proven to be accurate during the course of this study.

4.1. Results of the N_2-BET analysis

The nitrogen adsorption–desorption loops predicted the conceptual surface characteristics for the carbons studied. The analysis of the adsorption–desorption isotherms indicated strong micropore structures for Sorbonorite 4 and GAC 1240, and a smooth surface for amorphous carbons 1 and 3. However, a minor microporosity for amorphous carbon 2 was identified.

4.1.1. Analysis of the micropore volumes

Micropore volumes were estimated from N_2 adsorption isotherm data (de Boer's t-method) and DR plots. Based on the results, it was shown that the calculated DR micropore volumes were substantially higher than those estimated from N_2 analysis (de Boer's t-method) for disordered carbons. However, good agreement was observed between DR and N_2 (using de Boer's t-method) micropore volumes calculated for activated carbons. These results suggested that both DR and de Boer's t-method could be used to estimate the micropore volumes if adsorbents have a strong micropore structure.

4.1.2. Analysis of the fractal dimensions

The surface fractal dimensions estimated from the modified N_2-BET model were in good agreement with those calculated from the FHH model. In general, fractal analysis showed that Sorbonorite 4 and GAC 1240 (carbons with micropore structure) have fractal surfaces with dimensions ranging from 2.5 to 2.9. These models, however, suggested a non-fractal surface for amorphous carbons 1 and 3 (carbons with a low micropore structure).

For the fractal FHH model, we investigated both proposed fractal exponents, $(D_s - 3)/3$ and $(D_s - 3)$. The results of fractal analysis showed that the fractal exponent

$(D_s - 3)/3$ provides more realistic results. Fractal dimensions calculated from $(D_s - 3)$ were unrealistically high, especially for the disordered carbons with suggested smooth surfaces. These observations suggested that van der Waals forces dominate the interaction between the nitrogen and carbon surfaces. Finally, although the expression of the modified BET model was more complex, it provided more information about the surface properties of the carbons tested than the fractal FHH model. The fractal dimensions derived from the modified BET model agreed very well with observations of adsorption–desorption isotherm loops, and calculated isotherms from this model reasonably matched most of the experimental isotherms.

Two models proposed by Janoriec et al. [22] were used to demonstrate the extent of the linear relationship between the estimated relative adsorption, the amount adsorbed on the carbons and the surface fractal dimensions. Analysis of the plots constructed based on Eq. (10) showed that, if a carbon has a smooth surface ($D = 2$), the calculated relative adsorption $\Theta(A')$ is independent from the relative amount of gas adsorbed ($N(A')/N_{max}$) at the pressure range of 0.0 to 0.4 where the micropore filling is complete. However, the slopes of the lines increased as the values assumed for D increased (or when fractal surfaces were assumed). Based on these results, it was concluded that the slopes of the plots, or the extension of the relationships, are dependent on the surface characteristics, fractal dimensions, pore size range, and ultimately, the adsorption potentials of the surface.

4.2. Liquid-phase fractal analysis

To relate surface structure to liquid phase adsorption capacity, the phenanthrene adsorption isotherm data (room temperature experiment) were used to identify the fractal dimensions from the modified BET model. The predicted fractal dimensions of the carbons with smooth and irregular surfaces were all close to 2, suggesting that the phenanthrene adsorption interfaces for the liquid phase systems are smooth. Comparison between experimental and calculated phenanthrene monolayer adsorption capacities showed that the fractal dimension of amorphous carbon 2 was much smaller in the meso-scale than in the micro-scale system. For the other carbons, the fractal dimensions remained constant within the assumed scales.

Acknowledgements

Work at Argonne National Laboratory was performed under the auspices of the Office of Basic Energy Sciences; Division of Chemical Sciences, US Department of Energy, under contract number W-31-109-ENG-38.

References

[1] Kurk M, Jaroniec M, Gadkare KP. Nitrogen adsorption studies of novel activated carbon. J Colloid Interface Sci 1997;192:250–6.

[2] Bansal RC, Donnet BJ, Stoeckli F. Active carbon, New York: Marcel Dekker, 1988, Chapter 3.

[3] Kinoshita K. Carbon: electrochemical and physiochemical properties, New York: John Wiley, 1988, Chapters 1–3.

[4] Greg SJ, Sing KSW. Adsorption, surface area, and porosity, New York: Academic Press, 1982, Chapter 2.

[5] Wigmans T. Industrial aspects of production and use of activated carbons. Carbon 1989;27(1):13–22.

[6] Couto MS, Liu XY, Meekes H, Bennema P. Scanning tunneling microscopy studies on n-alkane molecules adsorbed. J Appl Phys 1994;75(1):627–9.

[7] Fryer JR. The micropore structure of disordered carbons determined by high resolution electron microscopy. Carbon 1981;19(6):431–9.

[8] Katz AJ, Thompson AH. Fractal sandstone pores: implications for conductivity and pore formation. Phys Rev Lett 1985;54(12):1325–8.

[9] Xu W, Zerda TW, Yang H, Gerspacher M. Surface fractal dimension of graphitized carbon black particles. Carbon 1996;34(2):165–71.

[10] Oshida K, Kogiso K, Matsubayashi K, Takeuchi K, Kobayashi S, Endo M, Dresselhaus MS, Dresselhaus G. Analysis of pore structure of activated carbon fibers using high resolution transmission electron microscopy and image processing. J Mater Res 1995;10(10):2507–11.

[11] Pfeifer P, Avnir D. Chemistry in noninteger dimensions between two and three. I. Fractal theory of heterogeneous surfaces. J Chem Phys 1983;79(7):3558–65.

[12] Avnir D, Farin D. Fractal scaling laws in heterogeneous chemistry: Part I: adsorptions, chemisorptions and interactions between adsorbates. N J Chem 1990;14:197–206.

[13] Segars R, Piscitelle L. Verification and application of a new adsorption model for fractal surfaces. Mater Res Soc Symp Proc 1996;407:349–54.

[14] Avnir D, Farin D, Pfeifer P. Chemistry in noninteger dimensions between two and three. II. Fractal surface of adsorbents. J Chem Phys 1983;78(7):3566–71.

[15] Avnir D, Farin D, Pfeifer P. Molecular fractal surfaces. Nature 1984;308:261–3.

[16] Lee CK, Chiang AST, Tsay CS. The characterization of porous solids from gas adsorption measurements. Key Eng Mater 1996;115:21–44.

[17] Montgomery JM. In: Water treatment principles and design, New York: John Wiley, 1985, pp. 177–9.

[18] Manes M, Hofer JE. Application of the Polanyi adsorption potential theory to adsorption from solution on activated carbon. J Phys Chem 1969;73(3):584–90.

[19] Drake JM, Yacullo LN, Levita P, Klafter J. Nitrogen adsorption on porous silica: model-dependent analysis. J Phys Chem 1994;98(2):380–2.

[20] Avnir D, Jaroniec M. An isotherm equation for adsorption on fractal surfaces of heterogeneous porous materials. Langmuir 1989;5(6):1431–3.

[21] Yin Y. Adsorption isotherm on fractally porous materials. Langmuir 1991;7(2):216–7.

[22] Jaroniec M, Kruk M, Olivier JP. Fractal analysis of composite adsorption isotherms obtained by using density functional theory data for argon in slitlike pores. Langmuir 1997;13(5):1031–5.

[23] Sandí G, Winans RE, Carrado KA. New carbon electrodes for secondary lithium batteries. J Electrochem Soc 1996;143:L95–98.

[24] Balaban A, Nenitzescu T. Chapter II, Scope and general aspects. In: Olah G, editor, Friedel-Crafts chemistry, Vol. 2, New York: John Wiley, 1973, p. 63.

[25] Copley JRD. The significance of multiple scattering in the interpretation of small angle neutron experiments. J Appl Crystallogr 1988;21:639–44.

[26] Thibodeaux LJ. In: 2nd ed, Environmental chemodynamics: movement of chemicals in air, water, and soil, New York: Wiley, 1996, p. 552.

Activated Carbon Compendium
H. Marsh (Editor)

Adsorption dynamics of carbon dioxide on a carbon molecular sieve 5A

S.W. Rutherford, D.D. Do*

Department of Chemical Engineering, The University of Queensland, St. Lucia, Queensland 4072, Australia

Received 5 August 1999; accepted 24 November 1999

Abstract

Measurement of batch adsorption of carbon dioxide with a carbon molecular sieve (commercially manufactured Takeda 5A) indicates no molecular sieving action but instead, micropore diffusion is shown to be rate limiting the adsorption dynamics. Permeation measurement through the same pellets is also performed and steady state analysis indicates there is negligible adsorbed phase transport along the pellet and that the gas phase diffusion process is a combined Knudsen and viscous mechanism. Batch adsorption and permeation methods are critically compared for their utility in determining which mass transfer processes are relevant and the conditions under which each technique is most useful are given. © 2000 Elsevier Science Ltd. All rights reserved.

Keywords: C. Adsorption; D. Adsorption properties, Diffusion, Transport properties

1. Introduction

The study of mass transport in porous adsorbent materials is important for its practical application in separation and purification technology. Whether the application involves chromatographic purification, separation via cyclic circulation through an adsorbent bed, or membrane separation, information concerning the nature of adsorption and structure of the adsorbent material is required for design purposes. The established view of the structure of many adsorbent materials manufactured in pellet form is that they are composed of crystalline or semi-crystalline grains in the order of micron size. These grains are surrounded by voids which allow fluid phase transport through the pellet [1].

When studying the transport of adsorbate through the pellet, it is necessary to identify the physical processes involved and relate such processes to the porous structure. These processes may include:

1. gas phase transport via Knudsen diffusion at low pressures in small pores, molecular diffusion for mixed gas diffusion in larger pores, or viscous flow in large pores if a pressure drop is applied.
2. adsorption into the grains via an adsorbed phase diffusion mechanism.
3. mass action mechanism for adsorbate to enter the grain as a result of pore mouth restriction sometimes generated through treatment of the micrograin. The mass action mechanism is useful in some molecular sieving materials which are produced especially to create a barrier resisting the adsorption of large molecules and preferentially allow smaller molecules to adsorb. In these materials, it can often be this barrier that is rate limiting the adsorption process [2,3].

The difficulty in the analysis of many dynamic adsorption processes lies in the determination of the role of the adsorption process itself. Ultimately it must be determined whether adsorption is fast or slow in comparison to the gas phase transport. This requires analysis of the timescales involved in the pellet scale diffusion and micropore diffusion processes. Generally, for species with high affinity diffusing through pellets of large macroscopic length, macropore diffusion controls the mass transfer process. This gives rise to analysis by the macropore diffusion control model, conveniently presented by Ruthven [1]. Alternatively, mass transfer of species of lesser affinity diffusing through pellets of small macroscopic

*Corresponding author. Tel.: +61-7-3365-4154; fax: +61-7-3365-2789.
E-mail address: duongd@cheque.uq.edu.au (D.D. Do).

length is rate limited by the micropore diffusion process. For this case, analysis by the micropore diffusion control model may be applicable [1]. A more general scenario may involve a combination of processes, a situation in which a more complex description is necessary for analysis of dynamics.

For purposes of experimental determination, it is useful to isolate each process and evaluate its contribution to the mass transfer process independently. This proves a more accurate means than evaluating parameters from optimisation when several processes are involved. Naturally, this is a more experimentally intensive path, but by this means it is possible to determine which of the possible mechanisms are relevant to the transport process under inspection. In this investigation, we employ batch adsorption and permeation methods to determine which of the processes are significant in the adsorption of carbon dioxide in a commercially supplied carbon molecular sieve. Analysis of dynamics is presented and discussed with the aim of further determining the conditions under which each technique is most useful.

2. Experimental

There are many established techniques for investigating dynamics of adsorption. Some are critically reviewed by Bulow and Micke [4] with discussion of their relative merits. Of these techniques, batch adsorption is well employed as it is used to simultaneously obtain dynamic information and the amount adsorbed at equilibrium. The batch adsorption experiment can be conducted gravimetri-

cally or volumetrically, with a carrier gas or single gas adsorption. Possibly the simplest method is to measure single gas batch adsorption by volumetric method as we have performed in this investigation. The procedure and apparatus have been described elsewhere [5] for differential operation of the batch adsorber. In this investigation the experiment is performed over a large initial pressure drop, making it an 'integral' procedure and nonlinear effects result in an averaged measurement being taken. Although the analysis of data obtained by this method is more complex, it is useful when only a small total amount is uptaken by the sample resulting in a small pressure change. For each of our measurements conducted in this fashion, the sample was prepared by outgassing for around 60 h at 10^{-6} Torr.

The material under analysis is a commercial carbon molecular sieve from the Takeda chemical company termed 5A, because it is capable of selectively adsorbing species of molecular size less than 5 Angstrom and excluding molecules of size greater than 5 Angstrom [6]. The sample under analysis is supplied in extruded pellet format of cylindrical shape with properties shown in Table 1.

Electron microscopy analysis of the material shows that the pellet is composed of an agglomeration of discrete

Table 1
Properties of material used in batch adsorption experiment

Sample	Weight	Diameter	Length
Takeda 5A	1.25 g	0.3 cm	0.3 cm

microparticles of the order of micron size [6]. Interstitial voids and microparticle porosity give the pellet its porous structure which has been probed using mercury intrusion on a Micromeritics Poresizer 9320. The pore size distribution is shown in Fig. 1, the average size (diameter) being calculated at 0.5 μm. The total macroporosity is measured at 0.28.

It is useful to couple the information obtained from batch adsorption with that obtained from a steady state technique such as permeation. For this reason, we have also measured permeation through a Takeda 5A pellet of dimensions shown in Table 2. The procedure and apparatus for the permeation setup has been described elsewhere [7].

2.1. Consideration of heat transfer

Because adsorption is an exothermic process, the dynamics of mass transfer can be affected by temperature build up within the particle. With a coupling of heat and mass transfer, it is necessary to

1. account for simultaneous heat and mass transfer, or
2. limit the pressure increment taken such that the maximum temperature rise is so small that the process is considered to be isothermal.

Because the mass transfer is of primary interest, it is advantageous to ensure that the process is isothermal. This requires an estimate of the maximum allowable pressure increment undertaken by comparing mass and heat transfer

Table 2
Properties of material used in permeation experiment

Sample	Length	Diameter
Takeda 5A	0.6 cm	0.3 cm

parameters as discussed in Ruthven [1]. The two parameters are defined as

$$\alpha = \frac{ha}{C_s} \frac{R_\mu^2}{D_\mu} \tag{1a}$$

$$\beta = \frac{\Delta H}{C_s} \left(\frac{\partial C_\mu}{\partial T} \right)_p \tag{1b}$$

For a 'worst case' calculation, the following correlation is valid under stagnant heat transfer conditions [8]:

$$h = \frac{k_f}{R} \tag{1c}$$

For carbon dioxide at low pressures ($k_f = 0.015$ W/m/K) we estimate a heat transfer coefficient in the order of 10 W/m^2/K. Typical values for heat capacity for coal can be obtained from [9] $C_p = 1306$ J/kgK and density 1400 kg/m^3. Later it is shown that for micropore diffusion of carbon dioxide the mobility is in the order of: $D_\mu / R_\mu^2 = 2 \times 10^{-3}$ s^{-1}. Furthermore, a typical value for heat of adsorption for carbon dioxide is around 20 kJ/mol [10] and in all experiments a maximum change in the adsorbed phase concentration of 0.45 mmol/ml is not exceeded. Under these conditions we have $\alpha \approx 6$ and $\beta \approx 0.1$. According to Ruthven [1], this makes the adsorption process

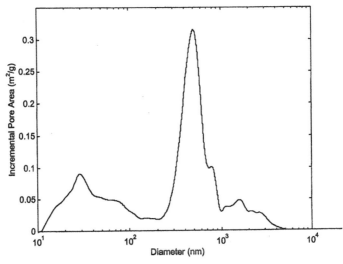

Fig. 1. Pore size evaluation by Mercury Intrusion.

clearly isothermal and coupled analysis of heat transfer is not warranted.

3. Steady state permeation and evaluation of gas phase processes

Because it provides a direct means for measurement of mass transfer, the permeation experiment provides an effective means for characterisation for many types of porous materials. Measurement can be performed by

(i) forcing liquid through the porous material under a hydrostatic head [11],
(ii) allowing diffusion of a species within a 'carrier gas' through the material utilising the concentration gradient [12] or
(iii) forcing gas through the material subject to a constant pressure drop (which we consider in this investigation).

In most instances the boundary conditions of the experiment can be maintained constant with respect to time, in which case a steady state of permeation is ultimately reached. A steady flow out of the material can be measured, the rate being proportional to the flux. By this means we are measuring the flux directly. If a constant pressure drop is used to drive the gas phase transport process (usually maintained by ensuring that the supply and receiving volumes are very large) then the pressure will rise linearly in the receiving volume with time. The slope of this linear rise (S) is evaluated from the data and the flux is given by

$$J = \frac{SV}{AR_g T} \tag{2a}$$

where A is the cross-sectional area of the media. After evaluating the flux directly, the diffusivity is often used to characterise the mobility of the transport process. Using Fick's law we obtain the diffusivity as a function of the measured slope and pressure drop (ΔP)

$$D_e = \frac{SVL}{A \, \Delta P} \tag{2b}$$

where D_e represents the combined effective diffusivity which will be the sum of all transport processes occurring in parallel along the direction of transport. For macro/microporous or bidispersed solids this may include a macroporous diffusion (Knudsen, viscous or molecular diffusion depending on the conditions) and possibly a coupled adsorbed phase diffusion process. The presence of the adsorbed phase diffusion increases the observed mobility beyond that which would be expected from the gas phase diffusion alone. Because both diffusion mecha-

nisms are coupled, an accurate means of determining the gas phase diffusion is vital for an accurate measurement of the adsorbed phase mobility. This is an important point to note because many methods of measuring the adsorbed phase flux simply rely on an estimate of the gas phase flux and hence the resulting accuracy of the adsorbed phase diffusivity can be heavily dependent upon this estimate. A better means for determining the gas phase contribution is by separate and independent measurement. Helium permeation is utilised for this process in many porous adsorbents. Transport within the macropores dominates the transport process for helium and the magnitude of the adsorbed phase flux is much less than that of the gas phase.

The gas phase diffusion process has been studied intensively and in the simplest case (single gas permeation) there are two mechanisms which dominate, Knudsen diffusion at low pressures and viscous permeation at higher pressures. For this case the diffusivity is described by

$$D_e = \frac{\varepsilon B_o}{\mu} P + \varepsilon D_P \tag{3}$$

where B_o is the viscous flow parameter, μ the viscosity of the penetrating gas, ε is the volume fraction of the material that is contributing to the flow and D_P is the pore diffusivity which is inversely proportional to the square root of molecular weight of the penetrating gas. It is obvious from Eq. (3) that a plot of effective diffusivity vs. pressure will be linear with gradient: $\varepsilon B_o/\mu$ and intercept: εD_P. If such a plot is linear, this does not exclude the presence of adsorbed phase diffusion accompanying gas phase transport. However, if the gradient of two differing species is inversely proportional to viscosity and the intercept inversely proportional to the square root of molecular weight, this ensures that gas phase flow dominates that of the adsorbed phase because the adsorbed phase flux has a more complex dependence on molecular properties of the diffusing species than a simple proportionality [13].

With the pellets described earlier in the experimental section, we have measured the steady state slope for carbon dioxide, helium, argon, oxygen and nitrogen permeation at 20°C for a range of pressures. The permeation process with helium, argon, oxygen and nitrogen reaches steady state very quickly and there is no measurable transient permeation phase. For carbon dioxide this is not the case and the dynamics can be monitored on a measurable timescale. This is due to the intrusion of adsorption dynamics in the permeation process which we discuss in a later section. Of interest is the steady state slope which we obtain from linear regression of the data at large times. The steady state slope for helium, oxygen, argon and nitrogen measurement is also obtained by linear regression. The correlation coefficient is high for all data sets, the lowest value of which is $R^2 = 0.98$. When plotted against pressure, the effective diffusivity, obtained from the steady state pressure rise using Eq. (2b), appears to be linearly related to pressure as is shown in Fig. 2. Linear regression shows that such a correlation is valid. For helium, argon, oxygen and nitrogen permeation at steady state we have determined the following relationships:

$$D_e^{He}(cm^2/s)7.0 \times 10^{-5}P(Torr) + 0.12 \quad (4a)$$

$$D_e^{O_2}(cm^2/s) = 6.0 \times 10^{-5}P(Torr) + 0.046 \quad (4b)$$

$$D_e^{N_2}(cm^2/s) = 7.3 \times 10^{-5}P(Torr) + 0.048 \quad (4c)$$

$$D_e^{Ar}(cm^2/s) = 5.5 \times 10^{-5}P(Torr) + 0.041 \quad (4d)$$

For carbon dioxide permeation at steady state we have:

$$D_e^{CO_2}(cm^2/s) = 8.4 \times 10^{-5}P(Torr) + 0.042 \quad (4e)$$

The viscous flow parameter (εB_0) and corrected pore diffusivity ($D_p\sqrt{M}$) are determined by equating these experimentally determined relationships with Eq. (3) for all five gases considered. These values are plotted against molecular weight and viscosity in Fig. 2f and 2g. The figures show that within $\pm 5\%$ error, the viscous flow parameter for all gases is independent of viscosity and the corrected pore diffusivity independent of molecular weight. We can conclude from this that Knudsen and viscous mechanisms are present within the gas phase flow and adsorbed phase flow if present, contributes negligibly to the mass transfer. For convenience we have included in Table 3, the viscosity of the gases used.

3.1. Evaluation of Tortuosity

It is well known that the structure of a porous material is fundamentally related to gas phase transport processes. A useful summary of early work in this area is given in the text of Carman [14]. For materials with a pore size distribution it is known that the permeability and Knudsen diffusivity can be related to the distribution of pore size through Darcy's law:

$$B_o = \frac{\overline{r^2}}{8\tau_V} \quad (4f)$$

where $\overline{r^2}$ is mean square pore size and τ_V is the viscous tortuosity factor and through the Knudsen diffusion relation:

$$D_P = \frac{2\overline{r}}{3\tau_K}\sqrt{\frac{8RT}{\pi M}} \quad (4g)$$

where $\overline{r}$ is mean pore size and τ_K is the diffusive tortuosity factor. The Takeda 5A sample used in this investigation has had pore size distribution measured using Mercury intrusion and diffusivity has been evaluated from permeation measurements. As a result, the tortuosity may be evaluated from Eq. (4f) and (4g). By this method we

obtain: $\tau_K = 4.2$ and $\tau_V = 5.8$. Approximately the same value for the diffusive tortuosity was obtained for diffusion within Ajax activated carbon [15] and similar values are reported by Ruthven [1]. The viscous tortuosity is also close to this value indicating that the pore network presents the same magnitude of resistance to both diffusional motion at low pressures and viscous transport at higher pressures.

4. Adsorption isotherm from batch measurement

The amount of carbon dioxide adsorbed at equilibrium by Takeda 5A pellets at 20°C has been determined by volumetric batch adsorption as discussed in the experimental section. Fig. 3 provides the isotherm graphically. It can be seen that the slope of the isotherm is larger at low pressure than it is at higher pressures typical of many isotherms of type 1. For comparative purposes, we have included isotherm data for other adsorbent materials, including an activated carbon fibre and a commercially available activated carbon extrudate which have similar adsorption properties to the Takeda 5A sample. This is a result of the fact that all samples share a generally related carbonaceous structure and therefore have similar affinity for gas adsorption. The comparison illustrated on Fig. 3 verifies that the measured isotherm is within an appropriate range for carbon dioxide adsorption.

The classic Langmuir isotherm equation is often utilised in the description of gas adsorption at low relative pressures where isotherm relations are generally of type 1. The mathematical form for such a relation is represented as

$$C_\mu = \frac{C_{\mu s}bC}{1 + bC} \quad (5a)$$

where b represents the affinity and $C_{\mu s}$ represents a saturation concentration. We can fit this isotherm equation to the data and in doing so we obtain $C_{\mu s} = 9.8 \times 10^{-4}$ mol/ml and $b = 3.3 \times 10^5$ ml/mol.

4.1. Adsorption dynamics

The steady state permeation result has proven useful in identifying the presence and absence of some common mass transfer mechanisms. We have shown that there is insignificant adsorbed phase flux of carbon dioxide, however consideration must be given to the dynamics of adsorption. Depending upon the conditions of measurement, the processes of macropore diffusion or micropore diffusion may dominate [1]. In molecular sieving materials such as the CMS used in this investigation, it is possible that a pore mouth resistance may influence mass transfer and cause molecular sieving [2,3]. However it has been shown that for Takeda 5A CMS, only molecules of size in the order of cyclohexane experience molecular sieving

290

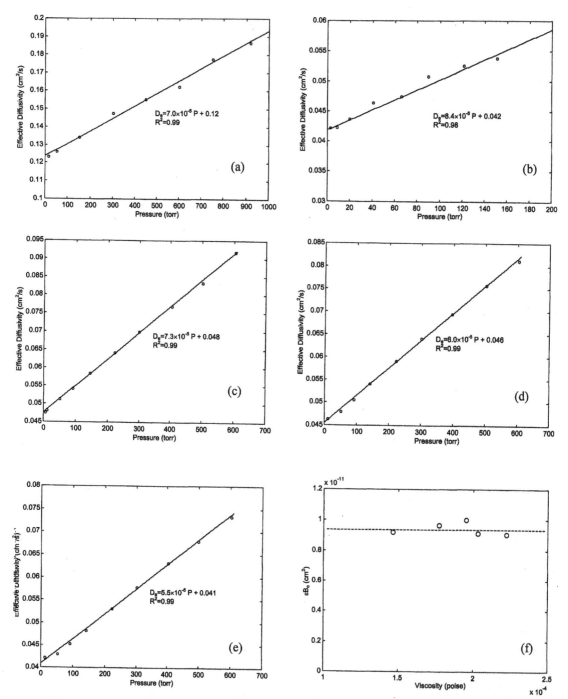

Fig. 2. (a) Effective diffusivity for helium at 293 K through Takeda 5A pellets; (b) Effective diffusivity for carbon dioxide at 293 K through Takeda 5A pellets; (c) Effective diffusivity for nitrogen at 293 K through Takeda 5A pellets; (d) Effective diffusivity for oxygen at 293 K through Takeda 5A pellets; (e) Effective diffusivity for argon at 293 K through Takeda 5A pellets; (f) The viscous flow parameter determined from gas permeation plotted against the viscosity; (g) The product of the pore diffusivity times the square root of molecular weight of the permeating gas plotted against the molecular weight.

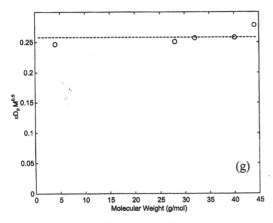

Fig. 2. (continued)

Table 3
Viscosity of gases at 293 K and 1 atm

Gas	Viscosity (poise)
Helium	1.95×10^{-4}
Nitrogen	1.77×10^{-4}
Argon	2.22×10^{-4}
Carbon Dioxide	1.46×10^{-4}
Oxygen	2.03×10^{-4}

[16]. Therefore we shall consider only micropore/macropore diffusion in the description of mass transfer of carbon dioxide.

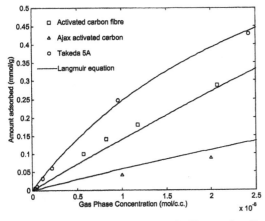

Fig. 3. Adsorption equilibrium of carbon dioxide on carbon fibre at 298 K (taken from [22]), on Ajax activated carbon (taken from [23]) and on carbon molecular sieve (CMS) (Takeda 5A in this work) at 293 K. Also included is the fit of the Langmuir isotherm Eq. (5a).

4.2. Mass balance

The general mass balance for considering single gas adsorption/diffusion process within our pellet must include a Knudsen and viscous flux and exclude any adsorbed phase flux. This can be represented as:

$$\varepsilon \frac{\partial C}{\partial t} + (1-\varepsilon)\frac{\partial \bar{C}_\mu}{\partial t} = \frac{1}{r^m}\frac{\partial}{\partial r}\left[r^m\left(\frac{\varepsilon B_\circ}{\mu}C + \varepsilon D_P\right)\frac{\partial C}{\partial r}\right]$$
(5b)

where C is the gas phase concentration and $\bar{C}_\mu$ is the volumetric average microparticle concentration. In most instances the permeation experiment is conducted with slab porous media implying that the shape factor m, in the mass balance is generally zero ($m=0$).

In the usual analysis of permeation data the adsorption process is often fast in comparison to the diffusion along the pellet scale. When this is the case, a change in the gas phase concentration makes an instant change in the average adsorbed phase concentration. Local equilibrium is established and this usually results in the 'pore diffusion' model being invoked to describe permeation dynamics and as has been useful in describing this situation [8]. However such a model can only be used when the timescale for macropore diffusion is much greater than the timescale for micropore diffusion. This condition has been conveniently quantified by the parameter γ and the condition [8]:

$$\gamma = \frac{C_\mu R_p^2}{C D_e}\frac{D_\mu}{R_\mu^2} \gg 1.$$

However, when micropore diffusion is slow (the microparticle is large or D_μ is low in magnitude), the pellet is small (R_p low in magnitude) or the gas phase diffusion process is fast (D_e large in magnitude), there is no local equilibrium between the sorbate and the microparticle. Diffusion within the microparticle must be considered and to account for this additional process within the pellet, uptake within the microparticle can be expressed using a parabolic approximation for the concentration profile, resulting in [17]:

$$\frac{\partial \bar{C}_\mu}{\partial t} = \frac{15 D_\mu(C_\mu)}{R_\mu^2}\left(C_\mu - \bar{C}_\mu\right)$$
(5c)

where C_μ is the equilibrium adsorbed phase concentration which is related to concentration through the isotherm relation and D_μ is the adsorbed phase diffusivity which normally adopts a dependence upon concentration described by the Darken relation, which for the Langmuir isotherm is:

$$D_\mu = \frac{D_{\mu 0}}{1 - C_\mu/C_{\mu s}}.$$
(5d)

Mathematically this system of equations (under linear

isotherm conditions), is similar to that solved by Good-knight and Fatt [18] who showed that a resistance to mass transfer perpendicular to the main direction of transport influenced the rate of pressure rise in the outgoing volume. It was found that a large resistance caused a faster initial rise than a lower resistance. In terms of the adsorption problem at hand, this would imply that a slower micropore diffusion would lead to a faster initial rise in the downstream pressure. In relation to the condition above, this would further imply that the lower the value of γ, the higher the initial rate of pressure rise. In the case where the maximum limit was approached when $\gamma \ll 1$, the downstream pressure rise would be dominated simply by the gas phase flow and rise at a fast initial rate. In the case where the other limit was approached when $\gamma \gg 1$ (local equilibrium), the downstream pressure rise could be described by the 'pore diffusion ' model and rise at an initially slow

rate. Hence inspection of the initial rate of pressure rise provides a useful visual identification for determining, based on the shape of the downstream pressure rise curve, whether there is local equilibrium between gas and adsorbed phases within the pellet.

As indicated earlier, the dynamics of permeation is displayed on a measurable timescale for carbon dioxide but not for the other gases. This is due to the influence of adsorption processes within the pellet effectively slowing the transient permeation rate on the pellet scale. Fig. 4 show the downstream pressure rise for carbon dioxide permeation for a number of upstream pressures and it is obvious that there is a relatively fast initial pressure rise, indicative of the fact that the micropore diffusion process is slow and that the parameter γ is not much greater than 1. Therefore, the simplified 'pore diffusion' description based on the assumption of local equilibrium is inappro-

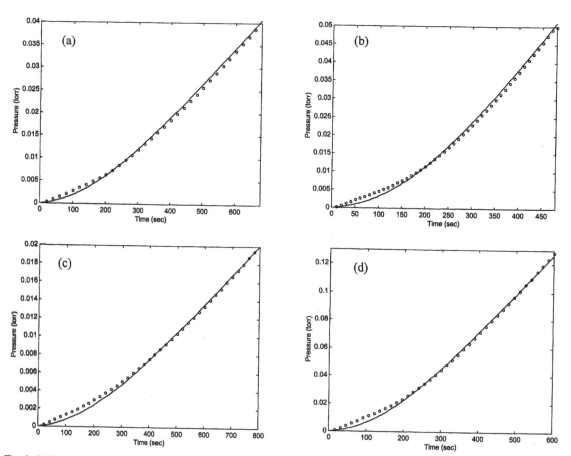

Fig. 4. (a) Downstream pressure rise in the permeation experiment of carbon dioxide at 293 K through Takeda 5A pellets at upstream pressure 19.623 Torr; (b) Downstream pressure rise in the permeation experiment of carbon dioxide at 293 K through Takeda 5A pellets at upstream pressure 40.86 Torr; (c) Downstream pressure rise in the permeation experiment of carbon dioxide at 293 K through Takeda 5A pellets at upstream pressure 8.197 Torr; (d) Downstream pressure rise in the permeation experiment of carbon dioxide at 293 K through Takeda 5A pellets at upstream pressure 65.5 Torr.

priate here. For this reason we must solve the general mass balance of Eq. (5) in order to obtain the parameter representing micropore diffusion $D_{\mu 0}/R_{\mu}^2$.

4.3. Solution of mass balance

In order to solve this mass balance, boundary conditions are required. The upstream boundary condition for the permeation experiment is

$$C(0,t) = C_0 \tag{5e}$$

and downstream, the continuity of mass implies that the amount entering the downstream volume is

$$x = L; \quad V \frac{\partial C}{\partial t} = A D_e \frac{\partial C}{\partial x}$$

In order to simultaneously solve Eq. (5) for the downstream pressure rise, it is necessary to use numerical techniques, because the isotherm and diffusivity relationships are non-linear. Here we opt to solve the equations simultaneously using the numerical method of lines. The kinetic parameter $D_{\mu 0}/R_{\mu}^2$ shall be evaluated by simultaneous fitting of the permeation curves which appear in Fig. 5 for the upstream pressures indicated. Other parameters have been evaluated previously and Table 4 summarises these values.

Fig. 4 contains the fit of Eq. (5) represented as continuous lines to the data which is represented as open circles. The prediction of the measured pressure rise against time is reasonable. The result of the optimisation is evaluation of the kinetic parameter, the limiting diffusivity, $D_{\mu 0}/R_{\mu}^2 = 1 \times 10^{-3}$ s^{-1}. This value is close to that presented by Kapoor and Yang [19] who obtained $D_{\mu 0}/R_{\mu}^2 = 9 \times 10^{-4}$ s^{-1} for carbon dioxide adsorption into a molecular sieving carbon at 298 K.

Table 4
Parameter values for solution of Eq. (5)

Constant	Value
ε	0.28
V	1100 ml
$\varepsilon B_o/\mu$	1.5×10^3 (ml cm^2/s/mol)
εD_P	0.04 (cm^2/s)
L	0.6 cm
A	0.707 cm^2

5. Analysis using time lag intercept

Fig. 4 show that the downstream pressure rise approaches a linear asymptote indicating steady state flow. The intercept of this asymptote with the time axis is known as the time lag and has been used as a general means for characterisation of transient permeation processes [20]. The relationship between time lag and the upstream pressure can reveal information about transport mobility and amount adsorbed. It can be shown by manipulation of the mass balance of Eq. (5b) according to the method of Frisch [21] that this relationship carries the form:

$$t_{\text{lag}} = \frac{L^2 \int\limits_0^{c_0} (\varepsilon C + (1-\varepsilon)C_\mu) H(u) \left(\int\limits_u^{c_0} H(w)\,dw \right) du}{\left(\int\limits_0^{c_0} H(u)\,du \right)^3} \tag{6a}$$

where the function H is given as

$$H = \frac{\varepsilon B_o}{\mu} C + \varepsilon D_P \tag{6b}$$

Having evaluated the isotherm relation from batch adsorption and obtained the time lag from linear regression of the steady state asymptote, it is possible to solve Eq. (6) by numerical integration and compare with the experimentally derived time lag. Fig. 5 shows the time lag for a series of upstream pressures represented as open circles and it is evident that the time lag decreases with upstream pressure for two reasons:

1. the slope of the isotherm decreases with increasing pressure
2. the gas phase mobility increases as a result of the intrusion of viscous permeation mechanism.

The solution of Eq. (6) is also shown on Fig. 5 and the fit is reasonable validating the accuracy of the time lag measurement and the time lag procedure as a method for characterisation for this case.

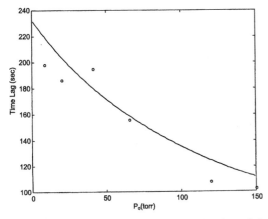

Fig. 5. The relationship between the measured time lag and the upstream pressure.

5.1. Shape of permeation curve and time lag

As we have discussed, the parameter γ, representing the ratio of the timescales for micropore diffusion and pellet scale diffusion processes, influences the shape of the curve of downstream pressure rise against time. This is evident in Fig. 6, which represents permeation of carbon dioxide with an upstream pressure of 19 Torr (curve B). The ratio of the macropore to micropore diffusion timescale, noted as γ, is in the order of 0.4 for permeation through the Takeda 5A pellet. According to Gray and Do [10] this implies that both macropore and micropore diffusion intrude into the permeation process a situation known as 'bimodal' diffusion. Also shown on Fig. 6 are two bounds, one being the limit of macropore diffusion control (curve C, $\gamma \gg 1$), the other being the limit of micropore diffusion control (curve A, $\gamma \ll 1$). It can be seen that curve A has a very small change in the gradient as time increases. However, all three curves should reach the same rate of pressure rise at steady state and according to Goodknight and Fatt [18], all three should have the same time lag for the advancing gas to penetrate the pellet. This would imply that for curve A, extrapolation of the linear steady state rise back to the time axis to obtain the time lag, would result in large error. This reinforces the need for other means of characterisation, such as batch adsorption, to be undertaken to compliment the time lag measurement when bimodal diffusion is observed.

5.2. Verification by batch dynamic measurement

Having characterised all the relevant mass transfer processes including gas phase permeation and micropore diffusion by permeation method, it is possible to verify

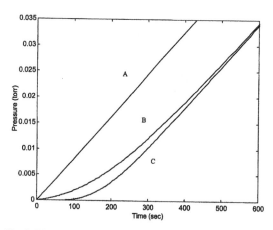

Fig. 6. The predicted downstream pressure rise of carbon dioxide with upstream pressure 19.62 Torr (as shown in Fig. 4a) noted as Curve B. Curve A represents the limit of slow adsorption (micropore diffusion control: $\gamma \ll 1$) and Curve C represents the limit of fast adsorption (local equilibrium: $\gamma \gg 1$).

these measurements using dynamic batch adsorption data. This dynamic data was obtained while monitoring pressure change in order to obtain the amount adsorbed at equilibrium (discussed earlier). This experiment was conducted within a large volume (2000 ml) for the purpose of keeping the maximum change in pressure between 5 and 10%. In this manner, the boundary condition can be considered effectively constant, but the pressure measurement is sensitive enough to monitor this relatively small transient change. The experiment was also conducted in an 'integral' fashion with outgassed pellets. As a result of the large total change in concentration within the pellet and the isotherm being non-linear over this range, the mass balance must be solved by numerical means.

The batch adsorption experiment was conducted with spherical particles of radius $R_p = 0.15$ cm for the purpose of measurement of the adsorption equilibrium isotherm. The dynamic pressure change was monitored with respect to time in order to determine when true equilibrium was reached. Under these measurement conditions the parameter γ is in the order of 0.1 implying that micropore diffusion is controlling the mass transfer. For this case, the mass balance will reduce to:

$$\frac{\partial C_\mu}{\partial t} = \frac{1}{r^2}\frac{\partial}{\partial r}\left(r^2 \frac{D_{\mu0}}{1 - C_\mu/C_{\mu s}}\frac{\partial C_\mu}{\partial r}\right) \tag{7a}$$

subject to the boundary and initial conditions of our 'integral' experiment:

$$r = R_\mu; \quad C_\mu = \frac{C_{\mu s}bC_0}{1 + bC_0} \tag{7b}$$

and to the initial condition

$$t = 0; \quad C_{\mu 7} = 0 \tag{7c}$$

and the fractional uptake (F) is given by

$$F = \frac{m_t}{m_{equ}} = \frac{(1 + bC_0)\left(\dfrac{3}{R_\mu^3}\displaystyle\int_0^{R_\mu} r^2 C_\mu \, dr\right)}{C_{\mu s}bC_0} \tag{7d}$$

The 5–10% change in concentration that is measured during adsorption will affect the shape of the uptake curve but to a negligible extent as shown by Ruthven [1]. Modification of the boundary condition (Eq. 7b) is therefore not necessary. Having already evaluated the necessary parameters, solution of Eq. (7) was undertaken and compared to the batch dynamic curves as is shown in Fig. 7. The continuous lines show the solution of Eq. (7) and the open circles represent data points. The fit is reasonable for all curves presented, showing good consistency between the dynamic methods of batch adsorption and permeation.

The batch adsorption experiment has been commonly used for determining kinetic data and proves more useful

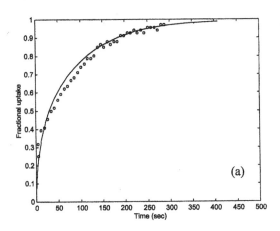

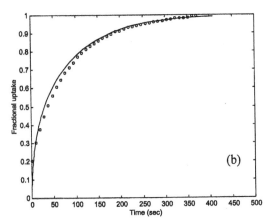

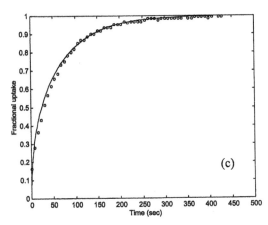

Fig. 7. (a) Batch uptake measurement of carbon dioxide at 293 K in outgassed Takeda 5A pellets at initial pressure 0.479 Torr; (b) Batch uptake measurement of carbon dioxide at 293 K in outgassed Takeda 5A pellets at initial pressure 4.66 Torr; (c) Batch uptake measurement of carbon dioxide at 293 K in outgassed Takeda 5A pellets at initial pressure 49.14 Torr.

than the steady state permeation experiment in the respect that the equilibrium isotherm can be determined simultaneously with dynamic data. However the flux cannot be determined directly in the batch experiment, often making estimation of the gas phase transport processes necessary, a procedure which can result in large error. This is an inherent weakness of the batch method. On the other hand, the permeation experiment can measure the flux accurately and when the two methods are coupled as we have used in this investigation, the results provide a valuable means of characterisation.

6. Conclusion

Analysis of equilibrium and kinetics of batch adsorption has shown that the micropore diffusion, characterised by the Darken relation, is rate limiting the dynamics of uptake in small Takeda 5A pellets. The permeation experiment undertaken with larger Takeda 5A pellets shows that bimodal diffusion control applies to the mass transfer process. Steady state diffusion measured by permeation experiment is shown to consist primarily of a combined Knudsen and viscous mechanism with negligible adsorbed phase flux along the pellet present. It is shown that a combination of both batch adsorption and permeation experiments prove to be a valuable method of characterisation.

Acknowledgements

This project is supported by the Australian Research Council.

References

[1] Ruthven D. Principles of adsorption and adsorption processes, New York: John Wiley, 1984.
[2] Koresh J, Soffer A. J Chem Soc Faraday Trans 1981;77(1):3005.
[3] Srinivasan R, Auvil SR, Schork JM. Chem Eng J 1995;57:137.
[4] Bulow M, Micke A. Adsorption 1995;1:29.
[5] Rutherford SW, Do DD. Chem Eng J (in press).
[6] Chihara K, Suzuki M, Kawazoe K. AIChE J 1978;24(2):237.
[7] Rutherford SW, Do DD. Ind Eng Chem Res 1999;38:565.
[8] Do DD. Adsorption analysis; equilibria and kinetics, London: Imperial College Press, 1998.
[9] Levenspiel O. Engineering flow and heat exchange, New York: Plenum Press, 1984.
[10] Gray PG, Do D. Gas Sep Purif 1989;3(4):201.
[11] Dullien FAL. Porous media, fluid transport and pore structure, New York: Academic Press, 1979.
[12] Wicke E, Kallenbach R. Die Oberflachendiffusion von Kohlendioxyd in activen kohlen. Kolloid-Z 1941;17:135.

[13] Hwang S, Kammermeyer K. Can J Chem Eng 1966;44:82.
[14] Carman PC. Flow of gases through porous media, London: Butterworth, 1956.
[15] Mayfield P, Do DD. Ind Eng Chem Res 1991;30:1262.
[16] Nakahari MH, Toshiaki O. J Chem Eng Data 1974;19:310.
[17] Yang RT. Gas separation by adsorption processes, Boston: Butterworths, 1987.
[18] Goodknight RC, Fatt I. J Phys Chem 1961;65:45.
[19] Kapoor A, Yang RT. Chem Eng Sci 1989;44(8):1723.
[20] Rutherford SW, Do DD. Chem Eng Sci 1997;52(5):703.
[21] Frisch HL. J Phys Chem 1957;61:93.
[22] Valenzuela DP, Myers AL. Adsorption equilibrium handbook, New Jersey: Prentice Hall, 1989.
[23] Gray PG, Do D. AIChEJ 1991;37:1027.

AUTHOR INDEX

Printed and bound by CPI Group (UK) Ltd, Croydon, CR0 4YY

03/10/2024

01040314-0017